3GPP-based Non-Terrestrial Networks in 5G and 6G

IEEE Press
445 Hoes Lane
Piscataway, NJ 08854

3GPP-based Non-Terrestrial Networks in 5G and 6G

Expanding the Frontiers of Wireless Communications

Nishith D. Tripathi
Virginia Tech
Blacksburgh, VA, USA

Jeffrey H. Reed
Virginia Tech
Blacksburgh, VA, USA

IEEE Press

WILEY

Published by John Wiley & Sons, Inc., Hoboken, New Jersey.

Library of Congress Cataloging-in-Publication Data
Names: Tripathi, Nishith D. author | Reed, Jeffrey H. author | John Wiley & Sons publisher
Title: 3GPP-based non-terrestrial networks in 5G and 6G : expanding the frontiers of wireless communications / Nishith D. Tripathi [and] Jeffrey H. Reed, Virginia Tech, Blacksburg, VA, USA.
Description: Hoboken, New Jersey : Wiley, [2026] | Includes index.
Identifiers: LCCN 2025025543 (print) | LCCN 2025025544 (ebook) | ISBN 9781394203383 hardback | ISBN 9781394203390 adobe pdf | ISBN 9781394203406 epub
Subjects: LCSH: Artificial satellites in telecommunication | 5G mobile communication systems | 6G mobile communication systems
Classification: LCC TK5104 .T75 2025 (print) | LCC TK5104 (ebook)
LC record available at https://lccn.loc.gov/2025025543
LC ebook record available at https://lccn.loc.gov/2025025544

Cover Design: Wiley
Cover Image: © aappp/Shutterstock

Set in 9.5/12.5pt STIXTwoText by Straive, Pondicherry, India
Printed and bound by CPI Group (UK) Ltd, Croydon, CR0 4YY

C9781394203383_221025

To Our Families

Contents

About the Authors

Dr. Nishith D. Tripathi is a research associate professor at Virginia Tech. Dr. Tripathi has 24 years of hands-on industry experience and is an expert on various aspects of commercial 3G, 4G, and 5G wireless networks including design, operations, testing, and optimization. At Virginia Tech, he has led several sponsored research projects on 5G, 5G-Advanced, and 6G in the areas such as 5G O-RAN testbeds, non-terrestrial network (NTN)/SpaceNet testbed, O-RAN xApps, O-RAN testing, enhanced security for 5G and 6G, NTN, V2X communications, geofencing, positioning, UAV/UAS, and smart warehouses. He has co-authored five books including the world's first multimedia book on 5G, a comprehensive textbook on cellular communications, a pioneering monograph on the RRM using AI, and two books on NTN and O-RAN. He has made more than thirty contributions toward the development of the 3GPP 5G specifications. As a wireless industry expert, Dr. Tripathi has contributed to organizations such as Federal Communications Commission (FCC), CTIA, Global System for Mobile communications Association (GSMA), Next G Alliance (NGA), NSC, Scientific American, FTC, EE Times University, and CNN. As an Academics Engagement Lead for the Next G Alliance (NGA), he has facilitated identification of 6G vision and requirements and made contributions to several NGA white papers on 6G. He is the founder of Aum Research and Consulting (ARC) that provides research and consulting services.

Dr. Jeffrey H. Reed is the Willis G. Worcester Professor of ECE. Professor Reed's research interests are wireless communications, wireless security, cognitive radio, software radio, telecommunications policy, and spectrum access. Dr. Reed has co-authored more than 500 articles and books. In addition, Dr. Reed co-founded several commercial companies, including Federated Wireless, which commercializes spectrum sharing; PFP Cybersecurity, which provides security solutions for IoT devices; and Cirrus360, which produces tools for rapid prototyping of O-RAN. He is the Founding Director of Wireless@Virginia Tech, a university research center, and co-founder of Virginia Tech's Hume Center for National Security and Technology, where he served as the Interim Director. He also served as the Interim Director of the Commonwealth Cyber Initiative and is currently its CTO. Dr. Reed is a Fellow of the IEEE for contributions to software radio and communications signal processing and for leadership in engineering education. Dr. Reed is a Fellow of the National Academy of Inventors.

Preface

Even with the deployment of five generations of cellular technologies, a significant portion of the planet's surface area and many people still do not have access to wireless communications. Non-Terrestrial Networks (NTNs) are becoming increasingly important in providing wireless communication services anywhere on the planet. The cost of sending a payload to space has reduced significantly recently, leading to a rise in the number of satellite launches. The NTN can also provide backhaul connectivity to the terrestrial radio network in remote areas, fallback connectivity when the terrestrial network infrastructure becomes damaged or unavailable, and supplemental connectivity to augment resource availability at the wireless device. This book explains key concepts of 3GPP[1]-based NTNs in ten chapters.

Chapter 1 illustrates a simplified NTN architecture and explains its key components and interfaces. The motivation for the NTN is discussed from the perspective of the use-case categories of service ubiquity, service continuity, and service scalability. The 3GPP roadmap on the NTN work is outlined to provide a historical perspective and summarize potential future work. The 3GPP started the NTN-related specifications work as part of 5G cellular technology and has continued enhancing the NTN in 5G-Advanced. The NTN is expected to play an even more important role in the sixth-generation (6G) cellular technology. Chapter 1 also provides a glimpse of the NTN's role in 6G.

Chapter 2 illustrates different types of NTN platforms. These NTN platforms include different types of satellites such as Geosynchronous Earth Orbit (GEO) satellites, Medium-Earth Orbit (MEO) satellites, Low-Earth Orbit (LEO) satellites, and High-Altitude Platform Stations (HAPSs). Chapter 2 also discusses the distinct characteristics of NTN platforms. The NTN platform, based on its capability and design, illuminates its target coverage area using a specific type of beam. Three types of beams are illustrated in Chapter 2: Earth-fixed beams, Earth-moving beams, and Quasi-Earth-fixed beams. Satellite systems that existed prior to the deployment of the 3GPP NTN deployments are also briefly described.

Chapter 3 provides a foundation of LTE-M, NB-IoT, and NR air interfaces by highlighting their key features and overall operations. The associated core networks, EPC and 5GC, are also briefly discussed. While this chapter provides the foundation of LTE-M and NB-IoT based on Release 13, enhancements in LTE-M and NB-IoT beyond Release 13 are also briefly summarized. Note that the 3GPP has two parallel work streams related to the NTN: NR-NTN and IoT-NTN. The NR-NTN utilizes an NR-based air interface. The NR-NTN utilizes gNBs in the radio network and the 5G Core

1 The Third Generation Partnership Project (3GPP) is the standards organization that has defined specifications of 3G, 4G, and 5G cellular communication systems. The 3GPP is also expected to define 6G specifications. See https://www.3gpp.org/ to learn about 3GPP activities.

(5GC) Network. The IoT-NTN supports LTE-M and NB-IoT air interfaces and uses the 4G Evolved Packet Core (EPC). The IoT-NTN utilizes pre-Release 17 LTE-M and NB-IoT specifications and leverages NTN-specific enhancements defined by the NR-NTN to determine NTN-related enhancements for LTE-M and NB-IoT specifications.

Chapter 4 describes NTN challenges and their implications for the design and operations of the NTN. The NTN poses distinct radio environment-related challenges compared to a Terrestrial Network (TN). For example, the NTN typically has a long propagation delay and often a significantly varying propagation delay, requiring timing synchronization and timer enhancements. A non-geostationary NTN platform leads to large and varying Doppler shifts, and suitable frequency synchronization adjustments are needed. While a TN cell is fixed, an NTN cell can move, requiring enhanced cell reselection, handover, and registration area management. An NTN may use Earth-fixed beams, quasi-Earth-fixed beams, or Earth-moving beams, requiring different mobility management approaches. The type of NTN payload, such as the transparent payload or the regenerative payload, affects the architecture, design, and operations of the NTN. Long-term signal strength characteristics differ between the TN and the NTN, requiring a different mobility management approach in the NTN. Special atmospheric effects such as the Faraday effect and scintillations may also need to be considered for the NTN. Chapter 4 also provides a high-level overview of the NTN solutions that the 3GPP has developed to address these NTN challenges.

Chapter 5 illustrates various candidate architectures of the NTN. While the 3GPP has defined an NTN architecture with a transparent payload in Release 17 and Release 18 NTN specifications, the 3GPP has also investigated other NTN architectures, such as an NTN with a regenerative payload and multi-connectivity NTNs. The regenerative payload is within the scope of the 3GPP Release 19 specifications. Furthermore, in the case of a multi-connectivity NTN, an NTN UE simultaneously communicates with (i) two NTNs (e.g., one GEO satellite-based NTN and another LEO satellite-based NTN) or (ii) one TN and one NTN. An example NTN architecture with non-terrestrial core and services networks is also illustrated. Enhanced Tracking Area management and enhanced QoS for an NTN are also described. A brief overview of optical communication in the context of an NTN is also given.

Chapter 6 aims to provide insights into NTN-specific aspects of RF Planning and Design (RPD) and RF propagation. An overall RPD framework is summarized. Potential NTN spectrum is discussed. Relevant propagation path loss models are specified, and NTN-specific propagation effects, such as atmospheric effects including scintillations, Faraday effect, and weather effects, are discussed. Various aspects of the NTN link budget are described, including the implications of different NTN platforms, types of beams, and device types. Capacity planning for an NTN is also discussed. NTN configuration guidelines based on the NTN RPD are outlined. Chapter 6 also summarizes 3GPP-estimated link budgets and UE throughput.

Chapter 7 explains pre-data transfer operations in the context of the NTN. Note that certain pre-data transfer operations occur between the NTN UE and the NTN radio and core networks before an NTN UE can exchange data with the radio and core network of the NTN. These operations include cell search, network acquisition, random access, Radio Resource Control (RRC) connection setup, registration or attach, and protocol data unit (PDU) session or EPS bearer setup. Chapter 7 discusses these operations for three radio access technologies: NR, LTE-M, and NB-IoT.

Chapter 8 focuses on data transfer in an NTN. The NTN reuses the TN's data transfer framework and makes enhancements to reflect NTN-specific radio channel challenges. The chapter illustrates various timing relationships in the NTN along with a new Timing Advance procedure for the NTN. The Key prerequisites for downlink and uplink data transfer in the LTE-M NTN and the NB-IoT-NTN are described. A procedure for managing Global Navigation Satellite System (GNSS) measurement gaps is also summarized.

Chapter 9 addresses mobility management. Mobility management principles for a TN are reused in an NTN along with NTN-specific enhancements. The NTN UE's RRC state influences mobility management. This chapter briefly overviews the NTN UE's RRC states, followed by a summary of major NTN mobility challenges and candidate solutions. Mobility management in the NR-NTN is addressed by discussing measurements and handover triggers, types of NTN handover, and activities of the NR-NTN UE in the RRC_IDLE and RRC_INACTIVE states. A detailed signaling flow for the Conditional Handover (CHO) in the NR-NTN is illustrated. Similarly, mobility management in the LTE-M NTN is addressed by discussing measurements and handover triggers, a detailed signaling flow for the CHO, and activities of the LTE-M NTN UE in the RRC_IDLE state. A detailed signaling flow for the CHO. Mobility management in the NB-IoT NTN is also described. Feeder link switchover, a unique aspect of the NTN, is explained. The issue of discontinuous coverage is also summarized.

Chapter 10 provides a brief overview of target NTN enhancements in Release 19. While the NTN-related specifications work carried out by the 3GPP is formally part of 5G and 5G-Advanced, the NTN is expected to play an even more key role in 6G. A brief introduction to the 6G vision and requirements is given, followed by the usage scenarios defined by the International Telecommunication Union (ITU) for 6G. The ITU-specified performance goals for 6G are also summarized. An overview of 6G technology enablers is also given. The NTN's role in realizing the vision of 6G is explained. Finally, Chapter 10 gives examples of potential NTN research directions.

Since we cannot escape acronyms in our wireless communications industry, we have included a list of acronyms in the front matter of the book to facilitate reading. References listed at the end of each chapter are quite helpful for digging deeper into NTN concepts. We have designed interactive exercises to reinforce learning of key NTN concepts and made them available at www.wiley.com/go/tripathi5g.

Our hope is that this NTN book will significantly increase your NTN knowledge and help you in your professional development. Please feel free to contact us via social media; we will be happy to strengthen our teacher–student relationship.

Happy NTN learning!
Well wishes,
Dr. Nishith D. Tripathi
Dr. Jeffrey H. Reed

Acknowledgments

We sincerely express our gratitude to our families for their patience and sacrifice of time while we worked on the book in the evenings, weekends, and holidays.

We thank our employer, Virginia Tech, where we have opportunities to perform non-terrestrial network (NTN) research. Our NTN research facilitated the development of this book. We also thank Samsung Research America for enabling Dr. Nishith Tripathi to make numerous contributions to the 3GPP 5G NTN specifications.

This book is supported by the research projects sponsored by the National Science Foundation (NSF) and the National Telecommunications and Information Administration (NTIA). Specifically, the book is partially supported by (i) the NSF's CCRI award number 2235139, (ii) the NSF's NeTS award number CNS-2312447, and (iii) the NTIA's Public Wireless Supply Chain Innovation Fund (PWSCIF) under the award number (51-60-IF007).

We thank our students who contributed directly or indirectly to this book. We thank Rahul Varma Chintalapati (Virginia Tech) for contributions to NTN and spectrum-sharing challenges. We thank Zac Martin (Virginia Tech) for developing nice diagrams for this book. We also thank Dr. Yash Vasavada of DAIICT (India) for his helpful feedback on the RF planning and design aspects of the NTN.

Finally, we sincerely appreciate the flexibility and guidance of our publisher, IEEE/Wiley.

Acronyms

Acronym	Expansion
16QAM	16-ary quadrature amplitude modulation
5GC	5G Core
5QI	5G QoS identifier
64QAM	64-ary quadrature amplitude modulation
AAA	authentication, authorization, and accounting
ACI	adjacent channel interference
ACK	acknowledgment
ADC	analog-to-digital converter
AF	application function
AiP	antenna in package
AL	atmospheric loss
AM	acknowledged mode
AMBR	aggregate maximum bit rate
AMF	Access and Mobility Management Function
AN	access network
AoD	antenna on display
API	application programming interface
APN	access point name
AR	augmented reality
ARP	allocation and retention priority
AS	access stratum
ATIS	Alliance of Telecommunications Industry Solutions
ATP	acquisition, tracking, and pointing
AUSF	Authentication Server Function
AWGN	additive white Gaussian noise
BIST	built-in self-test
BLER	block error rate
BPSK	binary phase shift keying
BR	bandwidth reduced
BSR	buffer status report
BWP	bandwidth part
CA	carrier aggregation
CAN	controller area network
CAPIF	common API framework

CBG	code block group
CCH	common control channel
CCI	co-channel interference
CDF	cumulative distribution function
CDMA	code division multiple access
CE	coverage enhancement
CE	control element
CGI	cell global identity
CHO	conditional handover
CI/CD	continuous improvement/continuous development
CIM	compute in memory
CINR	carrier-to-interference-plus-noise ratio
CL	clutter loss
CMOS	complementary metal-oxide-semiconductor
CN	core network
CNN	Cable News Network
CNR	carrier-to-noise ratio
CoAP	constrained application protocol
CP	control plane
CP	cyclic prefix
CPU	central processing unit
CQI	channel quality indicator
CRC	cyclic redundancy check
CRI	CSI-RS resource indicator
C-RNTI	cell-radio network temporary identifier
CRS	cell-specific reference signal
CSI-IM	channel state information-interference measurement
CSI-RS	channel state information-reference signal
CTIA	Cellular Telecommunications Industry Association
CTO	chief technology officer
CU	central unit
D2D	direct-to-device
DAC	digital-to-analog converter
DAPS	dual active protocol stack
DCI	downlink control information
DD	delay-Doppler
DID	decentralized identifier
DL	downlink
DLT	distributed ledger technology
DMRS	demodulation reference signal
DN	data network
DNN	data network name
DNS	domain name system
DPD	digital pre-/post-distortion
DPSK	differential phase shift keying
DRB	data radio bearer
DRX	discontinuous reception
DTC	direct-to-cell

DU	distributed unit
EAB	enhanced access barring
EARFCN	E-UTRA absolute radio frequency channel number
EASDF	edge application server discovery function
EC-GSM	extended coverage-Global System for Mobile Communication
EDA	electronic design automation
eDRX	extended discontinuous reception
EDT	early data transmission
EE	Electronic Engineering
EFC	earth-fixed cell
EIRP	effective isotropic radiated power
eMBB	enhanced mobile broadband
EMC	Earth-moving cell
eMTC	enhanced MTC
EN-DC	E-UTRA-NR dual connectivity
ENISA	European Network and Information Security Agency
ENOB	effective number of bits
EOC	edge of coverage
EPC	Evolved Packet Core
EPS	Evolved Packet System
E-RAB	E-UTRAN radio access bearer
ESIM	Earth station in motion
ETL	extract, transform, and load
E-UTRA	Evolved-Universal Terrestrial Radio Access
E-UTRAN	Evolved-Universal Terrestrial Radio Access Network
EVM	error vector magnitude
FCC	Federal Communications Commission
FDD	frequency division duplex
FFT	fast Fourier transform
FOV	field of view
FPC	flexible printed circuits
FPGA	field programmable gate array
FRF	frequency reuse factor
FSK	frequency shift keying
FSO	free space optical
FSPL	free space path loss
FSS	fixed satellite service
FTC	Federal Trade Commission
FTF	frequency translation factor
GBR	guaranteed bit rate
GDPR	general data protection regulation
GEO	geosynchronous/geostationary Earth orbit
gNB	next generation Node B
GPRS	General Packet Radio Service
GPS	Global Positioning System
GPU	graphics processing unit
GSM	Global System for Mobile Communication
GSMA	Global System for Mobile communications Association

GSO	geosynchronous orbit
GTP	GPRS tunneling protocol
GUAMI	globally unique AMF ID
GUTI	globally unique temporary identifier
GW	gateway
HAPS	high-altitude platform station
HARQ	hybrid automatic repeat request
HBF	holographic beamforming
HBT	heterojunction bipolar transistor
HEO	highly elliptical orbit
HFT	high frequency trading
HOM	high order modulation
HRLLC	hyper reliable and low-latency communication
HSS	home subscriber server
HTA	heavier than air
IAB	integrated access and backhaul
IE	information element
IETF	Internet Engineering Task Force
IFC	in-flight connectivity
IFFT	inverse fast Fourier transform
I-FSK	Impulsive FSK
IIC	incumbent informing capabilities
IIoT	industrial Internet of things
IMS	IP Multimedia Subsystem
IMSI	international mobile subscriber identity
IoT	Internet of Things
IoT-NTN	Internet of Things-Non-Terrestrial Network
IP	Internet Protocol
IPN	integrated phase noise
IPSM-GW	IP Short Message Gateway
I-RNTI	inactive-radio network temporary identifier
ISAC	integrated sensing and communication
ISL	inter satellite link
ISS	international space station
ITO	indium tin oxide
ITU	International Telecommunication Union
JCAS	joint communications and sensing
L1-RSRP	layer 1-RSRP
LAA	licensed assisted access
LBT	listen-before-talk
LCG	logical channel group
LCT	laser communication terminal
LDPC	low-density parity check
LED	light emitting diode
LEO	low Earth orbit
LHCP	left hand circular polarization
LI	layer indicator

LNA	low noise amplifier
LO	local oscillator
LOS	line of sight
LPF	low pass filter
LPWA	low-power wide-area
LTA	lighter than air
LTE	Long-Term Evolution
LTE-M	Long-Term Evolution-MTC
LTM	L1/L2 triggered mobility
LUT	look-up table
MAC	medium access protocol
MAPL	maximum allowable path loss
MBS	multimedia broadcast service
MCS	modulation and coding scheme
MEF	Metro Ethernet Forum
MEMS	micro-electromechanical system
MEO	medium Earth orbit
mHEMT	metamorphic high-electron-mobility-transistors
MIB	master information block
MIB-NB	master information block-narrowband
MICO	mobile initiated connection only
MIMO	multiple input multiple output
MIV	MPEG immersive video
ML	machine learning
MME	Mobility Management Entity
mMTC	massive machine type communications
MN	master node
MPDCCH	MTC PDCCH
MPEG	Moving Picture Experts Group
MQTT-SN	message queuing telemetry transport- sensor network
MR	mixed reality
MR-DC	multi-radio dual connectivity
MRSS	multi-RAT spectrum sharing
MSS	mobile satellite service
MSSA	Mobile Satellite Services Association
MT	mobile-terminated
MTC	machine type communication
MTC-IWF	Machine Type Communication-Inter Working Function
MU-MIMO	multi-user-multiple input multiple output
MV-HEVC	multiview high efficiency video coding
NACK	negative acknowledgment
NAS	non-access stratum
NB	narrowband
NBI	narrowband index
NB-IoT	narrowband IoT
NCGI	NR cell global identity

NDI	new data indicator
Near-RT RIC	near-real-time RAN intelligent controller
NEF	Network Exposure Function
NESAS	network equipment security assurance scheme
NF	Network Function
NF	noise figure
NFC	near-field communication
NG	next generation
NGA	Next G Alliance
NGAP	next-generation application protocol
NGC	Next-Generation Core
NGMN	Next-Generation Mobile Network
NG-RAN	Next-Generation Radio Access Network
NGSO	non-geostationary orbit
NIDD	non-IP data delivery
NIST	National Institute of Standards and Technology
NLOS	non-line of sight
NOMA	non-orthogonal frequency division multiple access
Non-RT RIC	non-real-time RAN intelligent controller
NPBCH	narrowband physical broadcast channel
NPCI	narrowband physical cell ID
NPDCCH	narrowband physical dedicated control channel
NPDSCH	narrowband physical downlink shared channel
NPRACH	narrowband physical random access channel
NPSS	narrowband primary synchronization signal
NPUSCH	narrowband physical uplink shared channel
NR	New Radio
NRF	Network Repository Function
NR-NTN	New Radio-Non-Terrestrial Network
NRS	narrowband reference signal
NSA	non-standalone
NSACF	Network Slice Admission Control Function
NSC	National Spectrum Consortium
NSSAAF	Network Slice-specific and SNPN Authentication and Authorization Function
NSSAI	network slice selection assistance information
NSSF	Network Slice Selection Function
NSSS	narrowband secondary synchronization signal
NTN	non-terrestrial network
NTN-GW	NTN-Gateway
OAM	orbital angular momentum
OAM/OA&M	operations, administration, and maintenance
OCC	orthogonal cover code
O-Cloud	O-RAN Cloud
O-CU-CP	O-RAN-Central Unit-Control Plane
O-CU-UP	O-RAN-Central Unit-User Plane
O-DU	O-RAN Distributed Unit
O-eNB	O-RAN eNodeB

OFDM	orthogonal frequency division multiplexing
OFDMA	orthogonal frequency division multiple access
OIA	Office of International Affairs
OLED	organic light emitting diode
OMA	Open Mobile Alliance
OOK	on–off keying
O-RAN	Open Radio Access Network
O-RU	O-RAN Radio Unit
OTA	over-the-air
OTDOA	observed time difference of arrival
OTFS	orthogonal time frequency space
PA	power amplifier
PAPR	peak-to-average power ratio
PBCH	physical broadcast channel
PCB	printed circuit board
PCC	Policy and Charging Control
PCEF	Policy and Charging Enforcement Function
PCF	Policy Control Function
PCI	physical cell ID
PCO	protocol configuration options
PCRF	Policy and Charging Rules Function
P-CSCF	Proxy-Call Session Control Function
PDB	packet delay budget
PDCP	packet data convergence protocol
PDN	packet data network
PDSCH	physical downlink shared channel
PDU	protocol data unit
P-GW	PDN-Gateway
PHR	power headroom report
PHY	physical
PLMN	Public Land Mobile Network
PMI	precoding matrix indicator
PQC	post-quantum cryptography
PRACH	physical random access channel
PRB	physical resource block
PSA	PDU session anchor
PSD	power spectral density
PSM	power saving mode
PSS	primary synchronization signal
PUCCH	physical uplink control channel
PUR	preconfigured uplink resource
PUSCH	physical uplink shared channel
QAM	quadrature amplitude modulation
QFI	QoS flow indicator
QoS	quality of service
QPSK	quadrature phase shift keying

RA	random access
RAB	radio access bearer
RACH	random access channel
RAN	radio access network
rApp	remote App
RAR	random access response
RAT	radio access technology
RBG	resource block group
RE	resource element
RF	radio frequency
RFIC	radio frequency integrated circuit
RHCP	right-hand circular polarization
RI	rank indicator
RIC	RAN Intelligent Controller
RIS	reflective/reconfigurable intelligence surface
RLC	radio link control
RLF	radio link failure
RNA	RAN notification area
RoHC	robust header compression
RoT	rise over thermal
RP	reference point
RPD	RF planning and design
RRC	radio resource control
RSRP	reference signal received power
RSRQ	reference signal received quality
RTD	round trip delay
RTT	round trip time
RU	resource unit
RV	redundancy version
SA	Service and System Aspects
SAE	System Architecture Evolution
SAN	Satellite Access Node
SBA	service-based architecture
SCEF	Service Capability Exposure Function
SC-FDMA	single-carrier frequency division multiple access
SCP	Service Communication Proxy
SC-PTM	single cell-point-to-multipoint
SCS	Service Capabilities Server
SCS	subcarrier spacing
SDAP	service data adaptation protocol
SDMA	space division multiple access
SDT	small data transmission
SEN	Societal and Economic Needs
SFBC	space frequency block coding
SFM	shadow fading margin
SFN	system frame number

S-GW	Serving Gateway
SI	system information
SIB	system information block
SIB1-BR	SIB1-bandwidth reduced
SIB1-NB	system information block 1-narrowband
SIB2-NB	system information block 2-narrowband
SiGe	silicon-germanium
SINR	signal to interference plus noise ratio
SIP	session initiation protocol
SI-RNTI	system information-radio network temporary identifier
SL	scintillation loss
SM	session management
SMF	Session Management Function
SMO	Service Management and Orchestration
SMS	short message service
SMTC	SSB measurement timing configuration
SN	secondary node
SNPN	standalone non-public network
SNR	signal-to-noise ratio
SOI	silicon-on-insulator
SON	self-organizing network
SOS	(Morse code distress signal)
SPI	serial peripheral interface
SPS	semi-persistent scheduling
SR	scheduling request
SRB	signaling radio bearer
SRS	sounding reference signal
SSB	SS/PBCH block
SSBRI	SSB resource indicator
SSC	session and service continuity
SSI	self-sovereign identity
SSS	secondary synchronization signal
S-TMSI	SAE-temporary mobile subscriber identity
SUCI	subscription concealed identifier
SU-MIMO	single user-multiple input multiple output
SUPI	subscription permanent identifier
TA	timing advance
TAC	tracking area code
TAI	tracking area identifier
TAR	timing advance report
TAU	tracking area update
TBS	transport block size
TCI	transmission configuration indication
TDD	time division duplex
TEE	trusted execution environment
TEID	tunnel endpoint ID

TLE	two-line element
TN	terrestrial network
TPC	transmit power control
TPM	trusted platform module
TPU	tensor processing unit
TTI	transmit time interval
UAS	uncrewed/unmanned aerial system
UAV	uncrewed/unmanned aerial vehicle
UCA	uniform circular array
UDM	Unified Data Management
UDR	Unified Data Repository
UDSF	Unstructured Data Storage Function
UE	User Equipment
UI	user interface
UL	uplink
ULI	user location information
UMTS	Universal Mobile Telecommunication System
UP	user plane
UPF	User Plane Function
URLLC	ultra-reliable low-latency communications
UTC	coordinated universal time
VGA	variable gain amplifier
VoNR	voice over NR
VR	virtual reality
VSAT	very small aperture terminal
WSSUS	wide-sense stationary uncorrelated scattering
WTFC	wideband time frequency coding
xApp	extended App
XR	extended Reality

About the Companion Website

This book is accompanied by a companion website:

www.wiley.com/go/tripathi5g

This website includes interactive exercises to reinforce the key concepts covered in all the chapters. References to the topics associated with exercise questions are given as part of hints.

1

Introduction to a Non-Terrestrial Network

Even with the deployment of five generations of cellular technologies, a significant portion of the planet's surface area and a large number of people still do not have access to wireless communications. Non-Terrestrial Networks (NTNs) are becoming increasingly important in providing wireless communications anywhere on the planet. This chapter defines an NTN and introduces different types of NTN platforms that help create an NTN. Since the NTN utilizes non-terrestrial platforms for radio coverage, a different network architecture is needed compared to a traditional terrestrial network (TN).

This chapter illustrates a simplified NTN architecture and explains its key components and interfaces. The motivation for the NTN is discussed from the perspective of three use-case categories—service ubiquity, service continuity, and service scalability. The Third-Generation Partnership Project (3GPP) has studied several NTN use cases, including user mobility between a TN and an NTN, satellite-based backhaul to support TN deployments in remote areas without the transport infrastructure, support for Internet of Things (IoT) devices, factories in remote areas, and offshore wind farms. The 3GPP roadmap on the NTN work is outlined to provide a historical perspective as well as potential future work. The NTN-related specifications work carried out by the 3GPP NTN is formally part of the fifth-generation (5G) and 5G-advanced cellular technologies. However, the NTN is expected to play an even more important role in the sixth-generation (6G) cellular technology. A glimpse of the NTN's role in 6G is also provided in this chapter.

1.1 Non-Terrestrial Networks: The Definition

An NTN is a wireless communication network that utilizes radio network equipment placed on an airborne or spaceborne vehicle. From this general perspective, if some radio network equipment is on an uncrewed aerial vehicle (UAV) or a drone, a network involving UAVs would qualify as an NTN. Indeed, in some literature, UAVs are considered to be part of an NTN. However, 3GPP NTN specifications consist of different types of satellites and high-altitude platform station (HAPS). A HAPS is an airborne vehicle that operates at an altitude between 8 and 50 km (3GPP 2024e). Examples of uncrewed/unmanned aerial systems (UASs) (3GPP 2021a) include lighter-than-air (LTA) UAS and heavier-than-air (HTA) UAS. Examples of the satellites include low Earth orbit (LEO), medium Earth orbit (MEO), and geosynchronous/geostationary Earth orbit (GEO) satellites. In general, LEO satellites operate at altitudes between 300 and

3GPP-based Non-Terrestrial Networks in 5G and 6G: Expanding the Frontiers of Wireless Communications, First Edition.
Nishith D. Tripathi and Jeffrey H. Reed.

Companion website: www.wiley.com/go/tripathi5g

1,500 km, MEO satellites operate at altitudes between 7,000 and 25,000 km, and GEO satellites operate at an altitude of 35,786 km.

The 3GPP formally defines the (NR) NTN to be "an NG-RAN consisting of gNBs, which provide non-terrestrial NR access to user equipments (UEs) by means of an NTN payload embarked on an airborne or spaceborne NTN vehicle and an NTN Gateway" (3GPP 2024e). The 3GPP-proposed NTN architecture is illustrated in Section 1.2.

1.2 Simplified 3GPP NTN Architecture

Figure 1.1 illustrates a simplified architecture of an NTN (3GPP 2024d,e). The NTN is one of the prominent 5G features defined by the 3GPP in Release 17. See Tripathi and Reed (2019) for details on 5G. While wireless communication is one of the most prominent features of a cellular system, the overall cellular system includes both wireless and wired links. A more detailed NTN architecture is described in Chapter 5 and 3GPP (2023).

An NTN UE communicates with an NTN payload using the radio interface protocol stack. The communication link between the NTN UE and the NTN payload is called the service link or the access link. A typical NTN UE is a handheld smartphone, an IoT device, or a very small aperture terminal (VSAT) device. The radio interface protocol stack may use technologies such as 5G new radio (NR), narrowband IoT (NB-IoT), and eMTC/LTE-M.[1] The NTN payload resides on an NTN platform. Examples of NTN platforms include satellites such as GEO satellites, MEO satellites, and LEO satellites and HAPSs such as stationary aircraft in the stratosphere.

The NTN payload communicates with an NTN-gateway (NTN-GW) using an implementation-specific radio interface. The communication link between the NTN payload and the NTN-GW is called the feeder link.

Figure 1.1 utilizes a transparent NTN payload, where the NTN payload does not carry out technology-specific baseband processing but performs conversion of signals between the service link

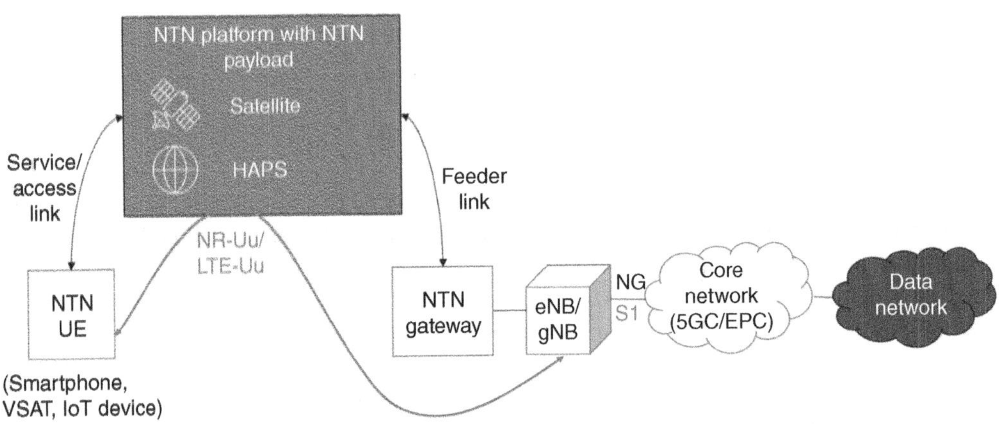

Figure 1.1 Simplified NTN Architecture.

1 eMTC is the enhanced MTC, MTC is the machine-type communication, and LTE-M is the long-term evolution-MTC.

and the feeder link. In the downlink from the gNB/eNB to the NTN UE, the NTN payload receives the signal from the NTN-GW and performs functions such as radio frequency (RF) filtering, frequency conversion, and amplification[2] and transmits the RF signals that the NTN UE can process using a suitable 3GPP-defined radio interface such as 5G NR, NB-IoT, and eMTC/LTE-M. In the uplink from the NTN UE to the gNB/eNB, the NTN payload receives the RF signal from the UEs in the cell on a suitable 3GPP-defined carrier frequency and transmits the RF signal to the NTN-GW on a different carrier frequency after suitable filtering and amplification.

The interface between the NTN-GW and the gNB/eNB is undefined in the 3GPP specifications. This interface can be internal to the gNB/eNB, where both the NTN-GW and the gNB/eNB are integrated into one entity. When this interface is external to the gNB/eNB, it can be a wired or wireless interface.

Similar to a TN, the gNB/eNB connects to the core network using 3GPP-defined interfaces. In general, the gNB connects to the 5GC in the case of NR-NTN, and the eNB connects to the EPC in the case of IoT-NTN. The core network connects to data networks such as the Internet, as in a TN.

1.3 Motivation for the NTN

A TN can provide good quality of service (QoS) to users when the users are in the radio coverage of the TN. Service providers carry out coverage and capacity planning to determine the number and locations of base stations in a target service area to meet the expected traffic demands. These base stations need to connect to other base stations to support features such as handover through a transport network. The base stations also need to connect to the core network via a transport network, because the core network interfaces with data networks such as the Internet.

Cost-effective deployment of the base stations and the associated transport network is not feasible in areas with sparse populations and remote and vast areas on land, such as mountains, deserts, and national parks. In cases such as large portions of oceans and remote islands, deployment of a terrestrial cellular network is challenging due to inadequate land infrastructure and the lack of a business case.

An NTN offers an attractive solution in the challenging environments mentioned earlier. Figure 1.2 shows three NTN use case categories identified by the 3GPP (2018b). These use case categories represent one of the key motivations behind the design and deployment of an NTN because TNs cannot adequately address the related use cases. Examples of use cases that belong to one or more of these categories are briefly described in Section 1.4.

Service ubiquity. An NTN platform can create large beams to provide radio coverage anywhere on the Earth. Remote land areas, homes, and IoT devices (e.g., devices in smart farms and manufacturing plants) in underserved or unserved rural areas with sparse populations, vast mountains, and oceans can all be illuminated by beams from spaceborne vehicles. Furthermore, the infrastructure of TNs may be damaged or destroyed due to natural disasters (e.g., earthquakes and floods) and armed conflicts such as wars. Through a suitable design, NTN platforms can connect to data networks on the Earth's surface via a gateway. Such a design obviates the need

2 Since the NTN platform carrying the NTN payload can be quite far away from the Earth's surface, a large propagation path loss is expected between the UE and the NTN payload and between the NTN payload and the NTN-GW. Hence, the NTN payload's amplification of the received signal before transmitting such signal is important to maintain a reliable communication link.

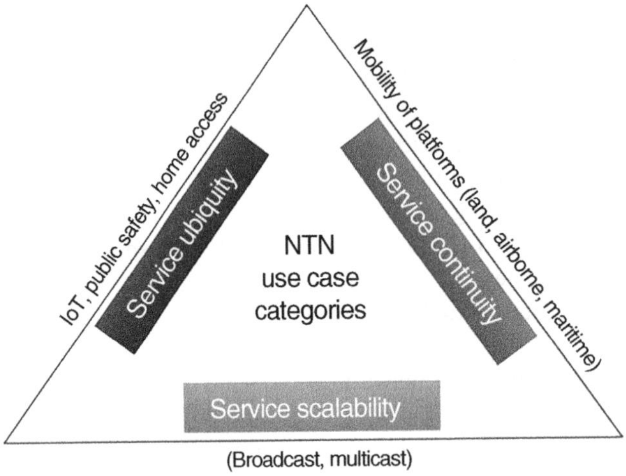

Figure 1.2 NTN Use Case Categories.

for the radio equipment to be near the geographic area that is being served. An NTN thus avoids the constraint a TN faces, where the radio equipment needs to be near or inside the area being served.

Service continuity. There are several cases in which the cellular subscriber moves out of the radio coverage of cellular networks. For example, moving out of a suburban area into a rural area, boarding a cruise ship, or during the takeoff in case of air travel often results in the loss of radio coverage and services. Cellular subscribers and IoT devices may be on moving land platforms (e.g., cars, trains, and trucks), airborne platforms (e.g., commercial or private aircraft), and maritime platforms (e.g., cruise ships and container or cargo ships). An NTN can provide service continuity in such cases by providing radio coverage in the areas that are not covered by a TN.

Service scalability. Cellular systems aim for high efficiency to optimize the network performance and the user experience while meeting resource constraints such as the available amount of spectrum. When the same service content needs to be sent to a target set of users in a large geographic area, all the base stations in such a geographic area need to consume precious radio resources. An NTN cell covers a much larger area compared to a TN cell, resulting in increased efficiency. Examples of the services that can benefit from large-area broadcasts include over-the-air software updates and selected entertainment programs, such as live games and prescheduled movies and shows.

A word of caution...all the target use cases may not be supported in initial NTN deployments. However, as NTN deployments become more capable and widespread, more use cases would become feasible.

1.4 An Overview of NTN Use Cases

An NTN can play an important role in numerous use cases. Table 1.1 lists the use cases identified by the 3GPP as part of a study on using satellite access in 5G (3GPP 2018b).

Table 1.1 Examples of NTN Use Cases.

Use Case ID	Brief Use Case Description
1	Roaming/movement between terrestrial and satellite networks
2	Broadcast and multicast with a satellite overlay
3	Internet of Things with a satellite network
4	Temporary use of a satellite component
5	Optimal routing or steering over a satellite
6	Satellite transborder service continuity
7	Global satellite overlay
8	Indirect connection through a 5G satellite access network
9	5G fixed backhaul between NR and the 5G core
10	5G moving platform backhaul
11	5G to premises
12	Satellite connection of the remote service center to the offshore wind farm

Roaming/movement between terrestrial and satellite networks. A device may move out of the TN coverage area. For example, a container ship may have containers with IoT devices to track the locations of the containers. On the land and near the shore, such IoT devices can connect to the TN. However, when the ship is away from the shore, these devices lose connectivity with a server that tracks the device/container locations. An NTN can be used to provide continuous location tracking. When two different operators own the TN and the NTN, a suitable roaming agreement is needed between the operators, just like two different TN operators. If the same operator owns both the TN and the NTN, a roaming agreement is not needed, and the mobility management is relatively simple.

Broadcast and multicast with a satellite overlay. A variety of digital content, including videos, can be broadcast or multicast to numerous devices simultaneously using a standalone receive-only mode or as a complement to a two-way communication mode. Some content may even be free. A TN may be congested, or a subscriber to the TN operator may be temporarily out of coverage of the TN. In such cases, the NTN can provide broadcast/multicast services to UEs. A UE may also receive services simultaneously from both the TN and the NTN.

Internet of Things (IoT) with a satellite network. IoT devices in a remote area may not have radio coverage from a TN. In such cases, the NTN can provide coverage to the IoT devices. One possible application is location tracking, which is similar to the container ship example discussed above. In smart farms, sensors can be connected to an NTN. GEO satellites or LEO satellites can be used to serve these IoT devices. LEO satellites can be helpful if the transmit power of an IoT device is a challenge. In contrast, a GEO satellite can cover a large area without requiring any significant mobility management-related processing by IoT devices.

Temporary use of a satellite component. In case of a crisis, such as an earthquake, flood, or war, the TN infrastructure in a given geographic area may be partially or fully destroyed. An NTN can be used to provide communication services so that people needing help can be contacted and medicines, food, and other supplies can be delivered. Services such as logistics and security can also be facilitated by the NTN.

Optimal routing or steering over a satellite. Factories in remote areas may not have adequate coverage from a TN. These factories can be connected with each other and a central command or management center via an NTN. A variety of data, including videos and control signaling, can be transported via a satellite. LEO satellites can be used to reduce end-to-end delays and support higher data rates.

Satellite transborder service continuity. This use case involves the use of the NTN for users moving between countries. One TN operator may serve areas in one country, while another TN operator may serve areas in another country. Hence, when a user goes from one country to another on a business or personal travel, an NTN can be used to provide services across countries regardless of which TN operator the user has subscribed to, as long as there are business agreements between TN operators and the NTN operator.

Global satellite overlay. A multinational company typically has offices in different countries with potentially hundreds or thousands of kilometers between the offices. An NTN can be used to connect such offices with custom QoS in support of target applications. Example applications include high-frequency trading (HFT), banking, or corporate communications. For low-latency applications, LEO satellites can be quite attractive.

Indirect connection through a 5G satellite access network. In areas without radio coverage of a TN, some UEs may not have the capability to communicate with an NTN (e.g., legacy pre-Release 17 UEs). Such UEs cannot communicate with far-away UEs or a web server on the Internet. In such cases, an NTN can act as a relay and transport packets between non-NTN-capable UEs and the NTN, enabling even non-NTN UEs to access communication services.

5G fixed backhaul between NR and the 5G core. It is economically challenging to provide TN coverage in remote areas such as villages due to the lack of a suitable transport network, such as a fiber network, and low population density. In one possible solution, a TN base station can connect to the 5GC using a satellite-based backhaul. Such an arrangement provides users in remote areas access to the Internet and the desired services.

5G moving platform backhaul. Trains are an important means of transportation in many countries. A train operator may want to offer entertainment services such as video streaming or other programs or basic Internet access. However, the train may not have coverage throughout its journey, preventing travelers from enjoying a seamless service experience. 5G gNBs can be placed on the train to provide 5G NR access to 5G UEs on the train. The gNBs can connect to the 5GC via a satellite.

5G to premises. In hard-to-reach areas such as remote vacation spots or remote communities, providing TN coverage to homes and businesses is not economically viable. An NTN can provide communications services to such areas without requiring any physical infrastructure in remote areas.

Satellite connection of the remote service center to the offshore wind farm. This use case targets a scenario where a wind farm is located offshore. There is a need to connect this farm to an inland service center. Since the wind farm is offshore, there is no TN available. An NTN can connect the wind farm to a remote service center. Such NTN use cases empower businesses to increase efficiency and economic output.

1.5 3GPP NTN Roadmap

The 3GPP specified the NTN as a new feature in Release 17 and has been enhancing the NTN for Release 18 and beyond. However, the preparatory work on the NTN started before Release 17 (3GPP 2021a). See 5G Americas (2022) for an overview of pre-Release 17 and Release

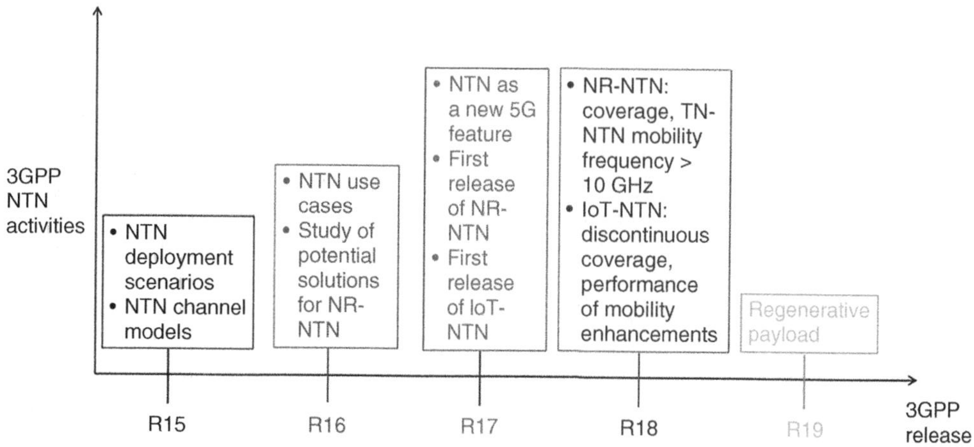

Figure 1.3 3GPP NTN Roadmap.

17 3GPP activities. Figure 1.3 summarizes key 3GPP activities and accomplishments related to the NTN.

1.5.1 Release 15 (R15)

The 3GPP started the work on the NTN in 2017, with a Study Item in Release-15 in 3GPP RAN WG1 that focused on NTN deployment scenarios and channel models (3GPP 2020). Since an NTN possesses unique propagation characteristics compared to a TN, new NTN-specific channel models are needed for the NTN analysis. Details of the NTN channel models can be found in 3GPP (2020).

1.5.2 Release 16 (R16)

The 3GPP Service and System Aspects (SA) working groups discussed use cases for a satellite-based NTN as part of the study item on satellite access in 5G (3GPP 2017). The study identified three main categories of use cases for satellite-based NTN service continuity, service ubiquity, and service scalability (3GPP 2018b), which are discussed in Section 1.2.

The work in R16 also included a comprehensive study in the radio access network (RAN) working groups on solutions for the NR to support an NTN to determine the necessary features to enable the NR support for the NTN (3GPP 2019). These studies provided a baseline for the normative work on the NR-NTN in R17 for the transparent payload (3GPP 2023).

1.5.3 Release 17 (R17)

The 3GPP introduced the NTN as a new feature in R17 by developing specifications for the NR-based NTN (3GPP 2024c). The R17 NTN specifications create a strong foundation for future enhancements in subsequent 3GPP releases. R17 NTN specifications assume UEs with a global navigation satellite system (GNSS) capability and a transparent NTN payload. R17 NTN enhancements to the basic NR specifications are related to synchronization, timing adjustments, system information, protocol timers, hybrid automatic repeat request (H-ARQ), timing and frequency precompensation by UEs, tracking or registration area management, handover, QoS, user location reporting, and feeder link management.

The SA Working Group 2 (SA2) created a study item on architecture aspects for using satellite access in 5G (3GPP 2018a) to identify key issues for satellite-based NTN in the 5G system architecture and provide solutions for both direct satellite access and satellite backhaul (3GPP 2021b). The SA2 normative work in R17 includes mobility management with large coverage areas and with moving coverage areas, delay in satellites, QoS with satellite access and with satellite backhaul, RAN mobility with non-geostationary orbit (NGSO) regenerative-based satellite access, and regulatory services with supernational satellite ground station. SA WG5 addressed management and orchestration aspects in their study (3GPP 2021d). The 3GPP Core Network and Terminals (CT) working groups created a study item to address aspects of the 5GC architecture for satellite networks. The Release17 work in CT working groups concentrates on aspects related to the public land mobile network (PLMN) selection (3GPP 2021c).

The 3GPP also worked on the IoT-NTN in R17. This work is based on the previously defined 3GPP Release 13 features of narrowband IoT (NB-IoT) and enhanced machine-type communication (eMTC) (also referred to as long-term evolution- machine type communication [LTE-M]) (3GPP 2021e). Enhancements to NB-IoT and eMTC specifications are made to support these features in an NTN. Since NB-IoT and eMTC are based on 4G LTE specifications, the NTN makes use of the 4G Evolved Packet Core (EPC) instead of the 5GC. The IoT-NTN work utilizes the studies carried out for NR-NTN and NR-NTN specifications as a baseline and adds enhancements unique to NB-IoT and eMTC.

1.5.4 Release 18 (R18)

The NTN enhancements continue in 5G-Advanced. 5G-Advanced refers to the 5G specifications defined in Release 18 and beyond. In Release 18, key NTN enhancements for the NR-NTN include coverage enhancement to support voice and low-rate data services, mobility and service continuity enhancements for mobility between the TN and the NTN, support for frequencies greater than 10 GHz, and network verification of the UE location. The IoT-NTN enhancements in Release 18 include discontinuous coverage enhancements (e.g., support for the UE movement across areas with and without NTN coverage), mobility enhancements (e.g., neighbor cell measurements), and performance enhancements (e.g., disabling of H-ARQ feedback and improved GNSS operations).

1.5.5 Release 19 (R19) and Beyond

The Release 19 workshop held by the 3GPP in June 2023 identified the NTN as one of the candidate areas for normative specifications in Release 19. The R19 work began in summer 2024 and is expected to complete in late 2025. The NR-NTN enhancements in Release 19 aim to enhance downlink coverage, enhance uplink capacity, support the multimedia broadcast service (MBS), support selected 5G network functions (NFs) on the NTN platform, including a regenerative payload, and support reduced capability (RedCap) UEs (3GPP 2024a). Two key enhancements targeted by the 3GPP for the IoT-NTN Phase 3 in Release 19 include the support for store and forward and uplink capacity enhancements (3GPP 2024b). See Chapter 10 for more details on Release 19 NTN enhancements.

1.6 Role of the NTN in 6G

While the 3GPP's NTN specifications for 5G provide a strong foundation for NTN deployments, 6G is expected to significantly expand the scope and applicability of the NTN in practice. 6G is anticipated to be deployed in the 2030s. While specifications for 6G are not yet being worked on, organizations worldwide, such as Alliance for Telecommunications Industry Solutions (ATIS) in North America, are defining a vision and requirements for 6G. Several organizations have published white papers describing their 6G vision and mentioned the NTN as one of the important aspects of 6G. See Chapter 10 for more details on 6G.

ATIS has an initiative called Next G Alliance that has defined six audacious goals for 6G systems (NGA 2022a). These goals include (i) trust, security, and resilience; (ii) enhanced digital world experience; (iii) cost-efficient solutions; (iv) distributed cloud and communications; (v) AI-native networks; and (vi) sustainability related to energy efficiency and the environment. The NTN can play an important role in achieving resilience in 6G by providing an additional mode of communications beyond terrestrial communications to users including public safety personnel. As part of distributed sensing and communications, the NTN can help bridge the digital divide by providing communications services to unserved or underserved geographical areas on the planet. In particular, the NTN can provide cost-effective coverage when it is too expensive to deploy a TN.

Indeed, the NGA has identified the NTN as one of the key technologies to meet the requirements of 6G (NGA 2022b). The NGA has identified numerous technologies for 6G and classified them into the following five technical areas: (i) component technologies, (ii) radio technologies, (iii) system and network architecture, (iv) OA&M and service enablement, and (v) trustworthiness. The NTN plays an important role in the technical areas of "Radio Technologies" and "System and Network Architecture." The NTN represents a new type of connectivity in the context of radio technologies to meet varying user traffic demands and extremely wide coverage area requirements. The NTN also fundamentally changes the typical RAN architecture of a TN because of the connectivity between the UE and the NTN payload and between the NTN payload and the NTN-GW. The alternative coverage provided by the NTN in 6G is important to ensure global coverage.

1.7 Major Takeaways

The NTN is expected to play an important role as a key component of 5G and 6G in advancing the reach of wireless communications. Here are the major takeaways from this chapter.

- Even with the deployment of five generations of cellular technologies, a significant portion of the planet's surface area and a large number of people still do not have access to wireless communications.
- An NTN is a wireless communication network that utilizes radio network equipment placed on an airborne or spaceborne vehicle.
- The 3GPP formally defines the NTN as "an NG-RAN consisting of gNBs, which provide non-terrestrial NR access to UEs by means of an NTN payload embarked on an airborne or spaceborne NTN vehicle and an NTN Gateway."
- The 3GPP considers platforms such as several types of satellites and HAPS as part of the NTN. A HAPS is an airborne vehicle that operates at an altitude between 8 and 50 km. Examples of the satellites include LEO, MEO, and GEO satellites. Hence, drones or UASs that operate below the 8 km altitude are not formally part of the 3GPP NTN activities.
- Key elements of the NTN architecture include the NTN-capable UE, the NTN payload on an NTN platform, the NTN-GW, the non-NTN processing of the gNB/eNB, and the 5GC/EPC.
- The motivation for the NTN stems from the use case categories of service ubiquity, service continuity, and service scalability. Service ubiquity refers to the coverage in geographic areas unserved or underserved by a traditional TN. Service continuity enables continuous wireless

connectivity for the user as the user moves between the TN and the NTN. Service scalability refers to a large number of users in a large geographic area for broadcast/multicast services, increasing the resource utilization efficiency.
- The 3GPP has studied several NTN use cases, including user mobility between a TN and an NTN, satellite-based backhaul to support TN deployments in remote areas without transport infrastructure, support for IoT devices, factories in remote areas, and offshore wind farms.
- The 3GPP started the work on the NTN in Release 15 and introduced the NTN as a new feature in Release 17. The 3GPP R15 identified NTN deployment scenarios and NTN-specific channel models. The NTN use cases were identified, and potential solutions for NTN–NTN were studied in the 3GPP R16. Both NR-NTN and IoT-NTN were specified as formal features in the 3GPP R17. The enhancements in the 3GPP R18 are related to coverage, TN-NTN mobility, frequencies above 10 GHz for NR-NTN, and discontinuous coverage and mobility for IoT-NTN. The 3GPP R19 makes additional NTN enhancements, including support for regenerative payloads.
- While the 3GPP NTN specifications are part of 5G and 5G-Advanced, the NTN is expected to play a vital role in 6G as well by providing alternative coverage and resiliency.

References

3GPP (2017). *New WID on study using satellite access in 5G (FS_5GSAT)*. SP-170788. https://www.3gpp.org/ftp/tsg_sa/tsg_sa/TSGS_77/Docs (accessed 4 August 2024).

3GPP (2018a). *Update to FS_5GSAT_ARCH: Study on architecture aspects for using satellite access in 5G*. SP-181253. https://www.3gpp.org/ftp/tsg_sa/TSG_SA/TSGS_82/Docs (accessed 4 August 2024).

3GPP (2018b). *Study on using satellite access in 5G*. TR22.822. https://portal.3gpp.org/desktopmodules/Specifications/SpecificationDetails.aspx?specificationId=3372 (accessed 4 August 2024).

3GPP (2019). *Study on solutions for NR to support Non-Terrestrial Networks (NTN)*. RP-190710. https://www.3gpp.org/ftp/tsg_ran/TSG_RAN/TSGR_83/Info_for_workplan/revised_SID_5/RAN3_1 (accessed 4 August 2024).

3GPP (2020). *Study on New Radio (NR) to support non-terrestrial networks*. TR38.811. https://portal.3gpp.org/desktopmodules/Specifications/SpecificationDetails.aspx?specificationId=3234 (accessed 4 August 2024).

3GPP (2021a). *Solutions for NR to support Non-Terrestrial Networks (NTN)*. RP-211557. 3GPP Work Item Description. 3GPP TSG RAN meeting #91-e, e-meeting, March 22–26, 2021, https://www.3gpp.org/ftp/tsg_ran/TSG_RAN/TSGR_92e/Docs (accessed 4 August 2024).

3GPP (2021b). *Study on architecture aspects for using satellite access in 5G*. TR23.737. https://portal.3gpp.org/desktopmodules/Specifications/SpecificationDetails.aspx?specificationId=3485 (accessed 4 August 2024).

3GPP (2021c). *Study on PLMN selection for satellite access*. TR24.821. https://portal.3gpp.org/desktopmodules/Specifications/SpecificationDetails.aspx?specificationId=3811 (accessed 4 August 2024).

3GPP (2021d). *Study on management and orchestration aspects of integrated satellite components in a 5G network*. TR28.808. https://portal.3gpp.org/desktopmodules/Specifications/SpecificationDetails.aspx?specificationId=3617 (accessed 4 August 2024).

3GPP (2021e). *Study on Narrow-Band Internet of Things (NB-IoT) / enhanced Machine Type Communication (eMTC) support for Non-Terrestrial Networks (NTN)*. TR36.763. https://portal.3gpp.

org/desktopmodules/Specifications/SpecificationDetails.aspx?specificationId=3747 (accessed 4 August 2024).

3GPP (2023). *Solutions for NR to support non-terrestrial networks (NTN).* TR38.821. https://portal.3gpp.org/desktopmodules/Specifications/SpecificationDetails.aspx?specificationId=3525 (accessed 4 August 2024).

3GPP (2024a). *Revised WID: Non-Terrestrial Networks (NTN) for NR Phase 3.* RP-240775. https://www.3gpp.org/ftp/tsg_ran/TSG_RAN/TSGR_103/Info_for_workplan/revised_WID_54/RAN2_5 (accessed 4 August 2024).

3GPP (2024b). *Revised WID on Non-Terrestrial Networks (NTN) for Internet of Things (IoT) Phase 3.* RP-240776. https://www.3gpp.org/ftp/tsg_ran/TSG_RAN/TSGR_103/Info_for_workplan/revised_WID_54/RAN2_5 (accessed 4 August 2024).

3GPP (2024c). *System architecture for the 5G system (5GS).* TS 23.501. https://portal.3gpp.org/desktopmodules/Specifications/SpecificationDetails.aspx?specificationId=3144 (accessed 4 August 2024).

3GPP (2024d). *Evolved Universal Terrestrial Radio Access (E-UTRA) and Evolved Universal Terrestrial Radio Access Network (E-UTRAN): Overall description.* TS36.300. https://portal.3gpp.org/desktopmodules/Specifications/SpecificationDetails.aspx?specificationId=2430 (accessed 4 August 2024).

3GPP (2024e). *NR and NG-RAN overall description.* TS38.300. https://portal.3gpp.org/desktopmodules/Specifications/SpecificationDetails.aspx?specificationId=3191 (accessed 4 August 2024).

5G Americas (2022). 5G and Non-Terrestrial Networks. White Paper. February 2022, https://www.5gamericas.org/wp-content/uploads/2022/01/5G-Non-Terrestrial-Networks-2022-WP-Id.pdf (accessed 4 August 2024).

NGA (2022a). Roadmap to 6G. Next G Alliance Report. February 2022. https://www.nextgalliance.org/wp-content/uploads/2022/02/NextGA-Roadmap.pdf (accessed 4 August 2024).

NGA (2022b). 6G Technologies. Next G Alliance Report. June 2022, https://www.nextgalliance.org/white_papers/6g-technologies/. https://www.nextgalliance.org/wp-content/uploads/2022/02/NextGA-Roadmap.pdf (accessed 4 August 2024).

Tripathi, N.D. and Reed, J.H. (2019). *5G Cellular Communications: Journey and Destination.* The Wireless University. https://thewirelessuniversity.com/ (accessed 4 August 2024).

2

Types of NTN Platforms

The NTN provides significant flexibility to operators by supporting different types of NTN platforms. Examples of NTN platforms include different types of satellites such as Geosynchronous Earth Orbit (GEO) satellites, Medium Earth Orbit (MEO) satellites, Low Earth Orbit (LEO) satellites, and High-Altitude Platform Stations (HAPSs). This chapter discusses the distinct characteristics of NTN platforms. The NTN platform, based on its capability and design, illuminates its target coverage area using a specific type of beam. Three types of beams are illustrated in this chapter: Earth-fixed beams, Earth-moving beams, and Quasi-Earth-fixed beams. Satellite systems that existed prior to the deployment of the 3GPP NTN deployments are also briefly described. Finally, the key takeaways of this chapter are summarized.

2.1 Types of NTN Platforms

Two main types of NTN platforms considered by the 3GPP are satellites and Uncrewed/Unmanned Aerial Systems (UASs), including HAPSs (3GPP 2023). Figure 2.1 shows examples of these NTN platforms.

Satellites are spaceborne vehicles that orbit the Earth using a specific orbit. A satellite can be classified as a geosynchronous orbit (GSO) satellite or a non-geosynchronous orbit (NGSO) satellite. A GEO satellite is an example of a GSO satellite. A LEO satellite, a MEO satellite, and a Highly Elliptical Orbit (HEO) satellite are examples of NGSO satellites.

Since the orbital movement of a GSO or GEO satellite matches the Earth's rotation, the GSO/GEO satellite appears to be fixed to a stationary observer on the Earth's surface. The GSO/GEO satellite is situated at an altitude of 35,786 km. Organizations such as Intelsat own and manage GEO satellites.

LEO satellites are situated at an altitude ranging from about 300 to 1,200 or 1,500 km. The International Space Station (ISS) and the Hubble Space Telescope utilize the LEO orbit. The LEO satellites orbit the Earth at a speed of approximately 7.8 km/s and take around 90 minutes to circle the Earth. There has been significant interest in LEO-based systems. Examples of such deployments are Kuiper by Amazon, Starlink by SpaceX, and OneWeb.

MEO satellites are situated at an altitude ranging from about 5,000 or 7,000 to 20,000 or 25,000 km. The US global positioning system (GPS) utilizes MEO satellites.

The orbit of a HEO satellite is highly elongated. Its altitude varies from about 400 to 50,000 km. Sirius Satellite Radio utilizes HEO satellites. The 3GPP has focused on LEO satellites and GEO satellites as two extremes while defining NTN specifications. Hence, HEO satellites have not been explicitly investigated by the 3GPP. As HEO satellites have significant variations in altitude, the

3GPP-based Non-Terrestrial Networks in 5G and 6G: Expanding the Frontiers of Wireless Communications, First Edition.
Nishith D. Tripathi and Jeffrey H. Reed.

Companion website: www.wiley.com/go/tripathi5g

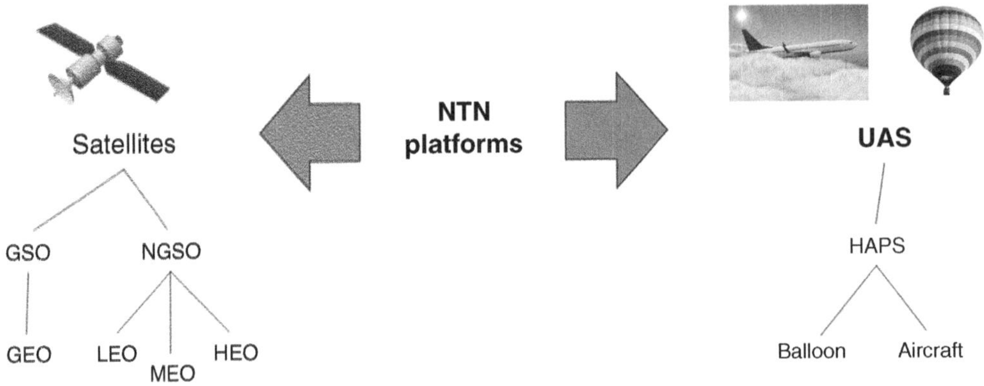

Figure 2.1 Types of NTN Platforms. *Source:* sdecoret / Adobe Stock Photos; littlestocker / Adobe Stock Photos.

propagation delays would have significant variations. Furthermore, the propagation path loss for HEO satellites can be expected to be smaller or larger compared to LEO satellites and much smaller compared to GEO satellites.

Operators can pursue multi-orbit satellite constellations to derive benefits from diverse types of satellites (OceanWeb 2023). For example, Inmarsat is promoting Orchestra as a communications network of the future that would combine GEO satellites, LEO satellites, and terrestrial 5G networks to realize an efficient and cost-effective solution. The European Space Agency initiated a study of multi-orbit satellite systems. Multi-orbit satellite systems also offer resiliency in military applications.

From the 3GPP NTN specifications perspective, a UAS platform is an airborne vehicle situated at an altitude ranging from about 8 to 50 km. As mentioned in Chapter 1, drones or uncrewed/ unmanned aerial vehicles (UAVs) are not within the scope of the 3GPP's NTN activities. For example, a HAPS may be located at an altitude of 20 km. Such a UAS platform maintains a fixed position and hence appears stationary to an observer on the Earth's surface. Examples of HAPS include balloons and aircraft, including solar-powered aircraft (Wikipedia 2024a). For example, Google experimented with balloons as part of the Loon project (Next G Alliance 2024). A HAPS is widely used in military applications. For example, the United States Air Force uses Northrop Grumman RQ-4 Global Hawk UAV.

The HAPS Alliance is an organization that aims to accelerate the commercial adoption of HAPS technologies and build a robust HAPS ecosystem. The HAPS Alliance includes companies from diverse industries such as telecommunications and aerospace. The HAPS Alliance promotes the use of HAPS in the stratosphere. The HAPS Alliance members advocate the advantages of HAPS to authorities in different countries, build a cooperative HAPS ecosystem, develop common product specifications, and promote the standardization of the HAPS network interoperability (HAPS Alliance 2024a).

2.2 Characteristics of Satellites

Different satellites possess different characteristics and hence have different implications on the system design and supported applications or services. Table 2.1 summarizes key characteristics of three widely used types of satellites (SES 2020; Stuhlfauth 2022).

Table 2.1 Characteristics of GEO, MEO, and LEO Satellites.

Characteristic/ Satellite Type	GEO Satellite	MEO Satellite	LEO Satellite
Altitude	35,786 km	5,000 or *7,000* km to 20,000 or *25,000* km	*300* or 500 km to 1,200 or *1,500* km
Appearance of the satellite to a fixed observer	Stationary	Slow moving	Fast moving
Types of beams	Earth fixed	Quasi-Earth fixed or Earth moving	Quasi-Earth fixed or Earth moving
Propagation delay	Large (119.3 ms)	Medium (23.3 to 83.2 ms)	Small (1 to 5 ms)
Propagation path loss	Large	Medium	Small
Coverage of Earth's surface	Large	Smaller than GEO systems	Smaller than GEO systems
Number of satellites for full Earth coverage	3	6	Hundreds
Number of Earth gateways	Very few	Few	Numerous
Types of supported services	Relatively low-rate and delay-tolerant services	Relatively medium-rate and medium-latency services	Relatively higher-rate and low-latency services
Life expectancy	Long	Medium	Short: Less than 10 years (e.g., five)
Size and weight	Large and heavy	Medium	Small and light

GEO satellites have a much higher altitude than MEO and LEO satellites. A GEO satellite orbits around the Earth at an altitude of 35,786 km. In contrast, LEO satellites orbit at altitudes ranging from 300 to 1,500 km, and MEO satellites orbit at altitudes ranging from 7,000 to 25,000 km. The altitude of an MEO satellite is much smaller than the altitude of an LEO satellite but significantly smaller than a GEO satellite's altitude.

A GEO satellite appears stationary to a fixed observer on the Earth's surface, while MEO and LEO satellites appear to be moving to a fixed observer on the Earth's surface. The relative movement between the satellite and the user device causes a Doppler shift. The velocities of MEO and LEO satellites usually dominate the typical velocities of user devices communicating with these satellites. An LEO satellite travels at about 7.78 km/s or 28,000 km per hour, while an MEO satellite travels at about 3.83 km/s or 13,800 km per hour (Stuhlfauth 2022). The orbital period of an MEO satellite ranges from 6 to 12 hours, and the orbital period of an LEO satellite is about 90 minutes (Stuhlfauth 2022).

The altitude of the satellite and the relative movement or velocity of the satellite relative to the user device influence several design aspects and operations of the NTN.

A GEO satellite utilizes Earth-fixed beams, while MEO and LEO satellites use quasi-Earth-fixed beams or Earth-moving beams. See Section 2.4 for details on these beam types.

The propagation delay between the user device and the satellite is much longer for GEO satellites than MEO and LEO satellites due to the much higher altitude of GEO satellites compared to MEO and LEO satellites. For example, the one-way propagation delay between the user device and a satellite is approximately 119.3 ms for a GEO satellite, 1 to 5 ms for an LEO satellite, and 23.3 to 83.3 ms for an MEO satellite.

The large distance between a user device on Earth and a GEO satellite leads to large propagation path loss. A TN has significant cluttering in the propagation path between the transmitter and the receiver, leading to a non-line of sight (NLOS) propagation environment. This results in significant attenuation of the signal. For example, the path loss exponent that influences distance-based path loss is quite high (e.g., n = 3.5) in a TN, leading to significant attenuation of the transmit signal. In contrast, the propagation path between the transmitter and the receiver tends to be line of sight (LOS) between the user device and the NTN platform, leading to a free-space path loss exponent corresponding to n = 2. Hence, from the perspective of the path loss environment, the NTN has a more benign radio environment than the TN. However, transmitter–receiver distances are much larger in an NTN compared to a TN, causing much larger path losses. A GEO satellite-based NTN experiences the largest path loss due to a very high altitude. The path loss is smaller for an MEO satellite-based NTN and even smaller for an LEO satellite-based NTN.

As LEO satellites are situated much closer to the Earth compared to MEO and LEO satellites, their coverage area is relatively smaller than GEO and MEO satellites. For example, LEO and MEO satellites have the largest dimension of the (elliptical) beam, ranging from 100 to 1,000 km.

Hence, to provide full coverage of the Earth's surface, hundreds or even thousands of LEO satellites are needed[1]. In contrast, six MEO satellites or three GEO satellites could cover the entire Earth. OneWeb's planned LEO satellite system is expected to consist of more than 600 LEO satellites; Starlink has applied for a license to deploy more than 40,000 (LEO) satellites (OceanWeb 2023). LEO satellites can provide suitable radio coverage in polar regions.

Satellites use Earth gateways to connect the user device to networks such as the Internet. Based on the coverage area of a satellite's beams, satellite altitude, and the mobility of satellites, fewer Earth gateways are adequate for GEO satellite-based NTN, while a large number of Earth gateways are needed for an LEO satellite-based NTN.

In general, GEO satellites are suitable for relatively low-rate and delay-tolerant services, and LEO satellites are suitable for relatively higher-rate and low-latency services. MEO satellites can support relatively medium-rate and medium-latency services.

LEO satellites typically have a lower life span (e.g., five years) than GEO satellites due to harsher environmental conditions such as radiation and atmospheric drag (OceanWeb 2023).

LEO satellites can be quite small and lightweight and hence are cost-efficient to produce. An LEO satellite's perimeter may be even less than a meter and even dozens of centimeters in the case of nanosatellites (Stuhlfauth 2022). Furthermore, multiple LEO satellites can be launched during a single trip of a large rocket (Sodders 2023). A high propulsion rocket is not required to launch LEO satellites, reducing the cost.

2.3 Characteristics of UASs of the NTN

As mentioned in Chapter 1, the UAS considered in this book as part of the NTN follows the 3GPP approach and does not consider low-altitude drones as part of the NTN. A HAPS is used as an example UAS in the discussions below.

Three categories of HAPS flight vehicles considered by the HAPS Alliance include fixed-wing, balloon, and airship (HAPS Alliance 2024b). The fixed-wing vehicle utilizes propulsion (i) to

1 LEO satellites can also provide suitable radio coverage in polar regions.

Table 2.2 Characteristics of HAPS-Based UAS.

Characteristic	Specifics
Altitude	8 to 50 km
Appearance of the HAPS to a fixed observer	Stationary
Types of beams	Earth fixed
Propagation delay	Comparable to TNs
Propagation path loss	Smaller than LEO systems
Type of coverage	On demand, area specific, and mission specific
Coverage area	Comparable to a TN and smaller than satellites
Data rate	Higher data rate and lower latency compared to LEO systems
Latency	Higher data rate and lower latency compared to LEO systems
Types of supported services	Telecommunications, high-resolution earth observation, and weather prediction and modeling
Endurance (service duration)	Tens of hours to a few days

generate aerodynamic lift and (ii) for lateral control. Control surfaces influence both attitude and altitude. In the case of a balloon, lighter-than-air gas is used to generate buoyant lift, and the vehicle may have altitude control by changing the net buoyancy. Lighter-than-air gas is used in an airship to generate buoyant lift. The airship has altitude control, and lateral control is achieved through propulsion.

Table 2.2 lists key characteristics of the HAPS-based UAS from the perspective of the NTN (Stuhlfauth 2022).

The HAPS typically operates in the stratosphere. The lowest altitude of the stratosphere ranges from 6 to 20 km, and the highest altitude of the stratosphere is approximately 50 km. The 3GPP focuses on the altitude between 8 and 50 km for the UAS, with the 20-km nominal altitude for the HAPS (3GPP 2023). The device-to-HAPS distance is much shorter than even the device-to-LEO satellite distance because of the HAPS altitude being much smaller than the LEO satellite altitude.

The HAPS maintains its position relative to the Earth's surface and hence appears as stationary to a fixed observer on the Earth's surface. The HAPS uses Earth-fixed beams like a GSO satellite.

As the device-to-HAPS distance is on the order of a few kilometers, the propagation delay for the HAPS-based NTN can be considered comparable to TNs. Furthermore, due to short distances, the propagation path loss for the HAPS systems is much smaller than that for LEO systems.

The HAPS coverage can be viewed as on-demand coverage, where the HAPSs are positioned in suitable locations for area-specific and mission-specific coverage. For example, in the case of temporary coverage of fairs and public events in large spaces, the HAPSs can provide coverage. The HAPS can also provide valuable coverage in support of Public Safety missions or military missions. The HAPS may also provide wireless access in rural and remote areas (e.g., coverage over the sea) (GSMA 2022).

A HAPS typically covers a larger area compared to a TN gNB and a smaller area than a satellite.

The radio environment in the HAPS-based systems is conducive to higher data rates and lower latency than even LEO systems.

In addition to supporting telecommunications services, the HAPS may also support high-resolution Earth observation and weather prediction and modeling.

A typical HAPS is expected to have a much shorter operational life than satellites. A typical HAPS has a service duration or endurance ranging from tens of hours to few days compared to multiple years of the operational life for satellites. Hence, to provide coverage continuously in a given geographic area, a new HAPS replaces an old HAPS when the endurance of an outgoing HAPS is reached.

The launch costs for the UAS, such as the HAPS, are generally much lower than satellite launch costs.

2.4 Types of Beams

Three types of beams can be used in an NTN to provide radio coverage: Earth-fixed beams, Earth-moving beams, and quasi-Earth-fixed beams (3GPP 2024). These beams are explained later. A given NTN payload utilizes one of such beams based on its capabilities and configuration. These beam types are also referred to as types of NTN cells or types of service links.

Figure 2.2 illustrates the concept of Earth-fixed beams.

An Earth-fixed beam covers a fixed region on the Earth's surface at all times. For example, the NTN platform, such as a GSO/GEO satellite or a HAPS, provides radio coverage to the same geographic area at all time instants t_1, t_2, ..., t_n and so on. These platforms are stationary in space or air and create fixed beams toward the Earth's surface.

Fixed beams significantly simplify operations and the design of the 5G NTN, such as mobility management, registration area management, and radio frequency (RF) planning and design. Fixed beams also reduce the overall processing requirements for the device and the radio and core networks.

Quasi-Earth-fixed beams are depicted in Figure 2.3.

A quasi-Earth-fixed beam covers a given geographic area X during the period from t_1 to t_2. Then, such beam covers a different geographic area Y during the period from t_2 to t_3. During the period from t_2' to t_3', a different quasi-Earth-fixed beam covers the geographic area X. The time overlap

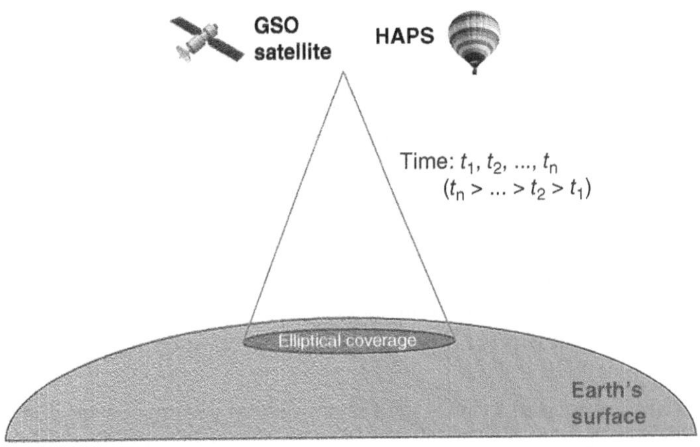

Figure 2.2 Earth-Fixed Beams in an NTN. *Source:* littlestocker / Adobe Stock Photos.

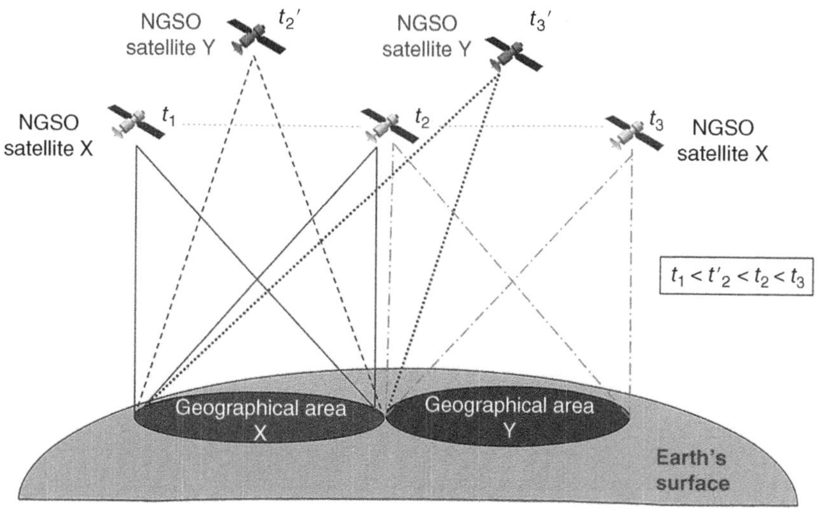

Figure 2.3 Quasi-Earth-Fixed Beams in an NTN.

between the outgoing and incoming beams (i.e., t_2' to t_2 in Figure 2.3) enables the transfer of user devices from one beam to the other. NTN platforms such as NGSO satellites (e.g., MEO and LEO satellites) may have beam steering capability such that the satellites can continue to illuminate the same geographic area[2] even as the satellites move from one location to another in their orbits. In summary, to provide coverage continuously in a given geographic area, an incoming quasi-Earth-fixed beam replaces an outgoing quasi-Earth-fixed beam.

Quasi-Earth-fixed beams increase the complexity of operations and the design of the 5G NTN and require significant changes in mobility management, registration area management, and RF planning and design compared to a TN. Quasi-Earth-fixed beams also increase the overall processing requirements for the device and the radio and core networks.

Figure 2.4 shows Earth-moving beams.

An Earth-moving beam covers a given geographic area X at the instant t_1. Then, such a beam covers a slightly different geographic area Y at the very next instant t_2. The beam covers a yet another geographic area Z at the next instant t_3. NTN platforms such as NGSO satellites (e.g., MEO and LEO satellites) may use fixed beams, and the movement of the NTN platform leads to different geographic areas being illuminated at different instants. While one Earth-moving beam ceases to cover a geographic area, another Earth-moving beam starts covering such a geographic area. For example, at the instant t_2, when NGSO satellite 1 covers the geographic area Y, NGSO satellite covers geographic area X′ to provide continuous and contiguous coverage. In summary, to provide coverage continuously in a given geographic area, an incoming Earth-moving beam replaces an outgoing Earth-moving beam.

Like quasi-Earth-fixed beams, Earth-moving beams also increase the complexity of operations and the design of the 5G NTN and require significant changes in mobility management, registration area management, and RF planning and design compared to a TN. Furthermore, Earth-moving

2 Note that the beams may not have identical coverage area from one elevation angle to another elevation angle due to the stretching of the elliptical beams for lower elevation angles. However, the beam covers approximately the same geographic area for a finite period. An elevation angle is the angle between the horizon of the user device and the line connecting the user device and the NTN platform.

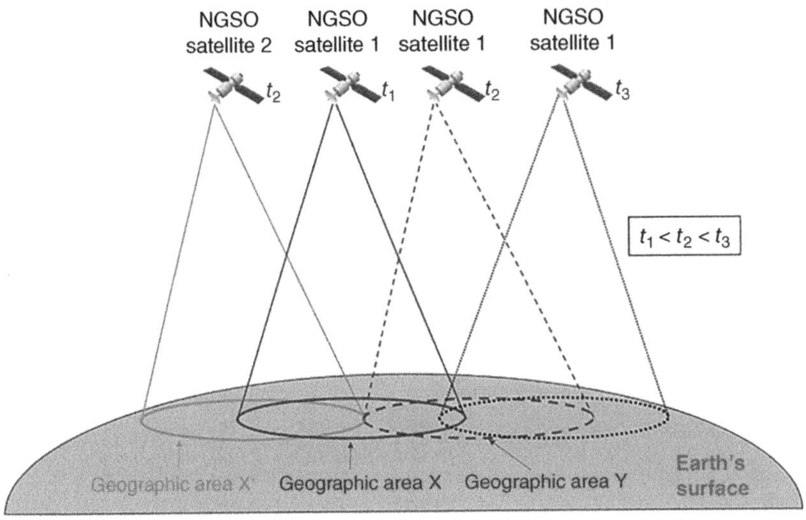

Figure 2.4 Earth-Moving Beams in an NTN.

beams increase the overall processing requirements for the device as well as the radio and core networks.

Note that the illustrations in Figures 2.2, 2.3, and 2.4 show continuous and contiguous coverage. In practice, such ideal coverage may not be feasible due to inadequate NGSO satellites, achievable beam patterns on the Earth's surface, and RF planning and design constraints.

2.5 An Overview of Pre-5G Satellite Systems

5G-based NTNs will be a common sight in the coming years. Both 5G-based NTNs and non-5G-based NTNs are expected to coexist. Several pre-5G-based satellite systems are already operating. Some of these systems may transition from the currently used (usually) non-5G systems to 5G-based systems due to the flexibility, efficiency, and ecosystem diversity of 5G. Figure 2.5 provides examples of pre-5G satellite constellations classified into established and emerging categories. Established constellations have been operating for a few decades while emerging constellations are

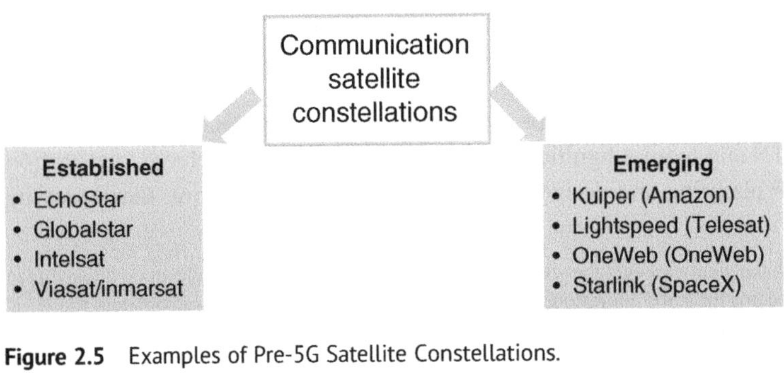

Figure 2.5 Examples of Pre-5G Satellite Constellations.

being deployed in 2020s. Note that the intent here is to show alphabetical examples of constellations; not all existing or planned constellations are mentioned.[3]

Many of the established constellations focused on GEO satellites earlier while emerging constellations have been focusing on LEO satellites. Examples of established satellite constellations include EchoStar, Globalstar, Intelsat, and ViaSat/Inmarsat. Kuiper, Lightspeed, OneWeb, and Starlink are examples of emerging satellite constellations. Established satellite constellations are discussed first, followed by emerging satellite constellations.

EchoStar provides satellite communication services through Hughes Network Systems and EchoStar Satellite Services business segments. Echostar and Dish Network merger was completed in January 2024. EchoStar provides satellite Internet service through HughesNet. Examples of Hughes satellites include GEO satellites such as Jupiter 1, Jupiter 2, and Jupiter 3 (Hughes 2024). These satellites offer connectivity for consumer broadband, enterprise networking, cellular backhaul, emergency communications, and aero and maritime mobility applications (GSMA 2022). EchoStar satellite services provides satellite communications infrastructure and solutions to media and broadcast organizations, enterprise customers, and US government and military service providers (Echostar 2024). As of early 2024, the EchoStar fleet includes 10 Ku-fixed satellite service (FSS), Ka-FSS, and S-band mobile satellite service (MSS) GEO satellites.

Globalstar offers three main types of satellite communication services—IoT services, messaging services, and voice/data services (Globalstar 2024). IoT devices can be used for asset tracking and monitoring. Messaging services support emergency services such as SOS alerts. Portable satellite phones can connect to Globalstar satellites for voice and data services. Globalstar satellites cover over 80% of the Earth's surface. Code Division Multiple Access (CDMA) is used by satellites. Twenty-four ground stations are used between LEO satellites and on-the-ground communications infrastructure on six continents. The IP Multimedia Subsystem (IMS) is used as the services network to enable low-cost local access for voice calls.

Intelsat offers services using GEO and LEO satellites and is also considering the deployment of MEO satellites (Intelsat 2024). Intelsat's satellites utilize C-, Ku-, and Ka-bands. Intelsat makes use of wide beams as well as spot beams. Intelsat is the largest provider of satellite communications to the US government. Intelsat also provides broadband connectivity to maritime passengers and crew. Furthermore, Intelsat provides in-flight connectivity (IFC) to more than 20 commercial airline partners and nearly 3,000 aircraft (Intelsat 2024).

Viasat offers services to consumers (e.g., home Internet), businesses (e.g., IFC) for aircraft, and the government (e.g., military applications) (Viasat 2024). Viasat provides coverage for 99% of US population, offering Internet access services even in hard-to-reach areas. Viasat also provides connectivity to unserved and underserved communities around the world. Viasat utilizes high-capacity GEO satellites in its GEO constellation and associated ground network to provide broadband services.

Project Kuiper by Amazon aims to provide satellite broadband Internet access to unserved or underserved communities in the United States and around the world (Kohnstamm 2023). Project Kuiper customers can obtain services such as Internet access, video calls, gaming, and high-definition streaming. Amazon obtained a license from the Federal Communications Commission (FCC) in July 2020 to deploy and operate Project Kuiper satellites. Amazon launched the first two prototype satellites on October 6, 2023. Project Kuiper expects to begin offering services to customers by the end of 2024. Project Kuiper carries out in-house design and development of ground infrastructure, satellites, and customer terminals. The initial satellite constellation design of Project Kuiper's includes 3,236 LEO satellites. These satellites will orbit at an altitude ranging from 590 to

3 Additional examples of existing constellations include Ligado Networks and Eutelsat.

630 kms. Target peak data rates are 100 Mbps for an ultracompact model, 400 Mbps for a standard model, and 1 Gbps for the largest model.

Telesat has developed a global network of 198 LEO satellites (Telesat 2024). Optical inter-satellite links (ISLs) are used to create a global mesh network. Multiple orbits are utilized to cover regions, including polar regions. Digital modulation, demodulation, and routing are carried out in space to achieve higher capacity and increased flexibility. Hopping beams are created using phased antenna arrays to dynamically focus capacity where needed. Telesat collaborates with terminals manufacturers that can connect to the Telesat Lightspeed satellite network. Metro Ethernet Forum (MEF) compliant interfaces are supported for seamless integration to the ground infrastructure network. Telesat's first LEO satellite was launched in January 2018. Telesat supports live demonstrations for various markets and applications.

Eutelsat combined its GEO satellite business with OneWeb's LEO satellite business in September 2023 to create Eutelsat OneWeb. Eutelsat OneWeb's LEO constellation design consists of more than 630 LEO satellites along 12 orbital planes at an altitude of 1,200 km (Eutelsat 2024). Each LEO satellite connects to the user terminals and the ground infrastructure. Eutelsat OneWeb LEO antennas on the ground can view multiple moving satellites at a given instant even in remote areas.

SpaceX's Starlink is the world's largest planned constellation of LEO satellites (Starlink 2024a). More than five thousand satellites have been operating as of August 2024, with a plan for 12,000 and possibly 42,000 satellites (Wikipedia 2024b). One of the famous uses of Starlink is satellite connectivity in Ukraine during the Russia-Ukraine war in the early 2020s. Examples of usage of Starlink include residential, connect-on-the-go coverage in unserved or underserved areas, and maritime communications (e.g., communications on boats). Starlink utilizes several technologies in its constellation of satellites (Starlink 2024b). A single solar array is used on the satellite to simplify the system design and facilitate manufacturing. Starlink satellites have a compact and flat-panel design, which facilitates the launch of satellites. Starlink satellites implement autonomous collision avoidance to avoid collisions with other satellites and any space debris. Custom-built navigation sensors are used to determine each satellite's location, altitude, and orientation so that beams can be focused in the areas where throughput is needed. Optical ISLs are used to provide global coverage. Each Starlink satellite has four-phased array antennas and two parabolic antennas. Krypton-powered thrusters are used by Starlink satellites for orbit maneuvering and end-of-life deorbiting.

2.6 Types of NTN Devices for the NTN

A variety of devices can connect to an NTN. Figure 2.6 shows examples of NTN devices. Characteristics of these terminals significantly influence the achievable data rates or throughput for a given satellite constellation.

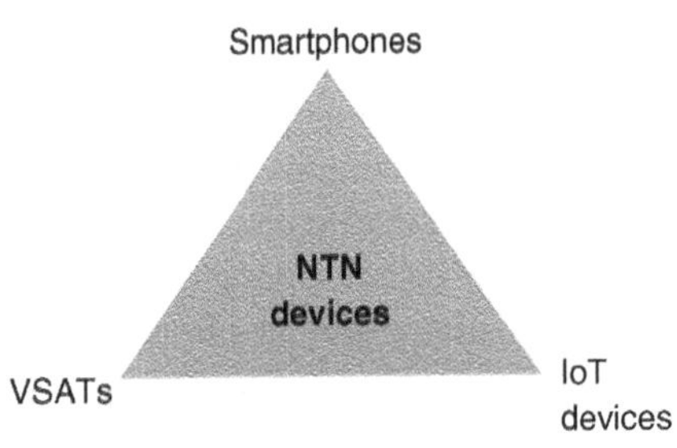

Figure 2.6 Example Types of NTN Devices.

Smartphones are expected to be a significant portion of user devices connecting to the NTN. In initial NTN deployments, data rates may be relatively small, and low- or medium-data rate services such as SMSs may be more common due to limited transmit power and low-gain antennas of smartphones. Certain iPhone models, including iPhone 14, iPhone 14 Pro, iPhone 15, and iPhone 15 Pro, can connect to a satellite to text emergency services, request roadside assistance, and share the phone location (Apple 2024). For a higher voice capacity per NTN cell, low-rate voice services may be used by trading off voice quality for increased cell capacity.

Low-rate and delay-tolerant services are adequate for many IoT devices. A massive number of such IoT devices scattered over a large remote area can be connected to a large NTN cell, leading to cost-effective coverage for such IoT devices.

Custom terminals with a high-gain antenna are quite common in pre-5G NTNs and would also likely be a significant portion of user devices connecting to an NTN. Examples of such devices include a very small aperture terminal (VSAT). A VSAT utilizes a high-gain antenna for a higher received signal strength and transmit power. Starlink, Project Kuiper, and Eutelsat OneWeb terminals are examples of such custom NTN terminals.

2.7 Trends in NTN Deployments

There has been a rising interest in the NTN, especially due to the reduction in the overall cost of the satellite launch. The cost of delivering a payload into orbit using space shuttles cost around $50,000 per kilogram in 2010s and 2020s. Such cost has been reduced by a factor of almost 2,500. For example, SpaceX's StarShip can cost about $20 per kg or less for delivering a payload (5G Americas 2023).

Different types of NTNs are expected to exist. An NTN may be a non-3GPP NTN supporting non-3GPP devices. For example, a Starlink terminal may communicate with a Starlink satellite via a Starlink-proprietary non-3GPP radio interface. A non-Starlink device cannot connect to a Starlink satellite in such an NTN. Another type of NTN is a Direct-to-Cell (DTC) or Direct-to-Device (D2D) NTN. Such DTC NTN supports unmodified 3GPP-based smartphones and IoT devices. Finally, an NTN may be a 3GPP-based NTN (which is the focus of this book!) that supports 3GPP NTN-compliant smartphones and IoT devices. Such smartphones and IoT devices support NTN specifications defined in Release 17 or later 3GPP releases.

Proprietary non-3GPP NTNs typically use VSAT devices with high-gain dish antennas and hence support broadband data rates. As of August 2024, Echostar, HughesNet, and Viasat provide broadband services via a non-3GPP NTN that uses GEO satellites, and Starlink and Eutelsat OneWeb provide broadband services via a non-3GPP NTN that uses LEO satellites (5G Americas 2023).

Examples of DTC NTN providers include Starlink and AST SpaceMobile. An industry alliance called Mobile Satellite Services Association (MSSA) also aims to be a DTC NTN provider (Mobile World Live 2024).

As mentioned earlier, SpaceX Starlink provides Internet access using numerous LEO satellites using a non-3GPP approach. Additionally, Starlink is using some satellites to support smartphones. T-Mobile is collaborating with Starlink for DTC NTN services. Specifically, T-Mobile's coverage above and beyond initiative aims to bring connectivity nearly everywhere in the US for T-Mobile customers by leveraging SpaceX's constellation of satellites with DTC technology and T-Mobile's wireless network (T-Mobile 2024). As part of Starlink DTC launch campaign, SpaceX launched 21 Starlink satellites with the DTC feature in January 2024 (Choudhury 2024). Starlink aims to provide seamless access to text, voice, and data for LTE phones across the globe. Starlink successfully transmitted and received the first text messages via T-Mobile's spectrum using DTC satellites on January 8, 2024. Furthermore, Starlink achieved a peak download speed of 17 Mbps from a satellite directly to an unmodified Samsung Android phone (Choudhury 2024). The larger second-generation SpaceX satellites can connect to the majority of T-Mobile's 4G and 5G phones over its 1,900 MHz spectrum.

AST SpaceMobile plans to support unmodified devices using over 240 planned LEO satellites in their constellation (AST SpaceMobile 2024). The UEs are regular smartphones used in terrestrial networks and do not need to support 3GPP-defined NTN specifications. AST SpaceMobile's test satellite, BlueWalker 3, is built with flat phased-array technology. The 693-square-foot array of the test satellite consists of thousands of antennas. AT&T and Verizon have invested in AST SpaceMobile and plan to use AST SpaceMobile to provide satellite-based connectivity to their customers. BlueWalker 3 is AST SpaceMobile's prototype satellite and a predecessor to planned commercial satellites called BlueBirds.

Five NTN providers, Viasat, Terrestar Solutions, Ligado Networks, Omnispace, and Al Yah Satellite Communications (Yahsat), formed the MSSA to create a global D2D ecosystem for connecting satellite services to mobile devices. MSSA was founded to utilize 100+ MHz of L-band (which ranges from 1 to 2 GHz) and S-band (which ranges from 2 to 4 GHz) allocated and licensed spectrum for MSSs. For example, Ligado Networks and Mavenir demonstrated an R17 NB-IoT NTN in the L-band using Ligado's SkyTerra satellite network and Mavenir's cloud-native RAN. MSSA plans to align its efforts with 3GPP specifications to enhance coverage for mobile operators and provide over-the-top Internet services.

A 3GPP Release 17 NTN using the sub-2 GHz spectrum can achieve the cell throughput of tens of Mbps in the downlink. A 3GPP R18 NTN using spectrum in the Ka band (ranging from 26.5 to 40 GHz) can support hundreds of Mbps to non-handheld devices using small dish antennas. The available cell throughput is shared among active users in the NTN cell. User-experienced throughput depends on factors such as the device capability, the spectrum, resource demands from active UEs, and antenna gains.

In response to the rising interest in NTNs, in April 2023, the FCC in the US launched the Space Bureau that will focus on policy and licenses for promoting satellite and space-based communications. The Office of International Affairs (OIA) will work with foreign regulatory authorities. The Space Bureau approved new rules on spectrum sharing and coordination between space-based operators to help new entrants while protecting incumbents.

2.8 Major Takeaways

The NTN is expected to play an important role as a key component of 5G and 6G in advancing the reach of wireless communications. Here are the major takeaways from this chapter.

- Two main types of NTN platforms considered by the 3GPP are satellites and UASs.
- Types of satellites include GEO satellites, MEO satellites, and LEO satellites.
- HAPSs are examples of an UAS.
- A GEO satellite operates at an altitude of 35,786 km, an MEO satellite operates at an altitude ranging from 5,000 to 25,000 km, an LEO satellite operates at an altitude ranging from 300 to 1,500 km, a HAPS operates at an altitude ranging from 8 to 50 km.
- To a fixed observer on the Earth's surface, a GEO satellite and a HAPS appear to be stationary, while MEO and LEO satellites appear as moving.
- A GEO satellite and a HAPS utilize Earth-fixed beams, while MEO and LEO satellites utilize quasi-Earth-fixed or Earth-moving beams.
- The HAPS experiences the least propagation delay and smallest propagation path loss, while the GEO satellite experiences the longest propagation delay and largest propagation path loss.

- The HAPS and the LEO satellite generally cover a relatively small geographic area, while a GEO satellite covers a relatively large area. Hence, three GEO satellites are adequate to provide global coverage, but hundreds of LEO satellites are needed for global coverage. The HAPS is usually intended for local coverage.
- The NTN involves three types of beams: Earth fixed, Earth moving, and quasi-Earth fixed.
- An Earth-fixed beam covers the same geographic area at all times, and an Earth-moving beam covers different geographic areas from one instant to the next. A quasi-Earth-fixed beam covers one geographic area for a certain period and a different geographic area for a certain period.
- Traditional pre-5G satellite systems include Echostar, Globalstar, Intelsat, and ViaSat/Inmarsat. Emerging pre-5G satellite systems include Kuiper Project by Amazon, Lightspeed by Telesat, OneWeb by Eutelsat OneWeb, and Starlink by SpaceX.
- Satellite systems that existed prior to the deployment of the 3GPP NTN deployments are also briefly described.

References

3GPP (2023). *Solutions for NR to support non-terrestrial networks (NTN)*. TR38.821. April 2023. https://portal.3gpp.org/desktopmodules/Specifications/SpecificationDetails.aspx?specificationId=3525 (accessed 8 August 2024).

3GPP (2024). *NR and NG-RAN Overall description*. TS38.300. July 2024. https://portal.3gpp.org/desktopmodules/Specifications/SpecificationDetails.aspx?specificationId=3191 (accessed 8 August 2024).

5G Americas (2023). Update on 3GPP Non-Terrestrial Networks. White Paper. July 2023. https://www.5gamericas.org/update-on-5g-non-terrestrial-networks/ (accessed 8 August 2024).

Apple (2024). *Connect to a satellite with your iPhone*. 8 December 2023. https://support.apple.com/en-us/105097 (accessed 8 August 2024).

AST SpaceMobile (2024). *AST SpaceMobile Announces First Five Commercial Satellites Completed Final Assembly and Ready for Shipment the First Week of August*. 25 July 2024. https://feeds.issuerdirect.com/news-release.html?newsid=5212606427775376 (accessed 8 August 2024).

Choudhury, R. (2024). *Starlink achieves peak download speed of 17Mb/s from satellite to cellphone: Elon Musk*. Interesting Engineering. 3 March 2024. https://interestingengineering.com/innovation/starlink-achieves-peak-download-speed-of-17mb-s-from-satellite-to-cellphone-elon-musk (accessed 8 August 2024).

Echostar (2024). *Services*. https://www.echostarsatelliteservices.com/services (accessed 8 August 2024).

Eutelsat (2024). *Eutelsat Group*. https://oneweb.net/ (accessed 8 August 2024).

Globalstar (2024). *Satellite Technology powered by The Globalstar Satellite Network*. https://www.globalstar.com/en-us/about/our-technology (accessed 8 August 2024).

GSMA (2022). *High Altitude Platform Systems: Towers in the Skies*. Version 2.0. February 2022. https://www.gsma.com/futurenetworks/resources/high-altitude-platform-systems-towers-in-the-skies-version-2-0/ (accessed 8 August 2024).

HAPS Alliance (2024a). *HAPS Alliance*. https://hapsalliance.org/ (accessed 8 August 2024).

HAPS Alliance (2024b). Guidelines for Payload Operation in the Stratosphere. White Paper. https://hapsalliance.org/wp-content/uploads/formidable/12/Guidelines_for_Payload_Operation_in_the_Stratosphere_WhitePaper_2022.pdf (accessed 8 August 2024).

Hughes (2024). *JUPITER GEO Satellites.* https://www.hughes.com/what-we-offer/satellite-services/jupiter-geo-satellites (accessed 8 August 2024).

Intelsat (2024). *Satellite Network.* https://www.intelsat.com/global-network/satellite-network/ (accessed 8 August 2024).

Kohnstamm, T. (2023). *Everything you need to know about Project Kuiper, Amazon's satellite broadband network.* 30 October 2023. https://www.aboutamazon.com/news/innovation-at-amazon/what-is-amazon-project-kuiper (accessed 8 August 2024).

Mobile World Live (2024). *NTN providers form satellite industry association.* 9 February 2024. https://www.mobileworldlive.com/network-tech/ntn-providers-form-satellite-industry-association/ (accessed 8 August 2024).

Next G Alliance (2024). 6G Radio Technology Part II: Basic Radio Technologies. NGA White Paper. May 2024. https://nextgalliance.org/white_papers/6g-radio-technologypart-ii-basic-radio-technologies/ (accessed 8 August 2024).

OceanWeb (2023). *A Guide: GEO, LEO and MEO Satellites.* 14 February 2023. https://www.oceanweb.com/a-guide-geo-leo-and-meo-satellites/#:~:text=Geo%20satellites%20have%20a%20much,satellite%20constellations%20complementing%20each%20other (accessed 8 August 2024).

SES (2020). *GEO, MEO, AND LEO.* May 2020. https://www.satellitetoday.com/wp-content/uploads/2021/02/Guide-GEO-MEO-LEO-1.pdf (accessed 8 August 2024).

Sodders, L. (2023). *LEO, MEO or GEO? Diversifying orbits is not a one-size-fits-all mission.* 18 July 2023. https://www.ssc.spaceforce.mil/Newsroom/Article-Display/Article/3462074/leo-meo-or-geo-diversifying-orbits-is-not-a-one-size-fits-all-mission-part-1-of (accessed 8 August 2024).

Starlink (2024a). *High-Speed Internet Around the World.* https://www.telesat.com/leo-satellites/ (accessed 8 August 2024).

Starlink (2024b). *Starlink Satellite Technology.* https://www.starlink.com/technology (accessed 8 August 2024).

Stuhlfauth, R. (2022). 5G NTN Takes Flight: Technical Overview of 5G Non-Terrestrial Networks. Rohde & Schwarz White Paper. Version 1.0. July 2022. https://www.rohde-schwarz.com/us/solutions/satellite-testing/landingpages/white-paper-5g-ntn-takes-flight-technical-overview-of-5g-non-terrestrial-networks_255919.html (accessed 8 August 2024).

Telesat (2024). *Telesat Lightspeed LEO Network.* https://www.telesat.com/leo-satellites/ (accessed 8 August 2024).

T-Mobile (2024). *First SpaceX Satellites Launch for Breakthrough Direct to Cell Service with T-Mobile.* 3January 2024. https://www.t-mobile.com/news/un-carrier/first-spacex-satellites-launch-for-breakthrough-direct-to-cell-service-with-t-mobile (accessed 8 August 2024).

Viasat (2024). *Unlimited High-Speed Satellite Internet.* https://www.viasat.com/ (accessed 8 August 2024).

Wikipedia (2024a). *High-altitude platform station.* https://en.wikipedia.org/wiki/High-altitude_platform_station (accessed 8 August 2024).

Wikipedia (2024b). *Starlink.* https://en.wikipedia.org/wiki/Starlink#:~:text=SpaceX%20started%20launching%20Starlink%20satellites,possible%20later%20extension%20to%2042%2C000 (accessed 8 August 2024).

3

Radio Interfaces of LTE-M, NB-IoT, and NR: A Concise Introduction

The 3GPP has two parallel work streams related to the Non-Terrestrial Network (NTN): NR-NTN and Internet of Things (IoT)-NTN.[1] As the name suggests, NR-NTN utilizes an NR-based air interface. The NR-NTN in Release 17 utilizes the 5G specifications defined in Release 16 as the baseline and introduces enhancements to the 5G specifications to address NTN-specific challenges. The NR-NTN utilizes gNBs in the radio network and the 5G Core (5GC) network. The IoT-NTN supports LTE-M and NB-IoT[2] air interfaces and uses the 4G Evolved Packet Core (EPC). The IoT-NTN utilizes pre-Release 17 LTE-M and NB-IoT specifications and leverages NTN-specific enhancements defined by the NR-NTN to determine NTN-related enhancements for LTE-M and NB-IoT specifications. This chapter provides a foundation of LTE-M, NB-IoT, and NR air interfaces by highlighting their key features and overall operations. The associated core networks, EPC and 5GC, are also briefly discussed. While this chapter provides the foundation of LTE-M and NB-IoT based on Release 13, enhancements in LTE-M and NB-IoT beyond Release 13 are also briefly summarized.

3.1 3GPP-Defined Wireless IoT Technologies and 5G New Radio: A Concise Introduction

5G is a transformational cellular technology with the potential to significantly influence several industries and segments of the economy. Flexibility and high performance of 5G can be exploited to offer three usage scenarios envisioned by the International Telecommunication Union (ITU) (Tripathi and Reed 2019): enhanced Mobile Broadband (eMBB), Ultra-Reliable Low-Latency Communications (URLLC), and massive Machine-Type Communications (mMTC). The eMBB usage scenario is characterized by high data rates. The URLLC usage scenario focuses on both high reliability and low latency. The massive Internet of Things (mIoT) usage scenario is characterized by support for many delay-tolerant IoT devices requiring only low data rates. Services such as high-definition video streaming, high-speed file downloads, augmented reality, virtual reality, self-driving cars, automated factories, remote surgery, and smart cities possess one or more service characteristics of these ITU usage scenarios.

---

1 3GPP stands for Third-Generation Partnership Project, NR represents New Radio, and IoT refers to Internet of Things.

2 LTE-M is Long-Term Evolution-Machine-Type Communications, and NB-IoT is Narrowband-Internet of Things.

3GPP-based Non-Terrestrial Networks in 5G and 6G: Expanding the Frontiers of Wireless Communications, First Edition.
Nishith D. Tripathi and Jeffrey H. Reed.

Companion website: www.wiley.com/go/tripathi5g

The 3GPP introduced a New Radio (NR) air interface for 5G in Release 15, when the first version of 5G, often referred to as 5G Phase 1, was defined. The 3GPP also defined a new core network for the 5G called 5G Core (5GC) or Next-Generation Core (NGC). The NR air interface and the 5GC play a key role in supporting services aligned with eMBB and URLLC characteristics. Section 3.1.2 provides a brief introduction to the NR air interface and the 5GC. More details of the 5G NR air interface are given in Section 3.4.

One of the key targets for the mMTC usage scenario is the support for a massive number of low-cost and delay-tolerant IoT devices. LTE-based technologies such as LTE-M and NB-IoT are more suited to support the mIoT usage scenario than 5G NR due to the lower cost and lower complexity of LTE-M and NB-IoT compared to NR. Hence, as of Release 18, the 3GPP has continued to evolve LTE-M and NB-IoT to cater to the mIoT usage scenario instead of modifying the NR air interface to support the mIoT usage scenario. Section 3.1.1 provides a brief introduction to the LTE, LTE-M, and NB-IoT air interfaces and the 4G EPC. More details of the wireless IoT technologies, LTE-M and NB-IoT, are provided in Sections 3.2 and 3.3, respectively. While the foundation of LTE-M and NB-IoT provided in this chapter is based on Release 13, enhancements in LTE-M and NB-IoT beyond Release 13 are briefly summarized in Section 3.5.

3.1.1 Wireless IoT Technologies and 4G Evolved Packet System

IoT technology connects an IoT device to one or more IoT application servers and potentially other IoT devices. IoT technology may use wireline connectivity or wireless connectivity to connect an IoT device to a network. Wireless connectivity offers more flexibility in IoT deployments and is quite popular. While the 3GPP defines some wireless IoT technologies, some are defined by organizations other than the 3GPP. Examples of 3GPP-defined wireless IoT technologies include LTE-M, NB-IoT, and Extended Coverage-Global System for Mobile Communication (EC-GSM). Examples of non-3GPP wireless IoT technologies include LoRaWAN,[3] Wi-Fi, ZigBee, and Bluetooth. The 3GPP has focused on LTE-M and NB-IoT for evolving wireless IoT technologies. Hence, this book also focuses on LTE-M and NB-IoT as key wireless IoT technologies.

A special type of wireless IoT is of considerable interest: low-power wide-area (LPWA) IoT. LTE-M, NB-IoT, and long range wide area network (LoRaWAN) can all be considered as LPWA IoT technologies. Example categories of LPWA IoT applications include utilities (e.g., smart gas and water meters), logistics (e.g., asset tracking), industrial (e.g., tank monitoring), smart cities (e.g., smart parking and streetlights), agriculture and environment (e.g., cattle tracking and soil monitoring), consumers and medical (e.g., patient monitors), and smart buildings (e.g., alarm systems) (Tripathi and Reed 2019).

LPWA IoT technologies aim to meet specific requirements, which are quite distinct from those for smartphones. For example, the LPWA IoT device cost must be low (e.g., only a few dollars). Furthermore, power consumption for the device must be low, and battery life must be long, often spanning multiple years. Additionally, deep indoor and subterranean coverage is desirable because some IoT devices may be on underground pipelines, and some IoT devices may be in basements of large buildings. The network needs to cost-effectively support a massive scale of LPWA IoT devices.

Another special type of wireless IoT that is gaining momentum is industrial IoT (IIoT). IIoT is expected to be important for 5G as well as 6G. Certain IIoT applications, such as those for automated manufacturing, often require extremely high reliability and high data rates. The URLLC

3 LoRaWAN stands for Long Range Wide Area Network. LoRaWAN defines a protocol stack that utilizes LoRa physical layer between the IoT device and the radio network.

usage scenario is better suited for such IIoT applications. Hence, the 3GPP focuses on the NR radio interface to support high-performance requirements associated with these IIoT applications.

The 3GPP has been enhancing LTE-M and NB-IoT with additional features and capabilities after their introduction in Release 13. Section 3.5 provides a brief overview of LTE-M and NB-IoT enhancements in Release 14 and beyond. The 3GPP Release 13 provides the foundation of LTE-M and NB-IoT. Figure 3.1 provides a high-level comparison among LTE, LTE-M, and NB-IoT based on Release 8 LTE, Release 13 LTE-M, and Release 13 NB-IoT (Tripathi and Reed 2019; Qualcomm 2018).

LTE, LTE-M, and NB-IoT share several common traits. For example, all of them utilize Orthogonal Frequency Division Multiplexing (OFDM) on the radio interface. Furthermore, LTE, LTE-M, and NB-IoT devices may use the maximum transmit power of 23 dBm or 200 mW. Additionally, all of them can use the same EPC with suitable enhancements for LTE-M and NB-IoT.

One of the key differences between LTE, LTE-M, and NB-IoT is the maximum channel bandwidth at a given carrier frequency. LTE devices typically support a maximum channel bandwidth of 20 MHz. LTE-M devices support the maximum transmission bandwidth of six physical resource blocks (PRBs) corresponding to 1.08 MHz. When the guard band is included, an LTE-M radio channel bandwidth becomes 1.4 MHz. NB-IoT devices support the maximum transmission bandwidth of one PRB corresponding to 180 kHz. The maximum channel bandwidth for NB-IoT devices is 200 kHz when the guard band is considered.

Another key difference between LTE, LTE-M, and NB-IoT is the device complexity. While typical LTE devices may have two or four antennas and associated radio frequency (RF) chains and baseband signal processing, LTE-M and NB-IoT devices are much simpler and have a single antenna to receive signals (and to transmit signals). Such design simplicity greatly reduces the cost and the size of LTE-M and NB-IoT devices.

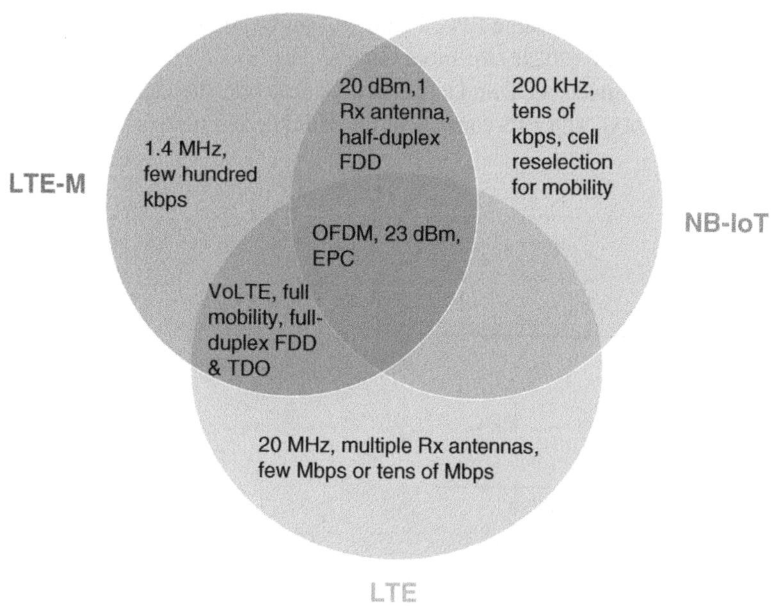

Figure 3.1 LTE, LTE-M, and NB-IoT: A Quick Comparison.

To increase battery life, LTE-M and NB-IoT make use of features such as extended discontinuous reception (eDRX) cycles and power saving mode (PSM). An eDRX cycle increases the time between successive control channel monitoring instants to reduce the amount of required processing for the UE. PSM enables the IoT UE to eliminate the monitoring of page messages to significantly reduce the processing requirements.

The constraint on the maximum channel bandwidth and other simplifications restricts the achievable data rates in LTE-M and NB-IoT. While LTE smartphones can often experience few Mbps or even tens of Mbps on one radio channel, LTE-M devices support a data rate of a few hundred kbps, and NB-IoT devices support a data rate of a few tens of kbps.

While LTE and LTE-M devices support full-duplex frequency division duplex (FDD) and time division duplex (TDD), NB-IoT devices support half-duplex FDD. FDD utilizes paired spectrum, where transmission on one portion of the spectrum and reception on a different portion of the spectrum occur at the same time. Half-duplex FDD also utilizes paired spectrum; however, transmission occurs on one portion of the spectrum at one instant, and reception occurs on a different portion of the spectrum at another instant. TDD utilizes unpaired spectrum, where the same portion of the spectrum is used for transmission at one instant and reception at another instant. Using half-duplex FDD instead of full-duplex FDD further simplifies the device design by eliminating the need for a duplexer and reducing the amount of required processing.

Delay-sensitive applications such as Voice over LTE (VoLTE) can be supported in LTE and LTE-M but not NB-IoT. Furthermore, both LTE and LTE-M support full mobility for the UEs in the RRC-CONNECTED state, where the UE provides handover-related measurements to the serving eNB, the serving eNB works with the target eNB to ensure the availability of radio resources in the target eNB, the serving eNB sends a handover command to the UE, and the UE establishes connectivity with the target eNB. Mobility management in NB-IoT is much simpler, reducing processing requirements for the NB-IoT device. The overall measurement and handover process are not used in NB-IoT; the NB-IoT performs cell reselection even when in the RRC_CONNECTED state while performing a handover from one cell/eNB to another.

LTE, LTE-M, and NB-IoT can all share the same EPC. Figure 3.2 shows the main EPC components (Tripathi and Reed 2019); see 3GPP (2024a) for more details.

The Evolved Packet System (EPS) connects a UE and a packet data network, such as the Internet and the IP Multimedia Subsystem (IMS). The EPS consists of the radio access network called the

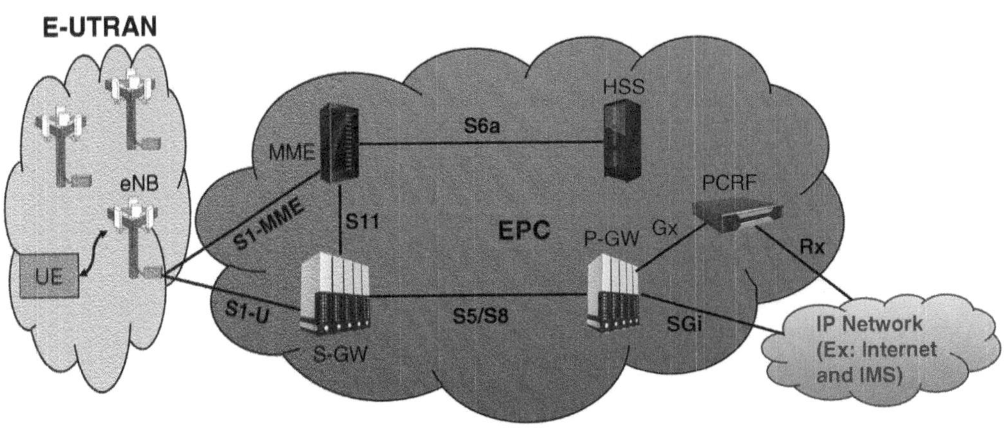

Figure 3.2 Traditional EPC Architecture. *Source:* Tatyana Drozdova / Adobe Stock Photo; DNY3D / Adobe Stock Photo; Andrey Popov / Adobe Stock Photo; cyberneticimages / Adobe Stock Photo.

Evolved-Universal Terrestrial Radio Access Network (E-UTRAN) and the EPC. The E-UTRAN consists of evolved Node Bs (eNBs). The UE and the E-UTRAN communicate using radio access technologies (RATs) such as LTE, LTE-M, and NB-IoT. Selected functions performed by the major network elements of the EPC are briefly discussed below. The EPC components together enable the service provider to manage various UE operations.

The Mobility Management Entity (MME) acts as a proxy or representative of the EPC for the UE and exchanges non-access stratum (NAS) signaling with the UE. The NAS signaling connection between the UE and the MME consists of (i) a signaling connection between the UE and the eNB in each LTE cell and (ii) a UE-specific S1-MME signaling connection. The MME obtains authentication parameters from the home subscriber server (HSS) and authenticates the UE during the mutual authentication process as part of the LTE attach procedure. The MME works with the eNB and EPC network elements to set up an EPS bearer that is necessary for the UE to be able to exchange user traffic with external packet data networks. The MME keeps track of the location of the idle UE so that a page message can be sent to such UE in case traffic for the UE arrives from an external packet data network. The serving eNB and the MME maintain a UE-specific connection on the S1-MME interface.

The Serving Gateway (S-GW) performs the function of a local mobility anchor. As the UE travels from one eNB to another, the S-GW maintains connectivity with the eNB that is serving the UE at a given instant. Such connectivity enables the serving eNB and the S-GW to send and receive user traffic via the UE-specific S1-U interface. The S-GW also buffers user traffic arriving from an external packet data network when the UE is in the idle mode.

The Packet Data Network-Gateway (P-GW) allocates an IP address to the UE during or after the LTE attach procedure. The P-GW is a gateway for the UE toward the external networks. The user traffic from networks such as the Internet arrives at the P-GW, and the P-GW forwards such traffic to a suitable S-GW responsible for the UE.

The HSS helps the MME with the authentication process, as mentioned earlier. The HSS stores the subscriber database.

The 3GPP has defined the Policy and Charging Control (PCC) framework to facilitate the implementation of quality of service (QoS) in the LTE EPS and charging. Three main PCC entities include the Policy and Charging Rules Function (PCRF), Policy and Charging Enforcement Function (PCEF), and application function (AF). The AF resides in the IMS and is typically implemented by the Proxy-Call Session Control Function (P-CSCF) in the IMS acts as the AF. The AF extracts QoS information from the session initiation protocol (SIP) signaling exchanged between two endpoints, such as two smartphones. The AF provides such generic QoS information to the PCRF. The PCRF maps generic QoS parameters to LTE-specific QoS parameters. These LTE QoS parameters are conveyed from the P-GW to the S-GW, S-GW to the MME, and the MME to the eNB. A suitable EPS bearer between the UE and the P-GW is established to conform to these target QoS parameters. The P-GW acts as the PCEF to enforce QoS, such as limiting the data rate below the maximum data rate limit.

The EPC architecture shown in Figure 3.2 can transfer IoT traffic between the IoT device/UE and an external network. The 3GPP has also introduced new network elements to provide additional flexibility for IoT traffic transfer. Figure 3.3 shows ways IoT traffic can be transported through the network (Tripathi and Reed 2019). These ways provide flexibility to service providers in deploying IoT solutions.

The IoT traffic may be transported through the network using the user plane or the control plane. When the user plane is used, the IoT traffic is carried by an EPS bearer between the UE and the P-GW and hence passes through the eNB, the S-GW, and the P-GW. When the control plane is used,

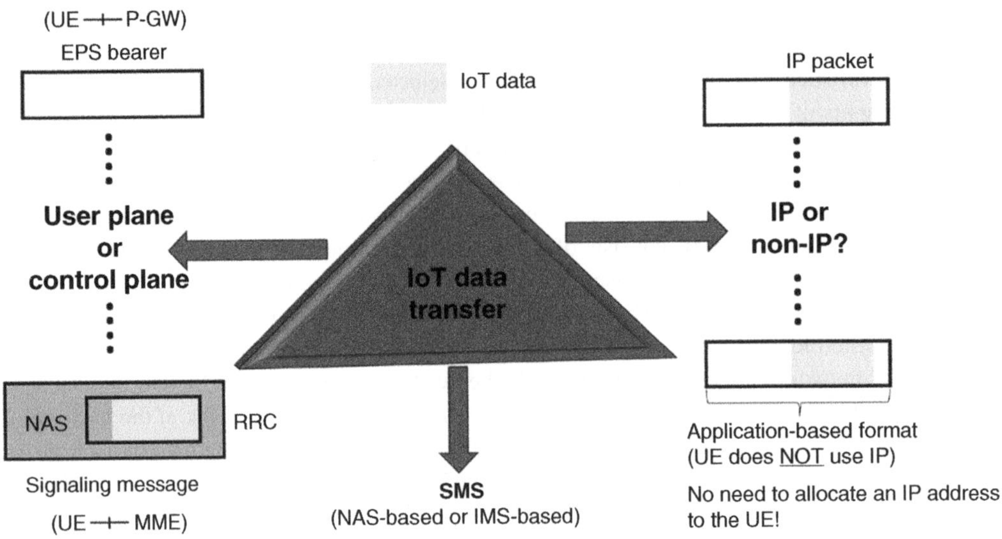

Figure 3.3 Methods for IoT Traffic Transfer.

the IoT traffic is carried by the NAS signaling message between the MME and the UE. The NAS message carrying the IoT traffic is encapsulated by an RRC signaling message. The user plane approach can be beneficial for more amount of data, while the control plane approach can be more efficient for lesser amounts of data.

The IoT traffic may be sent using SMS. LTE supports SMS delivery via NAS signaling or the IMS. When the NAS signaling is used, the IoT traffic is exchanged between the UE and the MME. When the IMS is used, the IoT traffic is placed inside an IP packet, and this IP packet is carried by an EPS bearer between the UE and the P-GW.

The IoT traffic may be transported with or without IP. Some IoT UEs are simple and may not have the IP protocol stack. In such cases, non-IP data delivery (NIDD) can be used to transport the IoT traffic. The network does not need to allocate an IP address to the UE in case of NIDD. In the case of the IP-based delivery, the IoT traffic is placed inside an IP packet, and this IP packet is transported through the LTE network using an EPS bearer, a NAS message, or an SMS.

While the LTE-M and NB-IoT can use the network elements of the EPC architecture shown in Figure 3.2, they can also use additional network elements for increased flexibility in IoT data transfer. Figure 3.4 shows the IoT-centric network architecture (Tripathi and Reed 2019; 3GPP 2024a). The roles of IoT-related functions of the network elements shown in Figure 3.4 are briefly explained.

The UE typically exchanges IoT traffic with an external server, such as the Machine-Type Communication (MTC) application server (AS). Note that the structure of the IoT traffic (e.g., sensor measurements) itself follows a certain protocol so that the UE and the IoT server such as the MTC AS can properly interpret the IoT data traffic. Examples of such IoT traffic protocols include constrained application protocol (CoAP) and Message Queuing Telemetry Transport-Sensor Network (MQTT-SN). Organizations such as oneM2M and open mobile alliance (OMA) play a key role in defining IoT standards/protocols.

There are two models for IoT services: a direct model and an indirect model. In the indirect model, the IoT traffic passes through the service capabilities server (SCS) in its journey between the P-GW and the MTC AS. In the direct model, the IoT traffic does not pass through the SCS

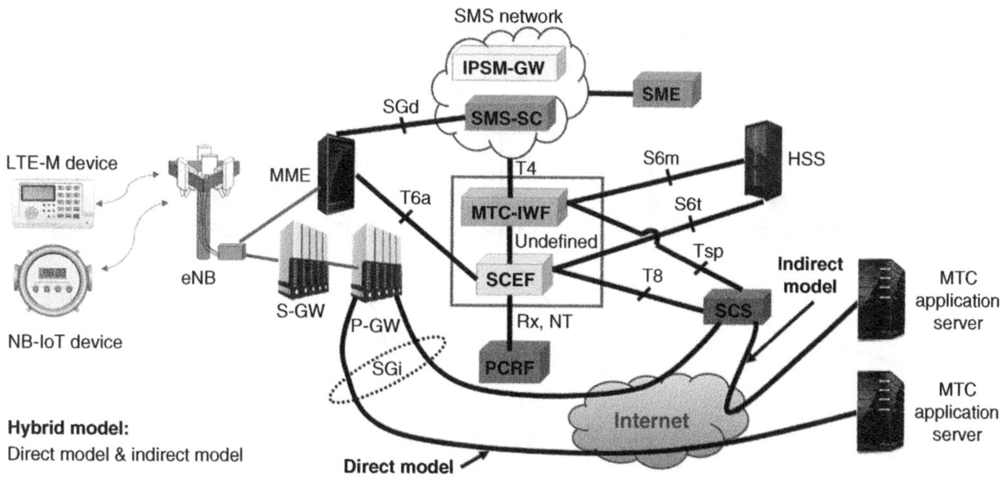

Figure 3.4 IoT-Centric Enhanced EPC Architecture. *Source:* DNY3D / Adobe Stock Photo; Art of Success / Adobe Stock Photos; Tatyana Drozdova / Adobe Stock Photo; Andrey Popov / Adobe Stock Photo; Jimmyi23 / Adobe Stock Photo.

in its journey between the P-GW and the MTC AS. The SCS, as the name suggests, provides value-added services. The SCS may be managed by the cellular service provider or a third party. An example of an SCS service is device triggering to facilitate mobile-terminated (MT) data transfer.

The Service Capability Exposure Function (SCEF) securely exposes the services and the capabilities of the LTE EPS to Application Servers via Application Programming Interfaces (APIs). The SCEF also carries out authentication and authorization of API consumers. Another key role of the SCEF is to support NIDD. The IoT UE and the MME exchange the IoT data via NAS signaling, and the SCEF transports the IoT data between the MME and the SCS. The SCS and the MTC AS can then exchange the IoT data through a packet data network such as the Internet. The SCEF is also responsible for policy enforcement through actions such as throttling.

The Machine Type Communication-Inter Working Function (MTC-IWF) authorizes SCSs. The MTC-IWF selects the delivery mechanism for the device trigger, such as (i) a suitable MME for the IoT traffic transport via NAS signaling or (ii) an IP Short Message Gateway (IPSM-GW) for IMS-based transport of the IoT traffic.

In support of IoT traffic transport, the HSS stores relevant information, such as information for device triggering and the identity of the serving node (e.g., MME and IP-SM-GW). The MME supports IoT traffic transport using NIDD and SMS via control plane signaling exchange with the IoT UE. For NIDD, the MME selects one of the two methods for IoT transport: the SCEF-based method and the SGi-based method. The IoT traffic passes through the SCEF using the SCEF-based method. The IoT traffic passes through the S-GW and the P-GW using the SGi-based method.

3.1.2 5G System

The 5GS includes the UE, the 5G access network (e.g., next-generation radio access network (NG-RAN) or a Non-3GPP access network), and the 5G Core Network (5GC). 5G defines a service-based architecture (SBA), where entities called network functions (NFs) provide services through service-based interfaces. The SBA provides significant flexibility and scalability, allowing NFs from different vendors to seamlessly interact with one another using service-based interfaces. Furthermore,

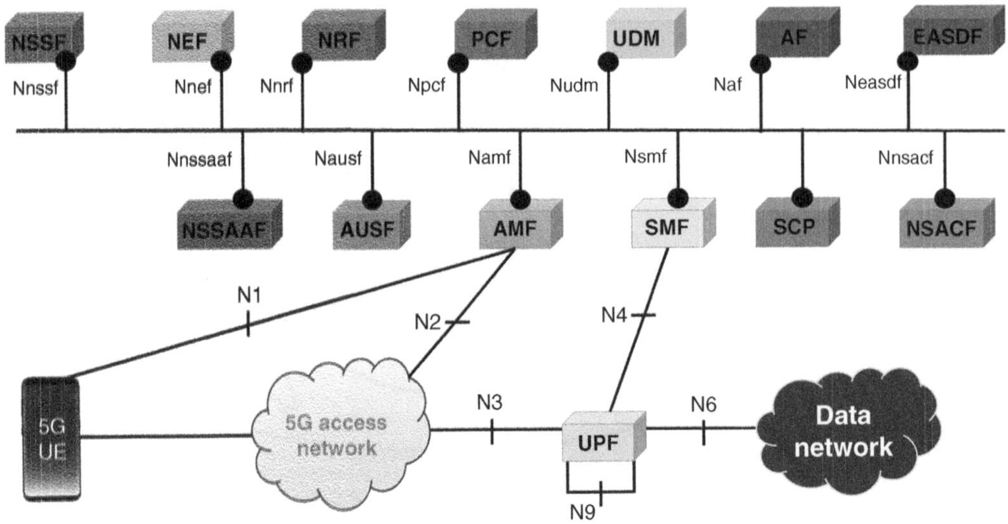

Figure 3.5 5G System Architecture. *Source:* Tripathi and Reed (2019) / The Wireless University; nadir_insan / Adobe Stock Photo.

the SBA can be implemented by exploiting virtualization and cloud technologies. Additionally, upgrades or enhancements to the NFs of the SBA can be made relatively easily.

Figure 3.5 illustrates major NFs in the SBA of a 5G System (5GS) (3GPP 2024b; Tripathi and Reed 2019); see 3GPP (2024b) for details of these and other NFs. The roles of the 5GC NFs shown in Figure 3.5 are briefly explained later. The NFs in the 5GC enable the service provider to manage a variety of UE operations.

The Access and Mobility Management Function (AMF) acts as a proxy or representative of the 5GC toward the UE and exchanges NAS signaling with the UE. The NAS signaling connection between the UE and the AMF consists of (i) a signaling connection between the UE and the gNB in a given 5G NR cell and (ii) a UE-specific N2 signaling connection. The AMF provides support for access authentication and access authorization by facilitating communications between the UE and the Authentication Server Function (AUSF). The AMF coordinates with the UE, the gNB, and the SMF to set up a 5G QoS flow for the UE, enabling the UE to exchange user traffic with external data networks. The AMF keeps track of the location of the idle UE so that a page message can be sent to the UE when traffic for the UE arrives from an external data network. The AMF, using its NAS signaling connection with the UE, facilitates the transport of session management (SM) messages between the UE and the SMF. The AMF manages the UE's registration, connection, reachability (e.g., UE tracking), and mobility (e.g., UE tracking and support for handover signaling).

The Unified Data Management (UDM) generates authentication credentials and provides those to the AUSF to support access authorization. The UDM also supports concealment of Subscription Permanent Identifier (SUPI), such as the Internal Mobile Subscriber Identity (IMSI) using Subscription Concealed Identifier (SUCI). The UDM manages subscriptions.

The AUSF obtains authentication credentials from the UDM and compares the expected authentication response with the response received from the UE during the mutual authentication procedure (which typically occurs during registration). The AUSF utilizes the help of the AMF to exchange authentication-related messages with the UE.

The SMF manages the UE's protocol data unit (PDU) sessions. The SMF exchanges SM signaling with the UE through the AMF (e.g., to set up, modify, and release a PDU session). The SMF allocates an IP address to the UE or works with the UPF, the UDM, and/or an external network to facilitate the allocation of an IP address to the UE. The SMF selects and controls a user plane function (UPF) for the UE.

The UPF acts as the mobility anchor for both intra-Radio Access Technology (RAT) mobility (e.g., UE mobility between two gNBs) and inter-RAT mobility (e.g., UE mobility between NR and LTE). The UPF also carries out UP policy enforcement and QoS handling (e.g., support for reflective QoS, where the uplink QoS is the same as the downlink QoS).

The Network Slice Selection Function (NSSF), when needed, determines allowed network slices for the UE and selects an AMF set for the target Network Slice. The NSSF also selects Network Slice instances for the UE when needed.

The AF interacts with the policy control function (PCF) for policy control. The AF is typically implemented by the P-CSCF in the IMS. The AF extracts QoS information from the SIP signaling exchanged between two UEs. The AF provides such generic QoS information to the PCF. The AF also influences traffic routing, where the user traffic may be routed to a central data network (DN) or a local DN. The service provider or a third party may deploy the AF.

The PCF maps the generic QoS information from the AF to 5G-specific QoS parameters. These 5G QoS parameters are conveyed from the PCF to the SMF, the SMF to the AMF, the AMF to the gNB, and the AMF/gNB to the UE. A suitable 5G QoS flow between the UE and the UPF is then established to conform to the target QoS parameters.

The network repository function (NRF) supports the service discovery function, where one NF can discover other NFs. The NRF also maintains profiles of NF instances, which includes information such as an IP address, network slicing specifics, and supported services.

The network exposure function (NEF) securely exposes NF capabilities and events for third party, AFs, and edge computing. The NEF also allows secure provisioning of information by external AFs. The NEF supports the Common API Framework (CAPIF) to facilitate exposure of the 5GC to third parties to enable development of applications for various verticals.

The network slice-specific and SNPN authentication and authorization function (NSSAAF) supports network slice-specific authentication and authorization by a AAA[4] server. Such a AAA server may belong to a third party. The NSSAAF supports access to the standalone non-public network (SNPN)[5] by interacting with a Credentials Holder or Default Credentials Server.

The service communication proxy (SCP) provides services such as (i) indirect communication among NFs/NF services and (ii) message forwarding and routing to destination NF/NF service. The SCP is useful when indirect communication between the consumer and the producer for the service. For example, the consumer sends a request to the SCP, and the SCP selects an NF service instance. In another example, the SCP performs service discovery instead of the consumer by interacting with the NRF.

The Edge Application Server Discovery Function (EASDF) supports edge computing. The EASDF is responsible for processing the Domain Name System (DNS) messages based on the guidelines from the SMF. The EASDF exchanges DNS messages with the UE and reports the related information to the SMF. The EASDF also buffers (and discards as needed) DNS messages received from the UE and the DNS server. The EASDF has direct UP connectivity with the UPF over the N6 interface for transmitting DNS messages exchanged with the UE.

The Network Slice Admission Control Function (NSACF) supports monitoring and controlling the number of registered UEs per network slice. The NSACF also supports monitoring and control

4 AAA stands for Authentication, Authorization, and Accounting.
5 The SNPN can be viewed as a private network such as an enterprise network.

of the number of established PDU sessions per network slice. The NSACF facilitates event-based network slice status notification and reports to a consumer NF.

5G has also defined data storage NFs such as unified data repository (UDR) and unstructured data storage function (UDSF). The UDR stores structured data, such as subscription and policy data, handled by the PCF and the UDM. Examples of structured data include registration state, RAT type, and IPv6 prefix. The UDSF supports storing, modifying, and retrieving unstructured data by an NF. Examples of unstructured data include UE contexts with information about PDU sessions.

3.2 LTE-M: A 3GPP-Defined Wireless IoT Technology

LTE-M is intended for a wide range of IoT services, supporting higher data rates compared to NB-IoT. Section 3.2.1 discusses the key features of LTE-M. Section 3.2.2 provides a brief overview of the LTE-M radio interface, including the radio protocol stack and the physical layer frame structure. Section 3.2.3 summarizes the main LTE-M operations.

3.2.1 Key Features of LTE-M

Figure 3.6 highlights the key features of LTE-M (Tripathi and Reed 2019). Several of these LTE-M features are also available in NB-IoT as discussed in Section 3.3.1.

3.2.1.1 Radio Interface

LTE-M, like LTE, utilizes an OFDM-based radio interface. The LTE-M downlink utilizes Orthogonal Frequency Division Multiple Access (OFDMA), while the LTE-M uplink utilizes Single-Carrier Frequency Division Multiple Access (SC-FDMA). The subcarrier spacing is 15 kHz in LTE-M for typical signals and channels, including the channels that carry IoT traffic, such as the Physical Downlink Shared Channel (PDSCH) and the Physical Uplink Shared Channel

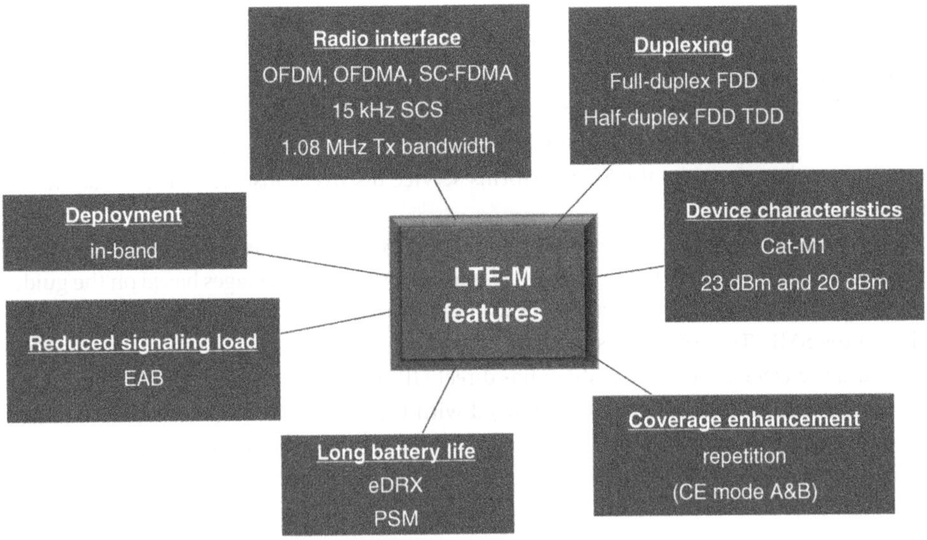

Figure 3.6 LTE-M Features.

(PUSCH). As the maximum number of PRBs in LTE-M is six and as each PRB occupies 180 kHz, the maximum transmission bandwidth for an LTE-M channel is 1.08 MHz.

3.2.1.2 Deployment

LTE-M is designed to be compatible with LTE. Hence, LTE-M utilizes in-band deployment, where a set of six contiguous PRBs is used as an LTE-M channel inside an LTE channel, and any remaining PRBs of the LTE channel are used for LTE UEs. Such in-band deployment of LTE-M significantly reduces the cost of LTE-M deployment.

3.2.1.3 Duplexing

An LTE-M device can support the duplexing methods that a typical smartphone UE supports, such as full-duplex FDD and TDD. Furthermore, an LTE-M device may support half-duplex FDD to simplify operations and reduce costs.

3.2.1.4 Receiver Design

The LTE-M UE receiver is designed to be much simpler than a typical LTE UE receiver. An LTE UE receiver supports multiple antennas and associated receiver chains consisting of components such as an analog-to-digital converter (ADC), a digital-to-analog converter (DAC), a filter, and a low noise power amplifier (LAN). In contrast, an LTE-M UE uses a single receive antenna and a single receiver chain.

3.2.1.5 Device Characteristics

Release 13 introduced a new category of devices for LTE-M: Category M1 (Cat-M1) devices. Cat-M1 devices are designed to operate with an LTE-M-capable eNB and possess characteristics such as the support for transmission in up to 1.08 MHz bandwidth. Cat-M1 devices may support Power Class 3 like LTE smartphones having the maximum transmit power of 23 dBm. Cat-M1 devices may also support a lower power level of 20 dBm, corresponding to Power Class 5, making an integrated power amplifier (PA) design feasible. Using an integrated PA reduces the space requirements and costs for an LTE-M device.

3.2.1.6 Coverage Enhancement

One of the key benefits of LTE-M is enhanced coverage compared to LTE. For example, the baseline LTE link budget is 140.7 dB, while the LTE-M link budget for Cat-M1 device is about 155.7 dB (Qualcomm 2018). Such an increase in the link budget in LTE-M can be attributed to repetitions, where the same information is transmitted multiple times to increase the reliability of reception through soft-combining of repeated signals at the receiver. Indeed, two coverage enhancement modes are supported: coverage enhancement (CE) Mode A and CE Mode B. CE Mode A has fewer repetitions, and CE Mode B has more repetitions. Specifically, CE Mode A allows up to 32 repetitions, and CE Mode B allows up to 2,048 repetitions. As CE Mode B supports more repetitions than CE Mode A, CE Mode B offers increased coverage compared to CE Mode A.

3.2.1.7 Long Battery Life

LTE-M introduces two features to increase the battery life—eDRX and PSM. A typical LTE UE in idle mode looks for a page message approximately once every 2.56 seconds. As an LTE-M IoT device is delay-tolerant, such frequent processing is unnecessary. Hence, LTE-M supports connected mode eDRX, such that the UE can process the downlink control information (DCI) approximately once every 10 seconds. In the idle mode, LTE-M supports an even longer eDRX cycle of approximately three hours. eDRX reduces the amount of processing that an LTE-M needs to perform, increasing battery life.

The PSM enables the LTE-M device to completely turn off the processing of LTE-M signals. When the PSM is activated, the network cannot reach the LTE-M UE because the LTE-M UE does not monitor LTE-M signals while in the PSM. The PSM supports device-initiated data transfer and periodic TAU.

3.2.1.8 Reduced Signaling Load

Many LTE-M devices are delay tolerant. Hence, when network congestion occurs on the radio interface, non-LTE-M devices can be prioritized. The feature of enhanced access barring (EAB) temporarily prohibits LTE-M devices of a target access class so that other higher priority devices can make use of the common physical random-access channel (PRACH). The System Information Block (SIB) 14 identifies the target access classes subjected to such access barring. An LTE-M device checks SIB14 before initiating a random-access procedure. When the expected signaling load is low, such access barring is removed through an indication on SIB14.

3.2.1.9 LTE-M/IoT Enhancements in the EPC

In addition to the radio interface features mentioned above, LTE-M supports IoT-centric core network-related features such as reduced tracking area update (TAU) signaling and EPC-Lite (Tripathi and Reed 2019). While LTE supports a periodic TAU timer of a few hours, LTE-M supports a much longer periodic TAU timer of many weeks. Such longer periodic TAU timers reduce the amount of TAU signaling generated by numerous IoT devices for a given MME, avoiding a Tsunami of signaling! If the service provider desires, a simplified EPC called EPC-Lite can be used for LTE-M, where MME, S-GW, and P-GW can be implemented in a single processor. EPC-Lite has limited functionalities (e.g., limited mobility support) compared to the EPC but can optimally support voice services.

3.2.2 LTE-M Air Interface: Protocol Stack and Physical Layer Frame Structure

3.2.2.1 LTE-M: Radio Protocol Stack

Figure 3.7 shows the radio interface protocol stack for LTE (Tripathi and Reed 2019). This protocol stack is used as a baseline in LTE-M and NB-IoT, and suitable modifications are introduced to support the unique characteristics of LTE-M and NB-IoT.

The radio protocol stack has three layers. Layer 1 is the Physical (PHY) layer. Layer 2 includes Medium Access Protocol (MAC), Radio Link Control (RLC), and Packet Data Convergence Protocol (PDCP). Layer 3 includes Radio Resource Control (RRC). User traffic enters the radio protocol stack as Layer 3 information. The PHY layer implements baseband and RF processing. For example, PHY layer functions are OFDM processing and multiple input multiple output (MIMO) processing. Example MAC functions include scheduling by the eNB to allocate radio resources to different UEs and controlling fast Hybrid Automatic Repeat reQuest (HARQ) retransmissions occurring at the PHY layer. The RLC protocol performs functions such as Automatic Repeat reQuest (ARQ) retransmissions. Examples of functions performed by the PDCP include Robust Header Compression (RoHC) and security mechanisms such as ciphering and integrity protection. The RRC layer carries access stratum (AS) signaling, such as mobility management-related signaling and configuration of Layer 1 and Layer 2 protocols.

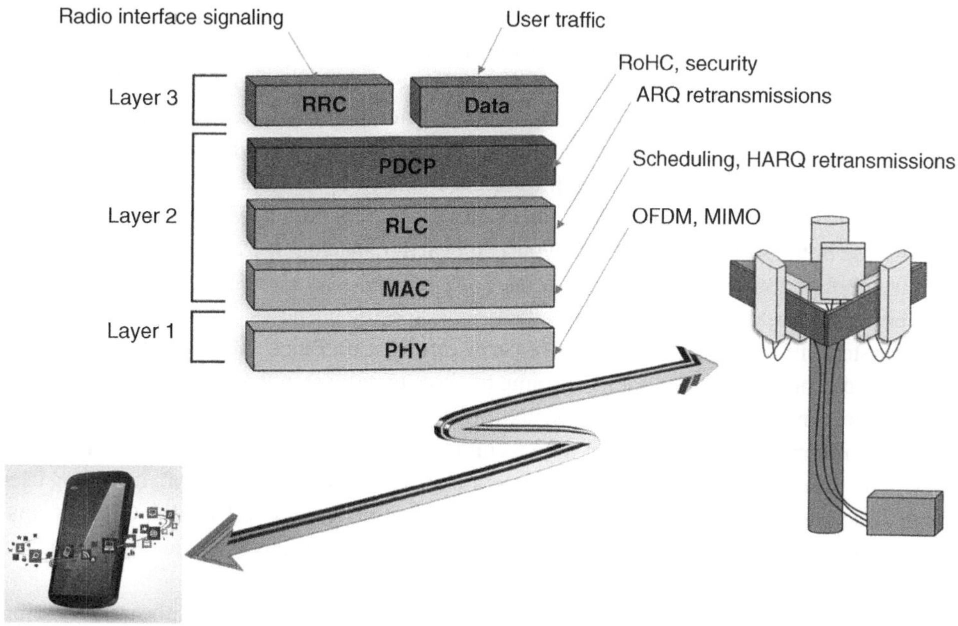

Figure 3.7 Radio Interface Protocol Stack for LTE, LTE-M, and NB-IoT. *Source:* Julien Eichinger / Adobe Stock Photo.

3.2.2.2 LTE-M: PHY Frame Structure

Figure 3.8 illustrates the time–frequency resources of the frame structure used by the LTE physical layer (Tripathi and Reed 2019). This frame structure is reused in LTE-M with a constraint on the maximum channel bandwidth. The frame structure along with the radio interface protocol stack enables the UE and the radio network to communicate successfully with each other.

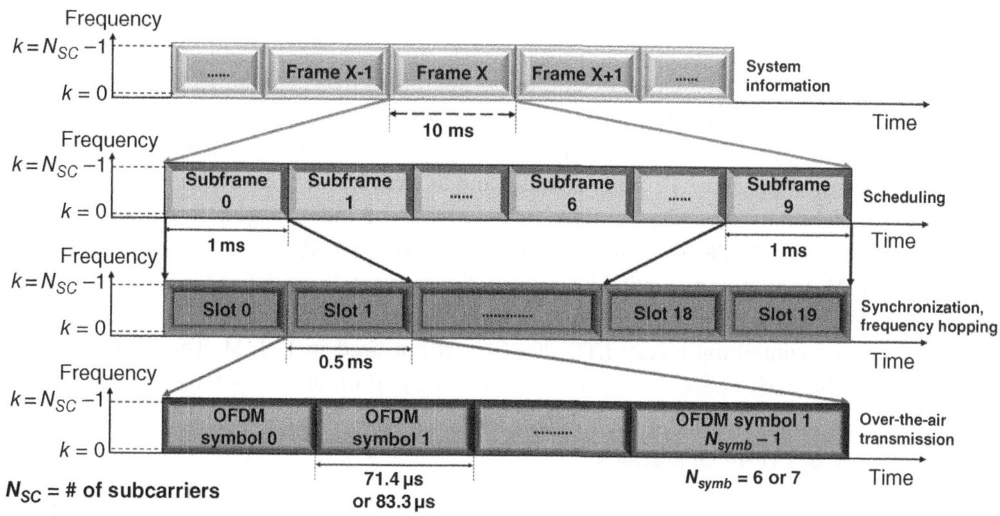

Figure 3.8 LTE Frame Structure.

LTE utilizes Type 1 frame structure for FDD, Type 2 frame structure for TDD, and Type 3 frame structure for operation in the unlicensed spectrum. Figure 3.8 shows Type 1 frame structure.

An LTE frame is 10 ms long and consists of 10 subframes. Each 1 ms subframe is divided into 10 0.5 ms slots. A slot consists of seven OFDM symbols when a normal cyclic prefix (CP) is used, and six OFDM symbols when an extended cyclic prefix is used. Cyclic prefix helps an OFDM system address the challenge of the Inter-Symbol Interference caused by multipath signals in the radio environment. The OFDM symbol includes CP and a useful symbol period, where the useful symbol period is inverse of the subcarrier spacing. Typical subcarrier spacing in LTE is 15 kHz, leading to the useful symbol period of $1/15\text{ kHz} = 66.67\ \mu\text{s}$.

The frame, subframe, slot, and symbol structures serve specific purposes. For example, the frame helps carry system information in the form of Master Information Block (MIB) and SIBs. The eNB executes a scheduling algorithm in each subframe to allocate radio resources to support user traffic and signaling transmission. Slots facilitate synchronization and frequency hopping. OFDM symbols transport information over the air and facilitate synchronization. Specifically, during an OFDM symbol, the subcarriers carry modulation symbols representing information, including user traffic, signaling messages, and physical signals (e.g., reference signals). In LTE, the OFDM transmitter utilizes fast Fourier transform (FFT) to combine modulation symbols carried by subcarriers and create a time-domain signal for transmission.

There are 600 subcarriers in the channel bandwidth of 10 MHz. As the typical subcarrier spacing is 15 kHz, a total of ($600\text{ subcarriers} * 15\text{ kHz} = 9\text{ MHz}$) spectrum is used as transmission bandwidth. The remaining ($10 - 9 = 1\text{ MHz}$) spectrum is the total guard band with a 0.5 MHz guard band on either side of the transmit signal. The combination of one subcarrier and one OFDM symbol is called a resource element (RE). One RE carries one modulation symbol. The RE is the smallest resource unit (RU) in LTE. A PRB in LTE is defined as a dozen consecutive subcarriers for the duration of a slot (i.e., 0.5 ms). Resource allocation in LTE occurs with the time granularity of a subframe. When an eNB decides to allocate radio resources to a UE, the minimum allocation is one PRB in each of the two slots of a 1 ms subframe. Various physical channels and signals are created using the time-frequency resources of the physical layer frame shown in Figure 3.8. The roles of these channels and signals are mentioned below while discussing LTE-M operations.

Figure 3.9 illustrates the physical layer concepts of narrowbands and subframes for LTE-M (Tripathi and Reed 2019).

As shown in the top-left corner of Figure 3.9, an LTE-M radio channel can be placed inside a regular LTE channel. The 3GPP has defined the concept of a narrowband for LTE-M, where each narrowband consists of six consecutive PRBs corresponding to an LTE-M radio channel. The bottom-right corner of Figure 3.9 shows how narrowbands are identified using a narrowband index. As there are 50 PRBs in a 10 MHz LTE radio channel and as each LTE-M narrowband consists of six PRBs, a total of eight narrowbands, identified by a narrowband index ranging from 0 to 7, can be defined. The remaining ($50\text{ PRBs} - (8\text{ narrowbands} * 6\text{ PRBs/narrowband}) = 50 - 48 = 2\text{ PRBs}$) is not used in LTE-M. In summary, a suitable number of narrowbands are defined in a given channel bandwidth. Furthermore, in the time domain, as shown in the right portion of Figure 3.9, the first part of a subframe containing legacy LTE PDCCHs is not used in LTE-M. The MTC PDCCH (MPDCCH) and the PDSCH are created in the remaining portion of the subframe.

3.2.3 Summary of LTE-M Operations

Examples of critical operations or procedures in LTE-M include initial attach, EPS bearer setup, random-access procedure, RRC connection setup, DL data transfer, UL data transfer, handover, and idle mode activities (e.g., cell reselection and tracking area updates). These LTE-M operations are briefly discussed in Sections 3.2.3.1 to 3.2.3.7.

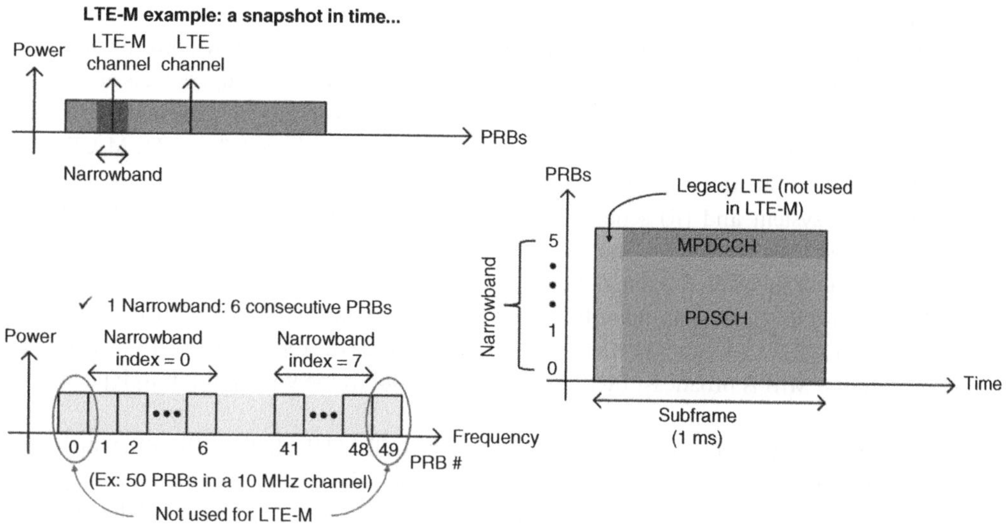

Figure 3.9 Narrowbands and Subframes in LTE-M.

3.2.3.1 Initial Attach and EPS Bearer Setup in LTE-M

Before an LTE-M UE can exchange IoT traffic, prerequisites such as initial attach and EPS bearer setup need to be met. The initial attach procedure makes the network aware of the UE's presence in the service area, and the EPS bearer setup procedure establishes suitable logical connections for the UE so that the UE can exchange user traffic via an EPS bearer. Figure 3.10 shows major steps of initial attach and EPS bearer setup procedures in LTE-M (Tripathi and Reed 2019).

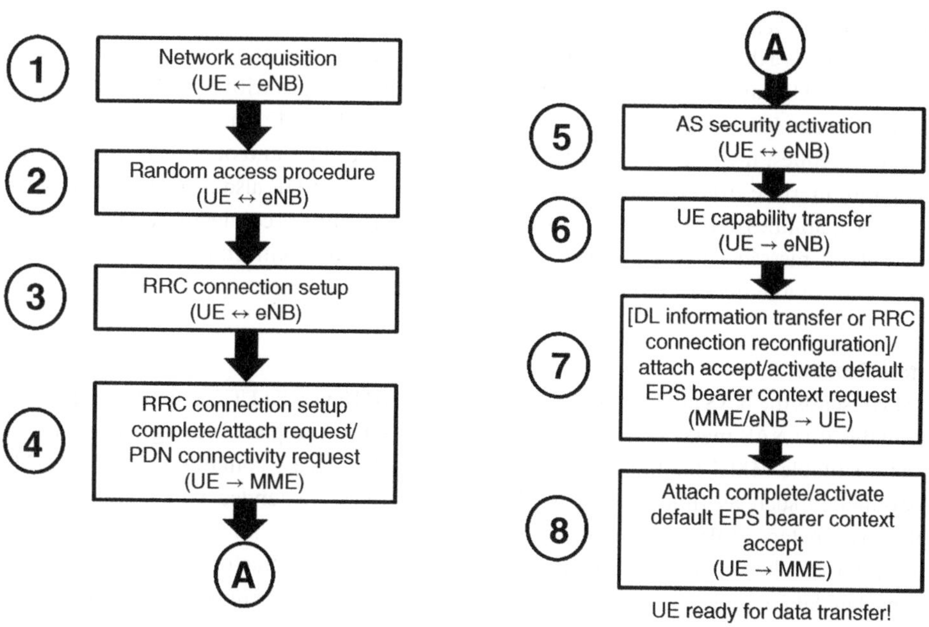

Figure 3.10 Initial Attach and EPS Bearer Setup in LTE-M.

Step 1: Network Acquisition

Network acquisition enables the UE to achieve downlink time and frequency synchronization and obtain critical information about the network. The LTE-M UE, upon power-up, detects an LTE-M cell by observing the Primary Synchronization Signal (PSS) and Secondary Synchronization Signal (SSS) and determines the Physical Cell ID (PCI) of the LTE-M-capable cell. The LTE-M UE then processes the Physical Broadcast Channel (PBCH) to obtain (i) the MIB containing critical information about the system and (ii) schedulingInfoSIB1-BR containing information about SIB1-bandwidth reduced (SIB1-BR). Note that the LTE-M UE can process the PSS, the SSS, and the PBCH of LTE because they are all located within six PRBs. However, an LTE-M UE cannot process LTE's SIB1, because SIB1 can span more than six PRBs. Hence, the 3GPP has defined SIB1-BR as something that an LTE-M UE can process in six PRBs. To increase the reliability of MIB reception at the LTE-M UE, the MIB is broadcast by the LTE-M cell in subframe 9 in addition to typical subframe 0.

After processing the MIB and schedulingInfoSIB1-BR, the LTE-M UE obtains SIB1-BR, which provides important LTE-M parameters such as the signal threshold for CE, the starting OFDM symbol number for the MPDCCH, and the schedule of other SIBs. The LTE-M UE learns about the random-access parameters by acquiring SIB2. For example, SIB2 contains Reference Signal Received Power (RSRP) thresholds for CE levels and PRACH configurations for different CE levels. SIB2 also provides information about the MPDCCHs to monitor for random-access response (RAR).

Step 2: Random-Access Procedure

After acquiring SIB2, the UE can initiate the random-access (RA) procedure. The RA procedure enables the UE to achieve uplink synchronization with a gNB and initiate communications with the radio network. The UE sends an RA preamble to the LTE-M cell and receives a RAR. Using the UL grant specified by the RAR, the UE initiates the establishment of a RRC signaling connection. See Figure 3.8 for details on the random-access procedure.

Step 3: RRC Connection Setup

The RRC connection is a dedicated signaling connection between the UE and the serving cell on the radio interface. The UE can exchange radio interface signaling messages with a gNB upon successful establishment of an RRC connection in a cell of a gNB. Like an LTE UE, the LTE-M UE sends an RRC connection request message toward the end of the RA procedure and obtains the configuration of the radio interface from the serving LTE-M cell. The LTE-M UE also provides some of its LTE-M related capabilities to the network. Furthermore, the UE initiates connectivity with the EPC by sending a NAS message in the last RRC message of the RRC connection setup procedure. Figure 3.9 provides more details of the RRC connection setup procedure.

Step 4: Attach Request

The LTE-M UE sends the NAS attach request message to the MME via the LTE-M eNB. The LTE-M UE encapsulates this NAS message in the RRC connection setup complete message. The LTE-M eNB selects a suitable MME for the UE. The attach request message contains several LTE-M-specific parameters such as timer T3324 for PSM, timer T3412 (extended value) for periodic TAU, eDRX parameters, RoHC parameters for IP-based control plane for IoT traffic transport, and an indication for the support of SMS for the IoT traffic. The NAS attach request message also contains a packet data network (PDN) connectivity request message for connecting to a PDN (e.g., Internet). The PDU connectivity request message includes typical protocol configuration option

(PCO) containing IP address request and DNS address request. The UE can indicate its preference for non-IP data transfer by indicating the PDU type= "Non-IP."

Mutual authentication between the LTE-M UE and the MME occurs. The MME obtains authentication-related parameters from the HSS in support of mutual authentication. The NAS security between the UE and the MME is also set up. The UE provides the access point name (APN) of the PDN through the EPS SM information transfer procedure. The MME coordinates with the S-GW and the P-GW to set up the S5/S8 bearer potion of the EPS bearer. The MME provides the NAS Attach Accept message to the eNB.

Step 5: Access Stratum Security Activation

The LTE-M UE and the LTE-M eNB exchange RRC messages to set up AS security to ensure security of RRC signaling messages on the radio interface. The RRC messages can now be both ciphered and integrity protected.

Step 6: UE Capability Exchange

The LTE-M UE conveys its LTE-M capabilities to the LTE-M eNB by sending the RRC UE Capability Enquiry message. For example, the LTE-M eNB Evolved—Universal Terrestrial Radio Access (E-UTRA) as RAT type in its UE Capability Enquiry messages to the LTE-M UE. The LTE-M UE replies with the UE Capability Information message and specifies its DL/UL UE category (e.g., M1 for Category M1). The UE also specifies its power class (e.g., power class 5 for 20 dBm). The capability exchange procedure enables the eNB to configure the UE with suitable radio interface parameters that are consistent with the UE's radio interface capabilities and the gNB's own radio interface capabilities.

Step 7: Attach Accept

The LTE-M eNB sends an RRC message, such as RRC Connection Reconfiguration or DL Information Transfer, which encapsulates the NAS Attach Accept message received from the MME in the S1AP[6] Initial Context Setup Request or DL NAS Transport message. The NAS Attach Accept message specifies, along with typical parameters such as Globally Unique Temporary Identifier (GUTI) and PCO, LTE-M-related parameters such as eDRX parameters, T3412 extended value for periodic TAU, T3324 for PSM, "Control Plane Only Indicator" (if applicable), and RoHC for IP-based CP. The Attach Accept message contains the Activate Default EPS Bearer Context Request message to set up a default EPS bearer with specific QoS parameters for the UE. The RRC Connection Reconfiguration message specifies the Data Radio Bearer (DRB) portion of the EPS bearer. The default EPS bearer helps carry user traffic between the UE and a network identified by an APN. If a new operator-aware application for a given APN requires QoS characteristics that is different from the QoS characteristics associated with the default EPS bearer for such APN, a separate dedicated EPS bearer is established to carry use traffic of that application.

Step 8: Attach Complete

The LTE-M UE replies to the Attach Accept/Activate Default EPS Bearer Context Request message with the Attach Complete/Activate Default EPS Bearer Context Accept message by encapsulating such NAS message in the RRC UL Information Transfer message. The network completes the S1-U/ access bearer portion of the default EPS bearer so that all three components of the default EPS

---

6 S1AP stands for S1 Application Protocol, which is a signaling protocol on the S1-MME interface between the eNB and the MME.

bearer are fully set up: (i) the DRB between the LTE-M UE and the LTE-M gNB in a cell, (ii) the S1-U or access bearer between the eNB and the S-GW, and (iii) the S5/S8 bearer between the S-GW and the P-GW.

3.2.3.2 Random Access in LTE-M

Figure 3.11 highlights LTE-M-specific enhancements to examine the RA procedure more closely (Tripathi and Reed 2019). As mentioned earlier, before the LTE-M UE can initiate the RA procedure, it completes network acquisition by synchronizing with the LTE-M cell using the PSS and the SSS and acquiring the MIB, SIB1-BR, and SIB2.

The LTE-M UE measures the RSRP of its serving LTE-M cell and determines its CE level. The CE level dictates the number of repetitions. If the measured RSRP is too low, the radio channel condition is poor. Hence, a CE level corresponding to more repetitions is used so that reliable communication becomes possible even under adverse channel conditions. The UE sends an RA preamble on the PRACH to the eNB using the PRACH configuration specified in SIB2 and uses the correct number of repetitions. The PRACH in LTE-M uses the subcarrier spacing of 1.25 kHz like LTE.

The LTE-M eNB responds with the RAR that contains typical configuration parameters for the UE, such as the Timing Advance Command to facilitate UL synchronization for the UE and Cell-Radio Network Temporary Identifier (C-RNTI). The LTE-M eNB also specifies LTE-M-related parameters such as (i) the Narrowband Index (NBI) associated with the UL grant and the PRBs allocated to the UE so that the UE can transmit an RRC message and (ii) information about MPDCCHs that the UE can monitor for future radio resource allocation. The channel bandwidth is divided into multiple narrowbands, with each narrowband consisting of six consecutive PRBs. Each narrowband is given an index.

Using the allocated UL grant, the LTE-M UE sends Message 3 (Msg 3), such as the RRC Connection Request message, containing its identity. If the UE has previously been allocated a GUTI by an MME, the UE includes S-Temporary Mobile Subscriber Identity (S-TMSI) in this Msg 3. S-TMSI is a short form of GUTI. If the UE does not have S-TMSI, it generates a random number and includes it in the RRC Connection Request message. As more than one LTE-M UE might have used the same RA preamble ID, more than one LTE-M UE may send the RRC Connection Request message. The UE ID (i.e., S-TMSI or a random number) enables the eNB to resolve the conflict among the UEs that might have used the same RA preamble. The LTE-M eNB sends the Contention

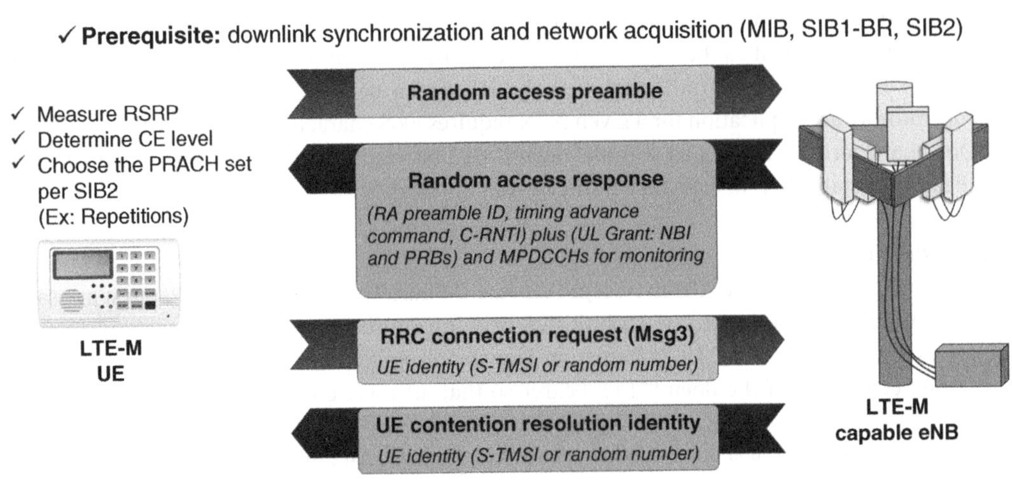

Figure 3.11 Random-Access Procedure in LTE-M. *Source:* Tripathi and Reed (2019) / The Wireless University; Art of Success / Adobe Stock Photos.

Resolution message (containing the identity of the UE) to the UE that the LTE-M eNB has successfully detected.

The successful RA procedure enables the RRC connection setup procedure to continue. In case an LTE-M UE's RA procedure fails, such UE retries the RA procedure by sending another RA preamble after a random wait time.

3.2.3.3 RRC Connection Setup in LTE-M

Figure 3.12 summarizes the key steps of the RRC connection setup for LTE-M (Tripathi and Reed 2019). A dedicated RRC connection between the LTE-M UE and the serving cell enables the LTE-M UE and the serving cell to reliably exchange UE-specific RRC and NAS signaling messages with each other.

Like LTE, LTE-M also utilizes a three-way handshake between the LTE-M UE and the LTE-M eNB to set up the RRC connection. As more than one UE may send the RRC Connection Request to the eNB on the same radio resources, the message is said to be using Signaling Radio Bearer (SRB) 0, which is common across multiple UEs. SRB0 uses a Common Control Channel (CCH) as the logical channel. As mentioned earlier, the RRC Connection Request message includes the UE ID, such as S-TMSI or a random number. The LTE-M UE indicates the purpose behind the message by specifying the cause of the RRC connection establishment (e.g., mobile-originated signaling).

The LTE-M eNB responds with the RRC Connection Setup, which also uses SRB0. The RRC Connection Setup message provides a configuration for a dedicated SRB called SRB1. SRB1 is specific to the UE and can specify a dedicated signaling configuration, such as the number of retransmissions. This message also specifies LTE-M-specific configuration, such as MPDCCH NBI and the number of MPDCCH repetitions, to facilitate the MPDCCH monitoring by the LTE-M UE.

Now that SRB1 is configured, the UE uses SRB1 to send the RRC Connection Setup Complete, which is the response to the RRC Connection Setup message. The RRC Connection Setup Complete message conveys certain capabilities useful for LTE-M operations, such as support for CE Mode B, attach without PDN Connectivity, and UP/CP Cellular IoT Optimizations. As discussed earlier in this section, the RRC Connection Setup Complete message also encapsulates the NAS Attach Request message, which in turn carries a PDN Connectivity Request message with a request to connect to a PDN.

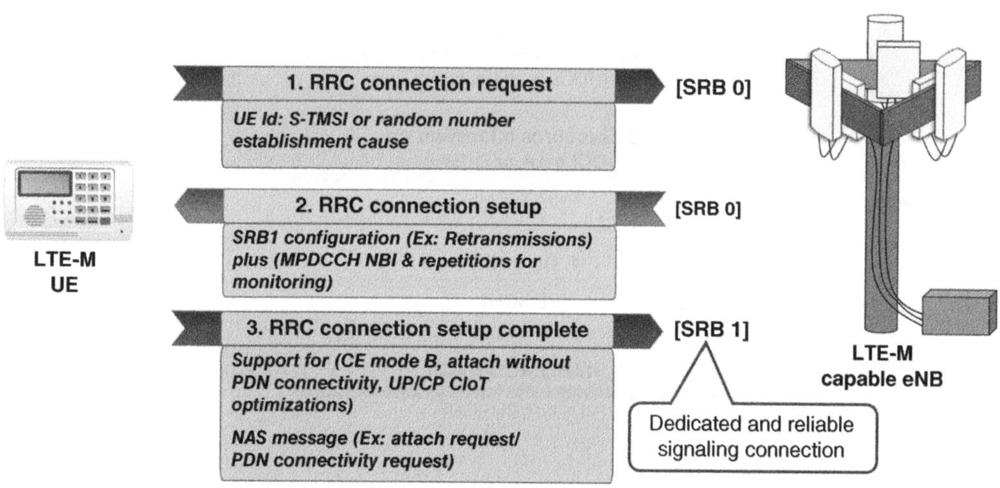

Figure 3.12 RRC Connection Setup in LTE-M. *Source:* Tripathi and Reed (2019) / The Wireless University; Art of Success / Adobe Stock Photos.

3.2.3.4 Example Downlink Data Transfer in LTE-M

After one or more UE-specific EPS bearers are set up, the LTE-M UE can exchange user traffic with the LTE-M eNB. Figure 3.13 illustrates the main steps of the DL data transfer on the radio interface in LTE-M (Tripathi and Reed 2019).

In *Step 1*, the LTE-M UE observes the Cell-specific Reference Signal (CRS) of the serving LTE-M cell to quantify the prevailing DL radio channel conditions. The CRS in LTE-M reuses the same structure as the CRS in LTE.

In *Step 2*, the LTE-M UE sends the channel condition report to the LTE-M eNB to enable the eNB to determine optimal PDSCH transmission parameters. This report contains information such as the Channel Quality Indicator (CQI), Precoding Matrix Indicator (PMI), and Rank Indicator (RI). As the LTE-M UE has a single receive antenna, RI cannot be greater than one in LTE-M. Furthermore, while LTE supports modulation schemes such as QPSK, 16QAM, and 64QAM, LTE-M supports QPSK and 16QAM only to reduce complexity. Additionally, while LTE supports Turbo coding for the PDSCH, LTE-M supports simpler convolutional coding for the PDSCH. The channel condition report is sent on the PUCCH or the PUSCH. If the UE has been allocated a PUSCH in the current subframe, the channel condition report is sent on the PUSCH along with other PUSCH payloads, such as IoT traffic. If no PUSCH is allocated to the UE for the current subframe, the UE utilizes the PUCCH to convey the channel condition report. For increased reliability, the channel condition report may be repeated.

The LTE-M eNB executes a scheduling algorithm that considers the channel condition report along with numerous factors such as the LTE-M UE capabilities (e.g., half-duplex FDD), the number of PRBs in the narrowband, the number of narrowbands, and QoS needs of LTE-M UEs to allocate DL radio resources. The LTE-M eNB then conveys allocated resources to the LTE-M UE using DCI Format 6-1A or 6-1B on the MTC PDCCH (MPDCCH) in *Step 3*. MTC is MTC. The MPDCCH itself may use repetitions for increased reliability. The MPDCCH may also use frequency hopping. The DCI format conveys PDSCH resource allocation parameters such as the allocated narrowband index, the number of PRBs assigned to the UE in the allocated narrowband index, MCS, and the

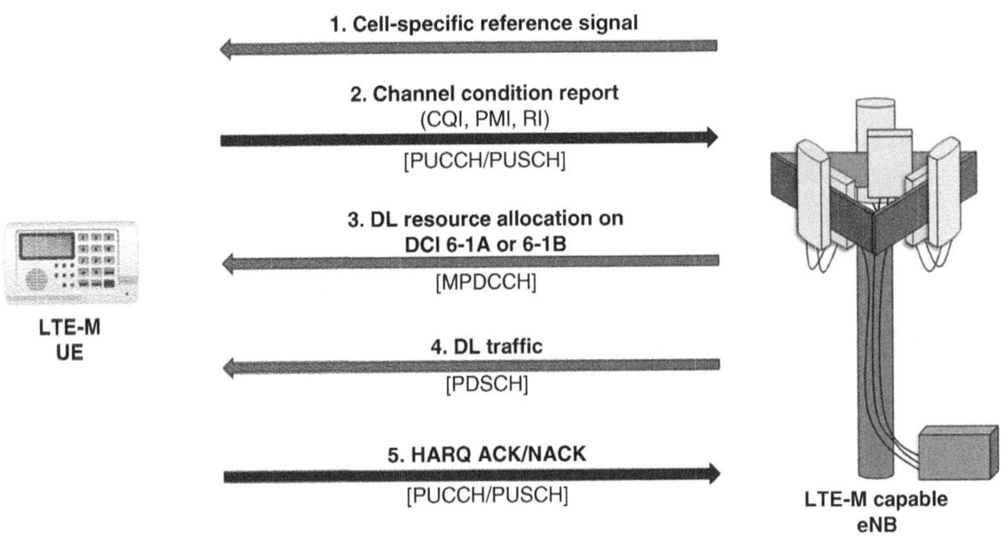

Figure 3.13 Downlink Data Transfer in LTE-M. *Source:* Art of Success / Adobe Stock Photos.

number of PDSCH repetitions. LTE-M utilizes cross-subframe scheduling, where the allocated PDSCH resources are used by the UE beginning subframe $(n + 2)$ if the DCI format has indicated the resource allocation in the subframe n. If the MPDCCH has been repeated, subframe n contains the last repetition of the MPDCCH.

The LTE-M retrieves the packet from the PDSCH in Step 4. Note that the PDSCH may be repeated as specified in the DCI Format in Step 3. The PDSCH utilizes QPSK or 16QAM and a single-layer transmission. Frequency hopping may be used on the PDSCH.

The LTE-M UE attempts to decode the packet on the PDSCH. The UE conveys the HARQ ACK or NACK[7] Feedback to the LTE-M eNB depends on whether or not the packet has been successfully decoded in Step 5. The UE sends this HARQ feedback on the PUCCH or the PUSCH, such as how it sends the channel condition report in Step 2. If the PDSCH (or the last repetition of the PDSCH if repetitions are used) is received in subframe m, the LTE-M UE sends the HARQ ACK/NACK in subframe $(m + 4)$. It is also possible to add additional delay to allow the LTE-M UE to send the HARQ feedback in a subframe after subframe $(m + 4)$. Furthermore, the HARQ ACK/NACK may be repeated for increased reliability. Additionally, HARQ-ACK bundling may be used, where the UE transmits a single HARQ-ACK feedback for multiple DL transport blocks (referred to as packets here).

The LTE-M eNB sends a new packet if it has received HARQ ACK or retransmits a previously sent packet if it has received a HARQ NACK, as long as the maximum number of HARQ retransmissions has yet to be reached.

3.2.3.5 Example Uplink Data Transfer in LTE-M

Key steps of the UL data transfer procedure on the LTE-M radio interface are summarized in Figure 3.14 (Tripathi and Reed 2019).

The main difference between the DL data transfer and the UL data transfer is the Buffer Status Report (BSR), which conveys the amount and priority of the data residing at the UE. In the DL, the

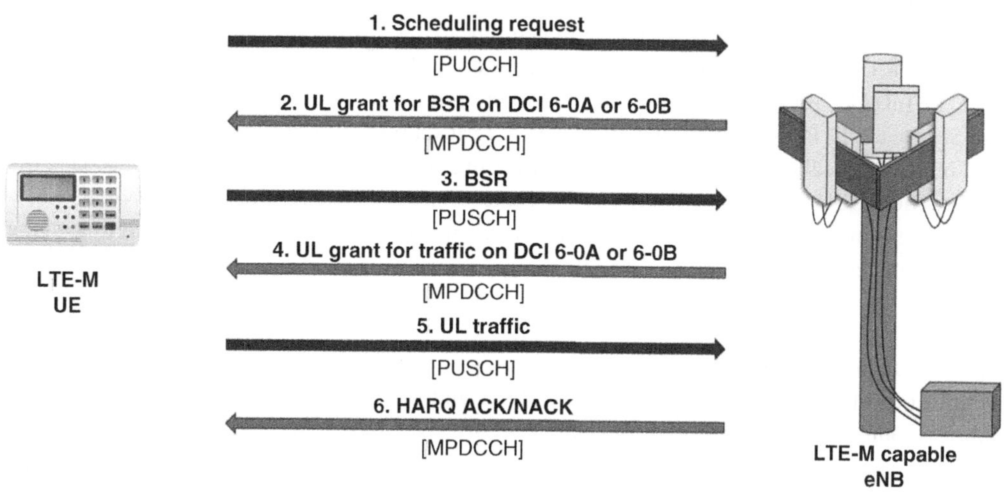

Figure 3.14 Uplink Data Transfer in LTE-M. *Source:* Art of Success / Adobe Stock Photos.

7 HARQ stands for Hybrid Automatic Repeat Request. ACK is acknowledgment and NACK is negative acknowledgment.

user data is in the buffer, and the eNB knows the amount and priority of the data. However, data resides at the UE for the UL. Hence, the BSR is needed for the UL data transfer so that the eNB can allocate a suitable amount of UL radio resources with appropriate priority to reflect the amount and priority of data.

In Step 1, when the LTE-M UE has data (e.g., a signaling message or IoT traffic) to send, it sends a Scheduling Request (SR) on the configured PUCCH. For increased reliability, the PUCCH can use repetition. Furthermore, frequency hopping is inherently supported on the PUCCH for enhanced reliability in the UL.

The LTE-M eNB executes a scheduling algorithm to allocate adequate UL resources to the UE so that the UE can send the BSR. The LTE-M eNB specifies the UL grant for the UE's BSR using the DCI 6-0A or 6-0B on the MPDCCH in Step 2. The MPDCCH itself may use repetitions for increased reliability. The MPDCCH may also use frequency hopping.

In *Step 3*, the LTE-M UE sends the BSR on the PUSCH to the LTE-M eNB using the allocated UL grant.

The LTE-M eNB learns about the UL resource requirements of the UL data waiting at the UE by processing the BSR. The eNB then executes a scheduling algorithm that considers a variety of factors such as the LTE-M UE capabilities (e.g., half-duplex FDD), UL channel conditions (estimated by the eNB based on the observation of the SRS), the number of PRBs in the narrowband, the number of narrowbands, the type and the amounts of data in the UE's UL buffer, and QoS needs of LTE-M UEs to allocate UL radio resources. The LTE-M eNB conveys allocated resources to the LTE-M UE using the DCI format 6-0A (for CE Mode A) or 6-0B (for CE Mode B) on the MPDCCH in Step 4. As mentioned in Step 2, the MPDCCH may use repetitions and frequency hopping. The DCI formats 6-0A and 6-0B include the resource allocation information, such as the narrowband index specifying the narrowband, the allocated PRBs in the narrowband, MCS, and the number of PUSCH repetitions.

The LTE-M UE transmits a packet on the PUSCH based on the received UL grant in Step 5. LTE-M utilizes cross-subframe scheduling, where the UE utilizes the allocated PUSCH resources to transmit a packet beginning subframe $(n + 4)$ if the DCI format has indicated the resource allocation in the subframe n. If the MPDCCH has been repeated, subframe n contains the last repetition of the MPDCCH. The UE transmits on a single antenna. The PUSCH in LTE-M can make use of QPSK or 16QAM. Frequency hopping is supported on the PUSCH in LTE-M.

In Step 6, the LTE-M eNB attempts to decode the packet on the PUSCH and conveys the HARQ ACK or NACK feedback to the LTE-M UE, depending on whether it has successfully decoded the packet. The LTE-M eNB sends this HARQ feedback on the MPDCCH. The LTE-M eNB also allocates UL resources for new packet transmission or retransmission on the PUSCH, depending on the HARQ feedback it sends to the LTE-M UE.

3.2.3.6 Handover in LTE-M

Handover in LTE-M is similar to handover in LTE. Handover enables the LTE-M UE to experience seamless communications even as it moves around in a given LTE-M service area. Figure 3.15 summarizes the key steps of the LTE-M handover (Tripathi and Reed 2019). Like LTE, LTE-M supports UE-assisted network-controlled handover, where the LTE-M UE in the RRC_CONNECTED state provides measurements to the serving LTE-M eNB that makes a handover decision and instructs the LTE-M UE to switch the RRC connection to a target cell by sending a handover command to the UE. The serving LTE-M eNB communicates with the target LTE-M eNB before issuing a handover command to the UE so that the target LTE-M eNB can reserve resources for the LTE-M UE.

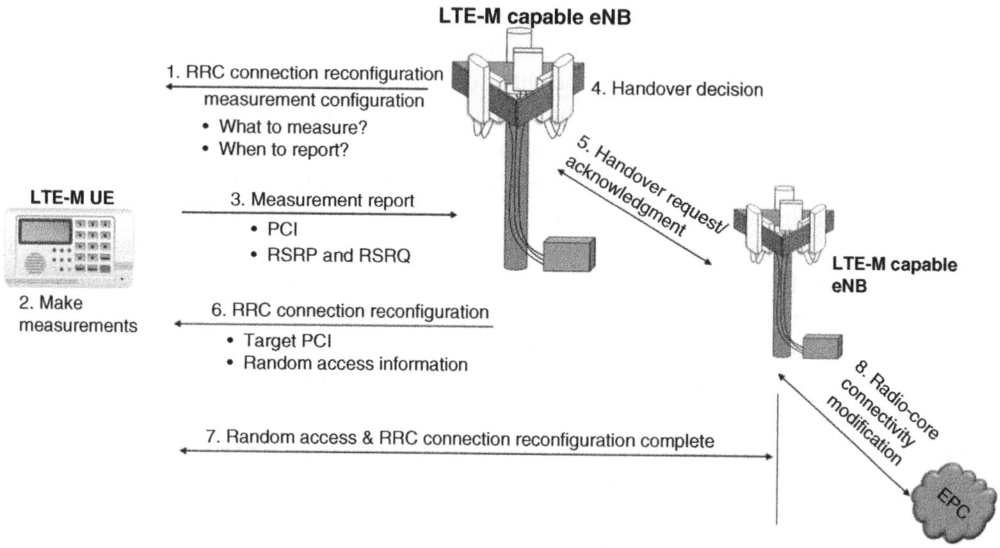

Figure 3.15 Handover in LTE-M. *Source:* Art of Success / Adobe Stock Photos.

The LTE-M eNB configures the UE with suitable measurements using the RRC Connection Reconfiguration message in *Step 1*. This configuration plays a key role in ensuring reliable handover and hence seamless communication experience for the LTE-M UE. The eNB specifies the carrier frequency or frequencies that the UE should measure. On a given carrier frequency, the UE autonomously detects cells (i.e., PCIs). The eNB also specifies what measurements to make, such as RSRP and Reference Signal Received Quality (RSRQ). Furthermore, the eNB configures the measurement reporting or handover events for which the UE needs to send a Measurement Report message. The eNB may specify the periodicity of measurement reports in case of periodic reporting.

In Release 13, when LTE-M was introduced as a feature by the 3GPP, six intra-RAT[8] handover events and two inter-RAT handover events were defined. Intra-RAT handover events facilitate handover of the LTE-M UE within the same (e.g., LTE-M) technology and are expected to be common in LTE-M deployments. Inter-RAT handover events are intended for handover between two different RATs. A1 to A6 are intra-RAT handover events, while B1 and B2 are inter-RAT handover events. Simplified and brief descriptions of these events are given later. See 3GPP (2024c) for exact definitions of these events; the exact definitions of these events have additional parameters such as hysteresis, offsets, and timers.

Event A1 corresponds to the case when a serving cell measurement (e.g., RSRP) exceeds a threshold, implying an improving radio channel condition between the UE and the serving cell. Hence, an impending handover can be avoided from the currently serving cell to a different cell.

Event A2 occurs when a service cell measurement drops below a threshold. This event implies a worsening radio channel condition. This event can be used to implement a blind handover to an inter-RAT cell when the inter-RAT cell measurement is unavailable or not configured.

Event A3 implies that a neighbor cell measurement exceeds the serving cell measurement by a certain amount. This is the most widely deployed handover event in commercial networks for intra-RAT handover.

8 RAT stands for Radio Access Technology. Examples of an RAT include LTE and NR.

Event A4 corresponds to the situation where a neighbor cell measurement exceeds a threshold. This event points to a potentially good radio channel condition between the UE and the neighbor cell, indicating that the handover to such neighbor cell may be desirable.

Event A5 corresponds to the case where the serving cell measurement is below a threshold and a neighbor cell measurement is above a threshold, pointing to a poor radio environment in the serving cell and a good radio environment in the neighbor cell, suggesting that the handover from the serving cell to the neighbor cell may be desirable.

Event A6 occurs when a neighbor cell measurement is better than a secondary cell measurement by a certain amount. This event can be used to replace a weaker secondary cell with a neighbor cell. This event is useful in the case of Carrier Aggregation (CA), where multiple carrier frequencies are used, with one carrier frequency providing a primary cell and other carrier frequencies providing secondary cells. Replacing a weak secondary cell with a better neighbor cell can increase throughput when CA is deployed.

Event B1 occurs when an inter-RAT neighbor cell measurement exceeds a threshold. This event indicates that the handover from the currently serving cell to the inter-RAT neighbor cell may be desirable.

Event B2 corresponds to the case where the serving cell measurement is below a threshold and an inter-RAT neighbor cell measurement is above a threshold. This indicates a poor radio environment in the serving cell and a good radio environment in the inter-RAT neighbor cell, suggesting that the handover from the serving cell to the inter-RAT neighbor cell may be desirable. When measurements of the serving cell and an inter-RAT neighbor cell are available, this event can be utilized for a reliable inter-RAT handover.

In *Step 2*, the LTE-M UE follows the measurement configuration specified by the LTE-M eNB and makes measurements such as RSRP and RSRQ of the currently serving cell and autonomously detects neighboring cells. On a given carrier frequency, the received signal contains signals from the serving cell and the neighboring cells, and the LTE-M UE can use the PCIs of cells to determine the RSRP and RSRQ of cells.

When the reporting criterion for the measurement reporting is met (e.g., the occurrence of an event such as A3), the UE sends the Measurement report message containing the cell identity (i.e., PCI) and RSRP and RSRQ of the serving and relevant neighbor cell(s) in *Step 3*.

In *Step 4*, the serving LTE-M eNB processes the UE's measurement report and executes a handover algorithm. If the serving eNB makes a handover decision, it exchanges handover signaling messages with the target eNB in Step 5.

In *Step 5*, the serving LTE-M eNB sends a Handover Request message to the target eNB[9]. The Handover Request message contains the target PCI and information related to the UE-specific S1-MME connection and the S1-U bearer. The E-RAB QoS parameters are also specified. The target LTE-M eNB executes the admission control algorithm and evaluates the QoS needs of the current E-RABs of the UE against the availability of radio resources. Suppose the target eNB decides to accept the handover request. In that case, it responds to the serving LTE-M eNB with the Handover Request Acknowledge, which contains parameters such as C-RNTI allocated to the UE in the new cell, a dedicated random-access preamble (to avoid contention with other UEs using the PRACH in the new cell), and system information. This message also includes the list of E-RABs

9 When the X2 connectivity between the eNBs is established when eNBs are activated, they exchange their cell configuration including PCIs that they manage. Hence, when the UE reports a PCI, the serving eNB knows which eNB is managing such reported PCI.

admitted by the target eNB. If packet forwarding is intended for the UE, the X2-U tunnel information is also part of this message.

In Step 6, the currently serving LTE-M eNB sends the RRC Connection Reconfiguration to the LTE-M UE. This message contains the handover command and the PCI of the target cell for handover. It also conveys random-access-related information, such as the dedicated random-access preamble for the target cell.

After receiving the handover command, the UE in *Step 7* initiates the random-access procedure toward the target cell of the new serving LTE-M eNB and utilizes a dedicated random-access preamble[10] as part of the contention-free random-access procedure. After the successful random-access procedure, the UE sends the RRC Connection Reconfiguration Complete message to the new serving LTE-M eNB in response to the RRC Connection Reconfiguration message.

In *Step 8*, the eNB-core network connectivity is modified. For example, the UE-specific S1-MME connection is changed so that the MME can exchange UE-specific signaling messages with the new serving eNB. Furthermore, the S1-U bearer is modified to enable the S-GW to exchange user traffic with the new serving eNB.

An LTE-M UE in the RRC_CONNECTED state has a dedicated radio connection with an LTE-M eNB via at least one cell. A UE-specific S1-MME connection exists between the serving LTE-M eNB and the serving MME. Furthermore, such UE typically has at least one EPS bearer that consists of a DRB on the LTE-M radio interface, an S1-U or Access Bearer on the S1-U interface between the eNB and the S-GW, and an S5/S8 Bearer between the S-GW and the P-GW. Suppose a UE does not need to exchange any traffic. In that case, keeping the UE in the RRC_CONNECTED state is inefficient because of the need for the UE to frequently process DCI, make measurements, and exchange handover signaling. The radio and core networks also need to carry out signaling to support UE mobility. Hence, the UE is asked by the eNB to enter the idle mode when appropriate. The idle mode reduces the amount of signaling and processing for both the UE and the network. Figure 3.13 summarizes the main operations related to the RRC_IDLE UE.

3.2.3.7 Activities of an LTE-M UE in the Idle Mode

The RRC_CONNECTED state enables the UE to actively exchange traffic with the network. If the UE needs to always operate in the RRC_CONNECTED state, there could be significant amount of unnecessary processing at the UE, the radio network, and the core network when the UE is not actively exchanging user traffic with the network. For example, even if there is no user traffic, the UE needs to monitor and process the PDCCH every Transmit Time Interval (TTI) for any resource allocation or directive from the network. The UE still needs to send frequent physical layer reports per configuration by the eNB. Furthermore, the UE needs to make measurements of the serving cell and neighboring cells and send a measurement report to the serving eNB when needed. The eNB has to look for the assigned PUCCH from the UE and process any reports and handover signaling messages from the UE. To avoid such extensive processing at the UE and the eNB in the absence of active user traffic exchange, RRC_IDLE state is utilized.

In the RRC_IDLE state, the dedicated UE-specific radio connection between the UE and the serving cell is released. If and when needed for DL or UL data transfer, the UE can quickly (e.g., within 100 ms or so) transition from the RRC_IDLE state to RRC_CONNECTED state.

In a typical scenario, in the absence of any data traffic for a few seconds (e.g., five seconds), the UE-specific inactivity timer expires. The LTE-M eNB sends the S1AP UE Context Release Request

10 If the UE does not receive a dedicated random-access preamble from the new serving eNB, it uses a common random access preamble as part of contention-based random access.

to the MME, and the MME works with the S-GW to release the S1-U/Access Bearer(s) by exchanging Release Access Bearers Request and Release Access Bearers Response messages. After releasing the S1-U/Access Bearer(s), the MME responds to the eNB with the UE Context Release Command. The LTE-M eNB then asks the LTE-M UE to enter the RRC_IDLE state by sending the RRC Connection Release command in *Step 1* of Figure 3.16 (Tripathi and Reed 2019). The LTE-M eNB removes the access stratum UE context. The MME still preserves the UE context with information such as the GUTI allocated to the UE and the location of the UE (e.g., a given tracking area to be used for any paging when needed). The IP address allocated to the UE also remains valid in the RRC_IDLE state of the UE.

In *Step 2*, the RRC_IDLE UE typically camps on the strongest cell to ensure a high probability of success in receiving any incoming page message and establishes a radio connection by initiating the random-access procedure for any UE-initiated data transfer. Hence, in an implementation-specific (i.e., proprietary) manner, the UE searches for strong cells and performs cell reselection to a neighbor cell; the neighbor cell can provide a better signal than the currently monitored cell.

In *Step 3*, the UE performs a TAU when necessary. For example, if the periodic TAU timer expires, the UE executes the TAU procedure. Furthermore, when the UE performs cell reselection, the UE checks if the new cell belongs to the tracking area that is part of the Tracking Area Identifier (TAI) list. If the new cell's TA is not in the UE's TAI list, the UE executes the TAU procedure.

There may be a few reasons for the UE to exit the idle mode in *Step 4*. For example, to execute the TAU procedure, the UE exits the idle mode, carries out the random-access procedure, establishes an RRC signaling connection, and exchanges non-access stratum TAU signaling with the MME. Specifically, the UE sends the TAU Request message to the MME, and the MME replies with the TAU Accept message. The UE sends the TAU Complete message to complete the TAU procedure. The MME (especially if it is an MME that has not allocated the current GUTI to the UE) allocates a new GUTI to the UE and provides an updated TAI list to the UE. The MME now becomes aware of the UE's latest location and updates the TAI list for the UE in its database. In future, when the MME needs to contact the RRC_IDLE UE, it sends a paging message to the eNBs that are part of the TAs

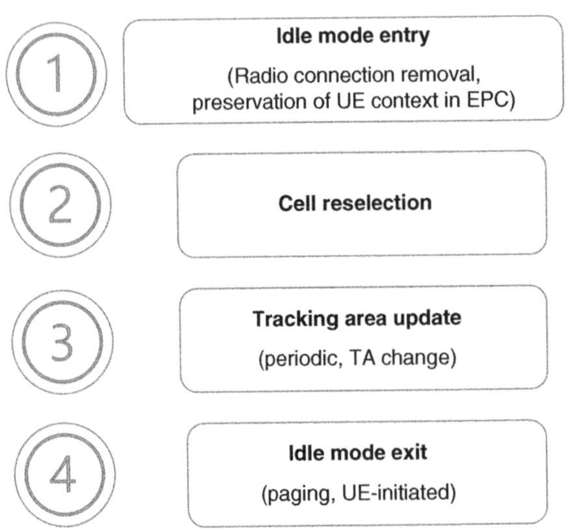

Figure 3.16 Cell Reselection and Tracking Area Update in LTE-M.

of the UE-specific TAI list. Upon completion of the TAU procedure, the UE usually re-enters the idle mode.

The UE also exits the RRC_IDLE mode when it needs to send data or to respond to an incoming paging message. To exit the idle mode, the UE carries out the random-access procedure and establishes an RRC signaling connection. The eNB establishes suitable UE-specific radio bearers, such as SRBs and DRBs, to facilitate the transfer of signaling messages and user traffic. For user traffic, the S1-U/Access Bearers are also established.

In support of paging, the UE monitors a specific paging occasion (which is a specific subframe) in a paging frame. The specific paging occasion and the paging frame that the UE monitors for paging is a function of the UE's ID (e.g., IMSI). The UE monitors the paging occasion during the period of the DRX paging cycle, which typically lasts for about two seconds. A UE wakes up every two seconds or so and looks for a page message in a specific subframe of a specific 10 ms paging frame based on its permanent identity (i.e., IMSI). The eNB sends the RRC Paging message during the paging occasion, which the UE is expected to monitor. The paging message typically includes S-TMSI as the identity of the UE.

3.3 NB-IoT: A 3GPP-Defined Wireless IoT Technology

NB-IoT is designed for IoT applications that can tolerate a long delay and IoT devices that are quite simple and inexpensive (e.g., a few dollars). Section 3.3.1 discusses key features of NB-IoT. Section 3.3.2 provides a brief overview of the LTE-M radio interface including the radio protocol stack and the physical layer frame structure. Section 3.3.2 summarizes the main NB-IoT operations.

3.3.1 Key Features of NB-IoT

NB-IoT and LTE-M share several common features, as discussed earlier in Section 3.2.1. Figure 3.17 lists such common features as well as NB-IoT-specific features (Tripathi and Reed 2019).

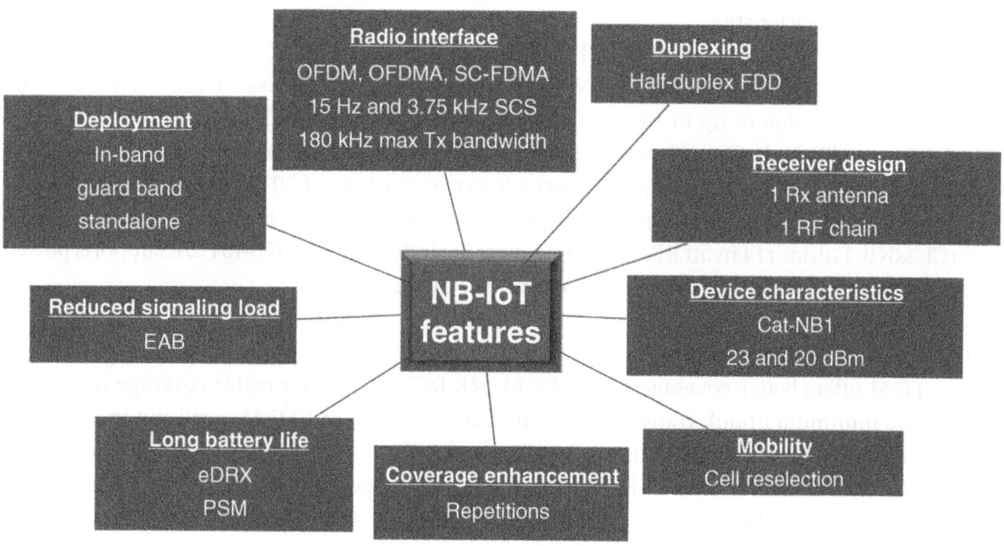

Figure 3.17 NB-IoT Features.

3.3.1.1 Radio Interface

NB-IoT, like LTE and LTE-M, also utilizes OFDM-based radio interface. NB-IoT utilizes OFDMA in the downlink and SC-FDMA in the uplink. The subcarrier spacings of 15 and 3.75 kHz are supported in NB-IoT for the channels carrying IoT traffic, such as the Narrowband Physical Downlink Shared Channel (NPDSCH) and the Narrowband Physical Uplink Shared Channel (NPUSCH). As the maximum number of PRBs in an NB-IoT channel is 1 in NB-IoT and as each PRB occupies 180 kHz for the subcarrier spacing of 15 kHz, the maximum transmission bandwidth for an NB-IoT channel is 180 kHz. While the minimum transmission bandwidth in the downlink is 180 kHz due to the minimum allocation of a PRB with 15 kHz subcarrier spacing, the minimum transmission bandwidth in the uplink could be as little as 3.75 kHz due to the minimum uplink allocation of a single subcarrier with 3.75 kHz subcarrier spacing.

3.3.1.2 Deployment

NB-IoT can be deployed in a variety of configurations. In the in-band deployment, NB-IoT is assigned one of the PRBs of the LTE channel. In the guard-band deployment, NB-IoT is assigned a 200 kHz spectrum in the guard band of an LTE channel. Out of 200, 180 kHz is occupied by the NB-IoT transmission, and the remaining 20 kHz is the NB-IoT guard band with 10 kHz on each side of the NB-IoT channel. In the standalone deployment, an NB-IoT channel is deployed by itself using a 200 kHz spectrum. NB-IoT can be deployed to make use of unused and leftover GSM[11] channels when the spectrum is transitioned to a new access technology.

3.3.1.3 Duplexing

An NB-IoT device is expected to be simple and inexpensive. To simplify operations and reduce costs, it uses half-duplex FDD.

3.3.1.4 Receiver Design

The NB-IoT UE receiver is designed to be much simpler than a typical LTE UE receiver and an LTE-M UE receiver. It uses a single receive antenna and a single receiver chain.

3.3.1.5 Device Characteristics

Release 13 introduced a new category of NB-IoT devices: Category NB1 (Cat-NB1) devices. Cat-NB1 devices are designed to operate with an NB-IoT-capable eNB and possess characteristics such as support for transmission in up to 180 kHz bandwidth. Cat-NB1 devices may support Power Class 3 like LTE and LTE-M devices, corresponding to a maximum transmit power of 23 dBm. Cat-NB1 devices, like Cat-M1 devices, may also support a lower power level of 20 dBm, benefiting from the space and cost savings associated with an integrated PA. The fields powerClassNB-20dBm-r13 and powerClassNB-14dBm-r14 in an RRC signaling message indicates if the NB-IoT UE supports power class 20 dBm and power class 14 dBm, respectively (3GPP 2024d).

3.3.1.6 Coverage Enhancement

While LTE-M offers better coverage than LTE-M, NB-IoT can provide better coverage than even LTE-M. The minimum uplink transmission bandwidth is 1 PRB in LTE-M, resulting in the minimum uplink transmission bandwidth of 180 kHz. In contrast, NB-IoT supports single-tone transmission. Using a 3.75 kHz subcarrier leads to increased power spectrum density (PSD) (expressed as dBm/Hz) compared to the case of power distribution across 180 kHz in LTE-M uplink. A higher

11 GSM stands for Global System for Mobile Communication. GSM is a second-generation cellular technology.

PSD in NB-IoT helps increase the link budget in NB-IoT compared to LTE-M. The expected link budget for NB-IoT is approximately 164 dB, which is significantly higher than the baseline LTE link budget of 140.7 dB and the LTE-M link budget of about 155.7 dB (Qualcomm 2018). Repetition also helps achieve a high link budget. The NPUSCH allows 128 repetitions, enhancing uplink coverage. While the link budget is often uplink-limited, enhanced downlink coverage is possible in NB-IoT due to many NPDSCH repetitions. The NPDSCH supports up to 2,048 repetitions.

3.3.1.7 Long Battery Life

Like LTE-M, NB-IoT also benefits from the features of eDRX and PSM, saving battery power and increasing battery life. NB-IoT can use connected mode eDRX and idle mode eDRX like LTE-M. Similarly, the PSM enables the NB-IoT device to completely turn off the processing of NB-IoT signals. When the PSM is activated, the network cannot reach the NB-IoT UE. However, the PSM supports device-initiated data transfer and periodic TAU.

3.3.1.8 Reduced Signaling Load

NB-IoT devices are expected to be delay tolerant. Hence, when network congestion occurs on the radio interface, non-NB-IoT devices can be prioritized. EAB's feature can be applied to NB-IoT in addition to LTE-M. EAB temporarily prohibits NB-IoT devices of a target Access Class so that other higher-priority devices can make use of the PRACH. An NB-IoT device checks SIB14 for any access class barring before initiating a random-access procedure.

3.3.1.9 Mobility

While LTE-M supports comprehensive mobility management like LTE, where the serving eNB uses UE measurements to make a handover decision, NB-IoT is much simpler to reduce the processing requirements on the NB-IoT device. Instead of measurement-based handover in the RRC_CONNECTED state, NB-IoT uses cell reselection as the mobility management mechanism. As the NB-IoT device travels from one NB-IoT cell to another, the radio connection with the serving NB-IoT cell is lost, and the UE performs cell reselection to a new serving cell.

3.3.1.10 NB-IoT Enhancements in the EPC

Like LTE-M, NB-IoT also supports IoT-centric core network-related features such as reduced TAU signaling and EPC-Lite (Tripathi and Reed 2019). For example, like LTE-M, NB-IoT supports a much longer periodic TAU timer of many weeks to reduce the amount of NB-IoT UE-MME signaling. A service provider may use EPC-Lite instead of a typical EPC to simplify the overall network design for NB-IoT.

As mentioned earlier in Section 3.2.2, like LTE-M, NB-IoT utilizes the LTR radio interface protocol stack as a baseline and makes enhancements to reflect the NB-IoT characteristics discussed in Section 3.3.1.

Regarding the frame structure, NB-IoT follows the LTE frame structure of frame, subframe, slot, and OFDM symbols with the constraint that the maximum transmission bandwidth on a given NB-IoT carrier frequency is one PRB. Due to the transmission bandwidth constraint, the overall physical channels and signals in NB-IoT are simplified compared to LTE and LTE-M. Figure 3.21 in Section 3.3.2 illustrates the NB-IoT physical signals and channels. The roles of these DL channels and signals are explained as part of the NB-IoT operations in Section 3.3.2. In essence, a given 1 ms subframe in NB-IoT contains only one physical channel. Furthermore, the Narrowband Reference Signal (NRS) is present or absent depending on the signal or channel transmitted in a given subframe.

In the NB-IoT uplink, there is no PUCCH. Two formats for the NPUSCH are supported. The NPUSCH Format 1 carries user traffic and signaling, such as the BSR. The NPUSCH Format 2 carries the HARQ feedback in response to a packet received in the NB-IoT DL. The NPUSCH resources are specified as RUs. The RU has a specific number of subcarriers, a total TTI, a subcarrier spacing (SCS), and a modulation scheme. For example, an RU may correspond to one subcarrier, a 32 ms TTI, the SCS of 3.75 kHz, and the modulation scheme of binary phase shift keying (BPSK) or quadrature phase shift keying (QPSK).

3.3.2 Summary of NB-IoT Operations

Examples of key operations or procedures in LTE-M include initial attach, EPS bearer setup, random-access procedure, RRC connection setup, DL data transfer, UL data transfer, mobility management. These LTE-M operations are briefly discussed in Sections 3.3.2.1 to 3.3.2.6.

3.3.2.1 Initial Attach and EPS Bearer Setup in NB-IoT

Before an NB-IoT UE can exchange IoT traffic, prerequisites such as initial attach and EPS bearer setup need to be met. If only the NIDD is used for data transfer for a given UE, the EPS bearer is not set up for such UE. Figure 3.18 shows the major steps of the initial attach and EPS bearer setup in NB-IoT (Tripathi and Reed 2019). Indeed, at a high level, these steps are like the ones discussed earlier for LTE-M. Certain details, however, differ between LTE-M and NB-IoT, as discussed later.

Step 1: Network Acquisition

The NB-IoT UE, upon power-up, detects an NB-IoT cell by observing the Narrowband Primary Synchronization Signal (NPSS) and the Narrowband Secondary Synchronization Signal (NSSS) and determines the Narrowband Physical Cell ID (NPCI) of the NB-IoT-capable cell using the NSSS.

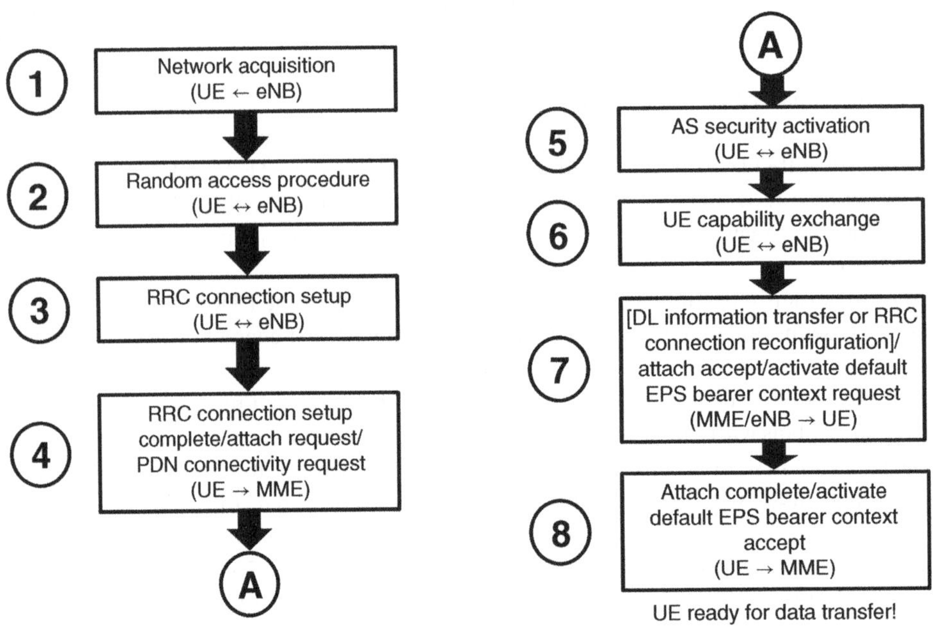

Figure 3.18 Initial Attach and EPS Bearer Setup in NB-IoT.

While synchronization signals in LTE and LTE-M occupy six PRBs, the NPSS and the NSSS are transmitted using a single PRB in NB-IoT because the maximum transmission bandwidth is one PRB in NB-IoT. The NB-IoT UE then processes the Narrowband Physical Broadcast Channel (NPBCH) to obtain the MIB-Narrowband (MIB-NB) containing critical information about the system such as (i) the System Frame Number (SFN) and hyper SFN and (ii) SIB1-Narrowband (SIB1-NB) scheduling information such as Transport Block Size (TBS) and number of repetitions, (iii) operation mode information (i.e., in-band, guard-band, and standalone), and (iv) access barring enabled or not. Note that NB-IoT defines all new physical signals and channels because LTE and LTE-M physical signals and channels cannot be contained in just one PRB. For increased reliability, the MIB-NB utilizes a long TTI of 640 ms, i.e., eight independently decodable blocks of 80 ms duration. The MIB-NB uses subframe 0 but skips the first three OFDM symbols.

After processing the MIB-NB, the NB-IoT UE obtains SIB1-NB that provides cell-related information such as the cell selection threshold (i.e., q-RxLevMin-r13), schedule of other SIBs, and frequency band indicator. SIB1-NB also utilizes a long TTI of 2,560 ms. SIB1-NB is broadcast in subframe 4 of every other radio frame in 16 consecutive frames.

After processing the SIB1-NB, the NB-IoT UE obtains SIB2-NB to learn about the random-access parameters. Specifically, SIB2-NB specifies RSRP thresholds for CE levels and PRACH configurations for different CE levels (e.g., subcarrier offset, number of subcarriers, and the maximum number of preamble attempts). SIB2-NB also specifies the search space for the Narrowband Physical Dedicated Control Channel (NPDCCH). The NB-IoT eNB can repeat SIB2-NB for increased reception reliability at the NB-IoT UEs.

Step 2: Random-Access Procedure

After acquiring SIB2-NB, the NB-IoT UE is ready to initiate the RA procedure. The UE sends an RA preamble to the NB-IoT cell and receives an RAR. Using the UL grant specified by the RAR, the UE initiates establishing an RRC signaling connection. See Figure 3.19 for details on the NB-IoT random-access procedure (Tripathi and Reed 2019).

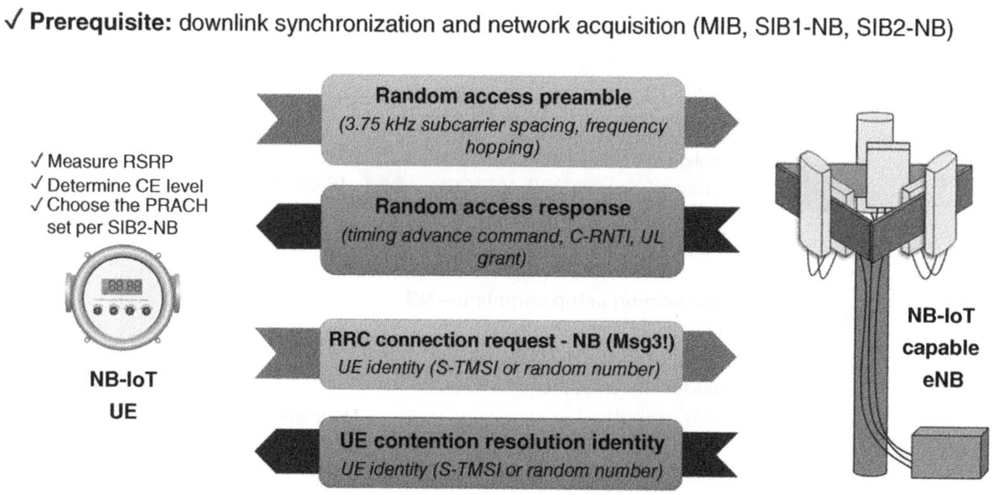

Figure 3.19 Random-Access Procedure in NB-IoT. *Source:* Tripathi and Reed (2019) / The Wireless University; Art of Success / Adobe Stock Photos.

Step 3: RRC Connection Setup

The NB-IoT UE sends an RRC Connection Request-NB message toward the end of the RA procedure to establish a dedicated RRC signaling connection with the NB-IoT eNB. The NB-IoT UE also provides some of its NB-IoT-related capabilities to the network. Furthermore, the UE initiates connectivity with the EPC by sending a NAS message in the last RRC message of the RRC connection setup procedure. Figure 3.20 provides more details of the RRC connection setup procedure in NB-IoT (Tripathi and Reed 2019).

Step 4: Attach Request

The NB-IoT UE sends the NAS Attach Request message to the MME via the NB-IoT eNB. The NB-IoT UE encapsulates this NAS message in the RRC Connection Setup Complete-NB message. The NB-IoT eNB selects a suitable MME for the UE. The Attach Request message contains several NB-IoT-specific parameters such as timer T3324 for PSM, timer T3412 (extended value) for periodic TAU, eDRX parameters, RoHC parameters for IP-based Control Plane for IoT traffic transport, and an indication for the support of SMS for the IoT traffic. The NAS Attach Request message also contains PDN Connectivity Request message for connecting to a PDN (e.g., Internet). The PDU Connectivity Request message includes typical PCOs containing IP address-request and DNS address-request. The UE can indicate its preference for non-IP data transfer by indicating the PDU type= "Non-IP."

Mutual authentication occurs between the NB-IoT UE and the MME. The MME obtains authentication-related parameters from the HSS to support mutual authentication. The NAS security between the UE and the MME is also set up. The UE provides the APN of the PDN through the EPS SM information transfer procedure. The MME coordinates with the S-GW and the P-GW to set up the S5/S8 bearer potion of the EPS bearer. The MME provides the NAS Attach Accept message to the eNB.

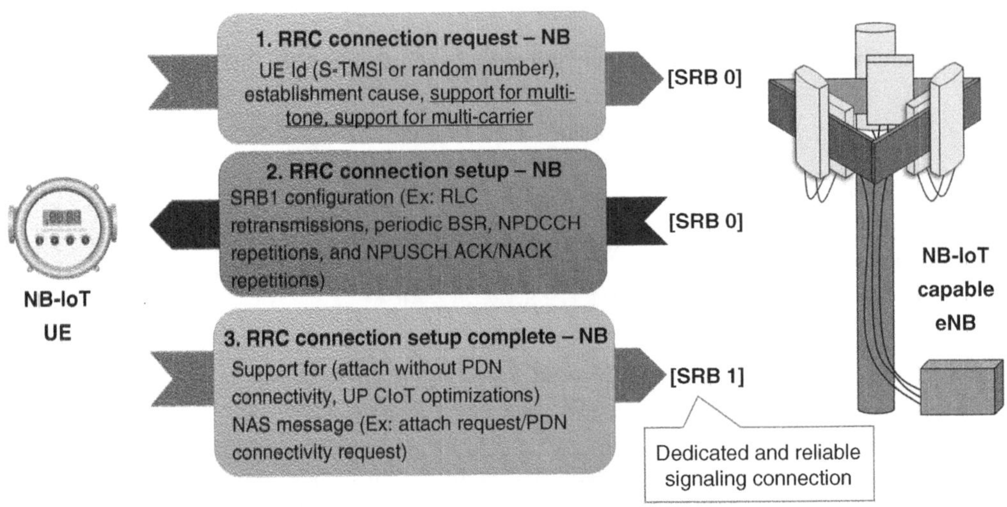

Figure 3.20 RRC Connection Setup in NB-IoT. *Source:* Art of Success / Adobe Stock Photos.

Step 5: Access Stratum Security Activation

The NB-IoT UE and the NB-IoT eNB exchange RRC messages to set up AS security. The RRC messages can now be both ciphered and integrity protected.

Step 6: UE Capability Exchange

The NB-IoT UE conveys its NB-IoT capabilities to the NB-IoT eNB by sending the RRC UE Capability Enquiry-NB message. For example, the NB-IoT eNB specifies E-UTRA as RAT Type in its UE Capability Enquiry-NB message to the NB-IoT UE. The NB-IoT UE replies with the UE Capability Information-NB message and specifies its DL/UL UE category (e.g., NB1 for Category NB1). The UE also specifies its power class (e.g., power class 5 for 20 dBm). Furthermore, the NB-IoT UE indicates its support for multitone transmission (if any) and multi-carrier operation (if any).

Step 7: Attach Accept

The NB-IoT eNB sends an RRC message such as RRC Connection Reconfiguration-NB or DL Information Transfer-NB that encapsulates the NAS Attach Accept message received from the MME in the S1AP Initial Context Setup Request or DL NAS Transport message. The NAS Attach Accept message specifies, along with typical parameters such as GUTI and PCO, NB-IoT-related parameters such as eDRX parameters, T3412 extended value for periodic TAU, T3324 for PSM, "Control Plane Only Indicator" (if applicable), and RoHC for IP-based CP. The Attach Accept message contains the Activate Default EPS Bearer Context Request message to set up a default EPS bearer with specific QoS parameters for the UE. The RRC Connection Reconfiguration-NB message specifies the DRB portion of the EPS bearer.

Step 8: Attach Complete

The NB-IoT UE replies to the Attach Accept/Activate Default EPS Bearer Context Request message with the Attach Complete/Activate Default EPS Bearer Context Accept message by encapsulating such NAS message in the RRC UL Information Transfer-NB message. The network completes the S1-U/access bearer portion of the default EPS bearer so that all three components of the default EPS bearer are fully set up: (i) the DRB between the NB-IoT UE and the NB-IoT gNB in a cell, (ii) the S1-U or access bearer between the eNB and the S-GW, and (iii) the S5/S8 bearer between the S-GW and the P-GW.

3.3.2.2 Random-Access Procedure in NB-IoT

Figure 3.19 takes a closer look at the RA procedure by highlighting NB-IoT-specific enhancements (Tripathi and Reed 2019). As mentioned earlier, before the NB-IoT UE can initiate the RA procedure, it completes network acquisition by synchronizing with the NB-IoT cell using the NPSS and the NSSS and acquiring the MIB-NB, SIB1-NB, and SIB2-NB.

The NB-IoT UE measures the RSRP of its serving NB-IoT cell and determines its CE level similar to an LTE-M UE. If the RSRP is too weak, the CE level corresponding to more repetitions increases reliability. The UE sends an RA preamble with the appropriate number of repetitions on the NPRACH to the eNB using the NPRACH configuration specified in SIB2-NB. The NPRACH uses the subcarrier spacing of 3.75 kHz, which is wider than the 1.25 kHz subcarrier spacing used in LTE-M and LTE.

The NB-IoT eNB responds with the RAR that contains (i) typical configuration parameters for the UE such as the Timing Advance Command to facilitate UL synchronization for the UE and C-RNTI and (ii) UL grant for the first RRC message from the UE.

Using the allocated UL grant, the NB-IoT UE sends Message 3 (Msg 3) such as the RRC Connection Request-NB message containing its identity (i.e., S-TMSI or a random number like an LTE-M UE). As more than one NB-IoT UE might have used the same RA preamble ID, more than one NB-IoT UE may send the RRC Connection Request-NB message. The presence of the UE ID such as S-TMSI or the random number enables the eNB to resolve the conflict among the UEs that might have used the same RA preamble. The NB-IoT eNB sends the Contention Resolution message containing the identity of the UE that the NB-IoT eNB has successfully detected.

The successful RA procedure enables the RRC connection setup procedure to continue. In case an NB-IoT UE's RA procedure fails, the UE retries it by sending another RA preamble after a random wait time.

3.3.2.3 RRC Connection Setup in NB-IoT

Figure 3.20 summarizes the key steps of the RRC connection setup for NB-IoT (Tripathi and Reed 2019).

Like LTE and LTE-M, NB-IoT utilizes a three-way handshake between the NB-IoT UE and the NB-IoT eNB to set up the RRC connection. As more than one UE may send the RRC Connection Request to the eNB on the same radio resources, the message is considered to be using SRB 0, which is common across multiple UEs. As mentioned earlier, for LTE-M, the RRC Connection Request-NB message includes the UE ID (i.e., S-TMSI or a random number). The NB-IoT UE indicates the purpose behind the message by specifying the cause of the RRC connection establishment (e.g., mobile-originated signaling). Based on its capabilities, the NB-IoT UE also indicates its support for multitone and multi-carrier transmission.

The NB-IoT eNB responds with the RRC Connection Setup—NB using SRB0. The RRC Connection Setup—NB message provides a configuration for the dedicated SRB1. SRB1 is specific to the UE and can specify dedicated signaling configurations such as the number of RLC retransmissions, periodic BSR, the NPDCCH configuration, such as the number of repetitions, and the NPUSCH configuration, such as the number of ACK/ANCK repetitions.

Now that SRB1 is configured, the UE uses SRB1 to send the RRC Connection Setup Complete-NB, which is the response to the RRC Connection Setup-NB message. The RRC Connection Setup Complete-NB message conveys additional capabilities useful for NB-IoT operations, such as support for attach without PDN Connectivity and UP Cellular IoT Optimizations. The RRC Connection Setup Complete-NB message also encapsulates the NAS Attach Request message, which in turn carries a PDN Connectivity Request message with a request to connect to a PDN.

After the dedicated RRC and NAS signaling connections and zero or more EPS bearers are set up, the NB-IoT UE can exchange traffic with the NB-IoT eNB. If the data transfer is to be carried out using only the CP, an EPS bearer is unnecessary. One or more EPS bearers are established if the UP is targeted for data transfer. In the case of the CP-based data transfer, the IoT traffic is placed inside an NAS PDU that is exchanged between the UE and the MME through the NB-IoT eNB. The CP-based data transfer requires the use of the NAS and RRC signaling. The "Dedicated Info NAS" structure containing the IoT traffic is encapsulated in the RRC signaling message, that is, the RRC Connection Setup-NB message or the DL Information Transfer-NB message. In the case of the UP-based data transfer, the IoT traffic is placed on an EPS bearer, and, hence, traffic is exchanged between the UE and the P-GW through the NB-IoT eNB and the S-GW. The UP-based IoT data transfer utilizes the DRB on the NB-IoT air interface. Figures 3.19 and 3.20 illustrate example methods by which DL and UL data transfer can occur in NB-IoT.

On the NB-IoT radio interface, the NB-IoT UE-specific IoT traffic from the NB-IoT eNB to the UE is sent on the NPDSCH. As there is a single PRB on an NB-IoT carrier frequency, physical signals and channels, including the NPDSCH, need to be time-multiplexed in different subframes, as shown in Figure 3.21 (Tripathi and Reed 2019).

The arrangement of physical signals and channels in NB-IoT depends on whether the 10 ms frame is an even frame or an odd frame. In the even frames (e.g., frame 0, 2, and so on), the NB-IoT transmits the NPBCH and the NRS in subframe 0, the NPSS in subframe 5, and the NSSS in subframe 9. Furthermore, in the even frames, the NB-IoT transmits the NPDCCH or the NPDSCH along with the NRS in subframes 1 to 4 and 6 to 8. In the odd frames (e.g., frames 1, 3, and so on), the NB-IoT transmits the NPBCH and the NRS in subframe 0 and the NPSS in subframe 5. Furthermore, in the even frames, the NB-IoT transmits the NPDCCH or the NPDSCH along with the NRS in subframes 1 to 4 and 6 to 9.

3.3.2.4 Example DL Data Transfer in NB-IoT

The main steps of the DL data transfer in NB-IoT on the radio interface are illustrated in Figure 3.22 (Tripathi and Reed 2019).

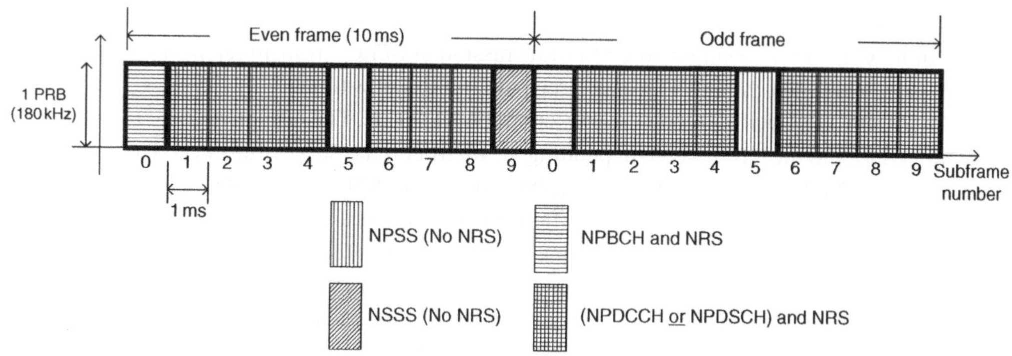

Figure 3.21 Downlink Physical Signals and Channels in NB-IoT.

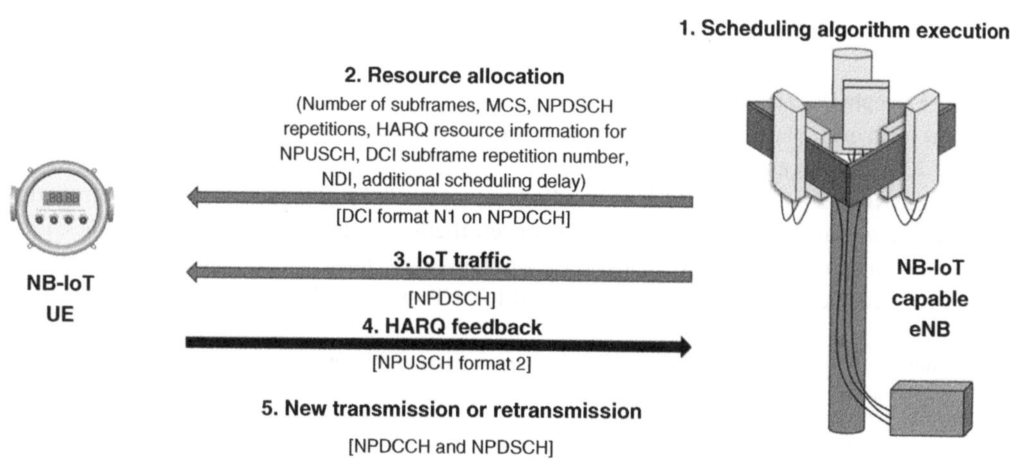

Figure 3.22 Downlink Data Transfer in NB-IoT. *Source:* Art of Success / Adobe Stock Photo.

In *Step 1*, the NB-IoT eNB executes an implementation-specific (i.e., proprietary) scheduling algorithm that considers various factors, such as historical HARQ feedback, the NB-IoT UE capabilities, such as cellular IoT optimization, the available radio resources on the anchor or nonanchor carrier, the amount of data in the buffer, and QoS requirements, to allocate the DL radio resources to the NB-IoT UE. Note that no explicit channel condition report contains information such as CQI in NB-IoT to reduce the processing requirements on the NB-IoT UE.

The NB-IoT eNB, in *Step 2*, conveys resource allocation to the NB-IoT UE using DCI Format N1 on the NPDCCH. Examples of the allocation information include the number of subframes allocated to the UE for transmission, MCS, the number of NPDSCH repetitions, HARQ Resource Information for the NPUSCH, DCI Subframe Repetition Number, and New Data Indicator (NDI)[12]. The NPDCCH itself may be repeated for reliability.

NB-IoT utilizes cross-subframe scheduling. The specified resource allocation usually applies to the NPDSCH transmission in subframe $(n + 5)$ if the associated NPDCCH has occurred in subframe n. If the NPDCCH has been repeated, subframe n contains the last repetition of the NPDCCH. Thus, the typical gap or the scheduling delay between the DL resource allocation by the NPDCCH and the associated NPDSCH transmission is four subframes. However, the NB-IoT eNB can also increase such scheduling delay beyond four subframes by specifying an additional scheduling delay.

The NB-IoT UE retrieves the packet from the NPDSCH in *Step 3*. Note that the NPDSCH may be repeated as specified in the DCI Format in Step 2. The NPDSCH utilizes QPSK and a single-layer transmission, which may be single-antenna transmission or SFBC[13] transmission. A single HARQ process is supported. The NPDSCH may be repeated up to 2,048 times.

The NB-IoT UE attempts to decode the packet on the NPDSCH. The UE conveys the HARQ ACK or NACK feedback to the NB-IoT eNB depending on whether it has successfully decoded the packet or not in *Step 4*. The UE sends this HARQ feedback using the NPUSCH Format 2. If the NPDSCH (or the last repetition of the NPDSCH if repetitions are used) is received in subframe m, the NB-IoT UE sends the HARQ ACK/NACK in subframe $(m + 13)$ or $(m + 21)$. It is also possible to add an additional delay to allow the NB-IoT UE to send the HARQ feedback in a subframe after subframe $(m + 13)$ or $(m + 21)$ based on the HARQ resource information specified by the NB-IoT eNB in the DCI Format N1. Furthermore, the HARQ ACK/NACK on the NPUSCH Format 2 may be repeated for increased reliability.

In *Step 5*, the NB-IoT eNB sends a new packet if it has received HARQ ACK or retransmits a previously sent packet if it has received a HARQ NACK if the maximum number of HARQ retransmissions has not been reached yet. Step 5 essentially repeats Steps 2 and 3 discussed earlier.

3.3.2.5 Example UL Data Transfer in NB-IoT

Key steps of an example approach for the UL data transfer procedure in NB-IoT are summarized in Figure 3.23 (Tripathi and Reed 2019).

When the NB-IoT UE has data (e.g., a signaling message or IoT traffic) to send, it must convey the need for UL resources to the NB-IoT eNB. However, there is no PUCCH in NB-IoT for the UE to send a scheduling request. NB-IoT relies on the random-access procedure for the NB-IoT to convey the UL resource needs. Hence, in *Step 1*, the NB-IoT UE executes the random-access procedure.

12 NDI indicates if this transmission is a new transmission or a retransmission. There is a single HARQ process in Release 13 NB-IoT.

13 SFBC stands for Space Frequency Blocking Coding. SFBC is a transmit diversity technique that utilizes both space diversity and frequency diversity, where the same information is sent on two different antennas and two different subcarriers.

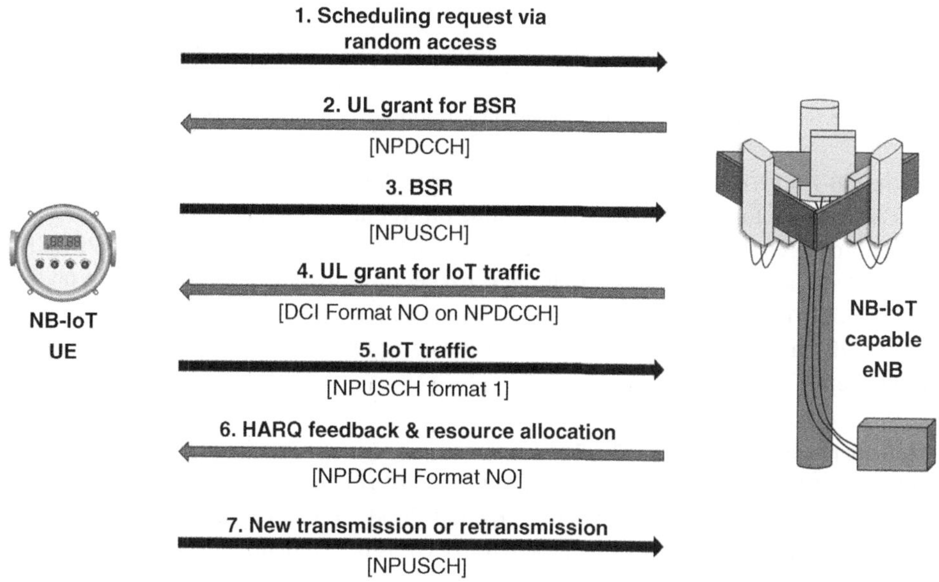

Figure 3.23 Example Uplink Data Transfer in NB-IoT. *Source:* Art of Success / Adobe Stock Photo.

In *Step 2*, the NB-IoT eNB executes a scheduling algorithm to allocate adequate UL resources to the UE so that the UE can send the BSR. The NB-IoT eNB specifies the UL grant for the UE's BSR using NPDCCH. The NPDCCH may be repeated for increased reliability.

In *Step 3*, the NB-IoT UE sends the BSR on the NPUSCH to the NB-IoT eNB using the allocated UL grant.

The NB-IoT eNB learns about the UL resource requirements of the UL data waiting at the UE by processing the BSR. The eNB then executes a scheduling algorithm that considers a variety of factors such as the NB-IoT UE capabilities (e.g., single-tone transmission and multi-carrier transmission), UL channel conditions (estimated by the eNB based on the observation of the DMRS[14] accompanying the NPUSCH; there is no SRS in NB-IoT), available radio resources, the amount of data in the UE's UL buffer, and QoS needs of NB-IoT UEs to allocate UL radio resources. The NB-IoT eNB conveys allocated resources to the NB-IoT UE using the DCI Format N0 on the NPDCCH in *Step 4*. As mentioned in Step 2, the NPDCCH may use repetitions. The DCI Format N0 includes the resource allocation information such as the allocated RUs, MCS, scheduling delay, and NDI. An RU is associated with a specific subcarrier spacing (SCS), a certain number of subcarriers, the TTI, and the supported modulation scheme. For example, an RU may use one subcarrier and a 32 ms TTI with an SCS of 3.75 kHz and may allow BPSK or QPSK. Another RU may use 12 subcarriers and the TTI of 1 ms for the SCS of 15 kHz and uses QPSK. The NPUSCH supports the SCS of 3.75 and 15 kHz. The NPUSCH supports 1, 3, 6, or 12 subcarriers. Specific combinations of the SCS and the number of subcarriers are used to define RUs, and the NB-IoT allocates RUs to the NB-IoT UE for the NPUSCH transmission.

The NB-IoT UE transmits a packet on the NPUSCH Format 1 along with the associated DMRS based on the received UL grant in *Step 5*. NB-IoT utilizes cross-subframe scheduling for the UL as well, where the UE utilizes the allocated NPUSCH resources to transmit a packet beginning

14 DMRS stands for Demodulation Reference Signal. The DMRS facilitates coherent demodulation.

subframe ($n+9$) if the DCI format has indicated the resource allocation in the subframe n and if additional scheduling delay has not been specified. If the NPDCCH has been repeated, subframe n contains the last repetition of the NPDCCH. The UE transmits on a single antenna. The NPUSCH in NB-IoT can make use of BPSK or QPSK. Frequency hopping is supported on the NPUSCH in NB-IoT.

In *Step 6*, the NB-IoT eNB attempts to decode the packet on the NPUSCH and conveys the HARQ ACK or NACK feedback to the NB-IoT UE, depending on whether it has successfully decoded the packet. The NB-IoT eNB sends this HARQ feedback using the DCI Format N0 on the NPDCCH. The NB-IoT eNB also allocates UL resources for new packet transmission or retransmission on the NPUSCH, depending on the HARQ feedback it sends to the NB-IoT UE.

In *Step 7*, the UE transmits a new packet or retransmits the previously transmitted packet on the NPUSCH like Step 5.

3.3.2.6 Mobility Management of NB-IoT UEs

While handover in LTE-M is similar to handover in LTE, NB-IoT does not support UE-assisted network-controlled handover that is supported in LTE and LTE-M. To reduce the complexity and the processing requirements on the NB-IoT UE, the NB-IoT UE is not configured by the eNB to make frequent measurements of neighbor cells. When the NB-IoT UE in the RRC_CONNECTED state moves from one NB-IoT cell to another, its radio link with the currently serving NB-IoT cell experiences a Radio Link Failure (RLF). The NB-IoT UE enters the RRC_IDLE state and performs cell reselection to the strongest neighbor cell. An RRC connection is established between the NB-IoT UE and the new NB-IoT cell (which may belong to the previously serving eNB or the new eNB). The NB-IoT can carry out any data transfer upon establishing suitable SRBs and DRB(s).

The NB-IoT UE performs idle mode activities such as cell reselection and TAU like an LTE-M UE, as shown earlier in Figure 3.13. For increased power savings, the PSM may be activated. Furthermore, the MME can configure a long periodic TAU timer value for the NB-IoT UE.

3.4 5G: A 3GPP-Defined Transformational Technology

5G is a high-performance transformational technology with significant flexibility. Section 3.4.1 discusses key features of 5G NR radio interface. Section 3.4.2 provides a brief overview of the 5G NR radio interface including the radio protocol stack and the physical layer frame structure. Section 3.4.3 summarizes the main 5G operations.

3.4.1 Key Features of 5G NR

5G NR is a flexible and high-performance air interface. Figure 3.24 summarizes key features of 5G NR (Tripathi and Reed 2019). These features include scalable and flexible OFDM, unified and flexible frame structure, support for a variety of spectrum, massive MIMO and advanced beamforming, advanced channel coding, and Multi-Radio Dual Connectivity (MR-DC).

3.4.1.1 Scalable and Flexible OFDM

Like 4G LTE, 5G NR also utilizes OFDM in the DL and the UL. However, 5G NR provides scalability and flexibility in the use of OFDM. For example, while LTE typically utilizes the fixed SCS of 15 kHz for various physical signals and channels, 5G NR in Release 15 supports the variable SCS of 15, 30, 60, and 120 kHz. The SCS values of 15, 30, and 60 kHz are used at lower frequencies, such as below 7 GHz, and the SCS values of 60 and 120 kHz are used in the millimeter wave (mmW) spectrum.

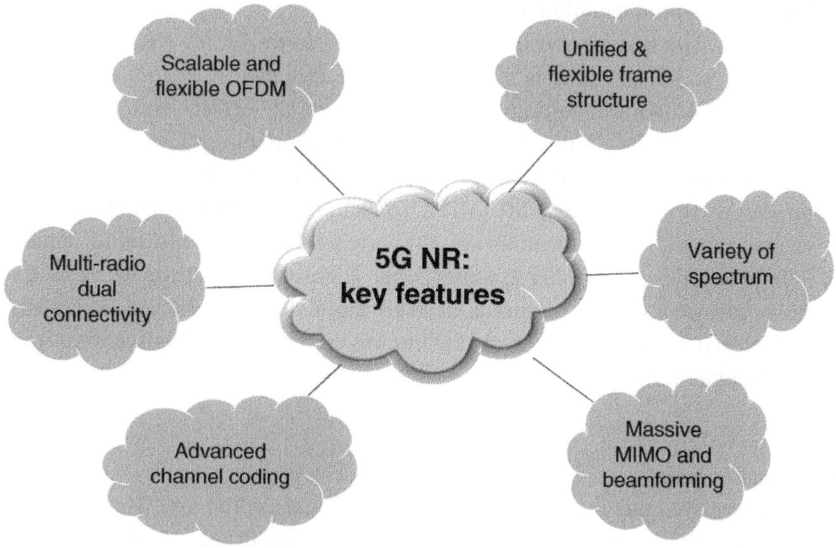

Figure 3.24 Key Features of 5G NR.

To support variable SCS, 5G NR defines the concept of an OFM numerology. The OFDM numerology defines a configuration associated with a specific SCS. Furthermore, such configuration also has an associated Cyclic Prefix (CP). All numerologies support the normal CP, while the numerology corresponding to the SCS of 60 kHz also supports an extended CP. OFDM numerologies reduce the complexity of the FFT processors. Different numerologies help efficiently meet the diverse requirements of services and deployment scenarios. For example, wider SCS leads to shorter TTIs, significantly reducing the radio interface latency and facilitating URLLC applications.

3.4.1.2 Unified and Flexible Frame Structure

LTE defines different frame structures for FDD, TDD, and unlicensed spectrum sharing. 5G NR defines a unified and flexible frame structure that supports FDD and TDD, asymmetric downlink and uplink channel bandwidths, and flexible time allocation to downlink and uplink for TDD. 5G NR defines a slot with fourteen OFDM symbols. The OFDM symbol duration is inversely proportional to the SCS, leading to different slot lengths for different SCS values.

3.4.1.3 Variety of Spectrum

In Release 15, 5G NR supports a variety of spectra such as low-band spectrum below 1 GHz, mid-band spectrum from 1 to 7 GHz, and high-band spectrum from about 24 GHz to about 53 GHz[15]. The low-band spectrum provides good 5G coverage due to small propagation path losses. The mmW spectrum enables 5G to achieve extremely high capacity per unit area and throughput due to large channel bandwidths and smaller cells resulting from large propagation path losses. The midband spectrum provides coverage and capacity between those of the low-band spectrum and the high-band spectrum.

15 The high-band spectrum includes the mmW spectrum near 28 and 39 GHz, which has been widely deployed in the United States.

3.4.1.4 Massive MIMO and Advanced Beamforming

MIMO techniques have been widely deployed in commercial LTE networks. 5G NR allows more advanced antenna techniques. 5G NR defines a comprehensive framework to support various antenna array structures and antenna techniques. 5G NR supports SU-MIMO and MU-MIMO along with advanced 3D beamforming[16]. Support for many antenna elements, grouping of antenna elements in subarrays, and multiple panels of antennas provide significant flexibility in implementing single-layer beamforming, SU-MIMO, and MU-MIMO, depending on the deployment scenarios. Furthermore, several types of beamforming are supported—digital beamforming, analog beamforming, and hybrid digital and analog beamforming. High-band spectrum such as the mmW spectrum enables numerous (e.g., hundreds or even more than a thousand) small-size antenna elements to be placed in an antenna array, facilitating the creation of high-gain narrow beams (often called "pencil beams").

3.4.1.5 Advanced Channel Coding

Two main channel coding techniques used in LTE are convolutional coding and Turbo coding. Convolutional coding is used for the PDCCHs to enable faster decoding in the UE. Turbo coding helps achieve a low error rate on the radio interface and is used on the PDSCH and the PUSCH. 5G NR essentially replaces convolutional coding with polar coding and Turbo coding with Low-Density Parity Check (LDPC) coding. Polar coding and LDPC coding provide better error performance than convolutional coding and Turbo coding, respectively. Polar coding is more efficient than convolutional coding in decoding. LDPC coding also reduces power consumption compared to Turbo coding.

3.4.1.6 Multi-radio Dual Connectivity (MR-DC)

MR-DC enables a UE to communicate simultaneously with two different RATs base stations. In MR-DC, the base station of one RAT acts as the Master Node, and the base station of the second RAT acts as the Secondary Node (SN). The MN provides the UE with NAS signaling connectivity to the core network. Depending on the MR-DC flavor, the core network may be the 5GC or the EPC. The MN and the SN carry out scheduling independently. The UE can obtain radio resources from both the MN and the SN. The user traffic for a given application or Service Data Flow can flow through one of the RATs or both RATs to the UE.

E-UTRA-NR Dual Connectivity (EN-DC) is a widely deployed 5G network architecture, where the LTE eNB acts as the MN, the 5G NR gNB acts as the SN, and the EPC is used as the core network. EN-DC is also known as non-standalone (NSA) architecture with the EPC or Option 3x. In one approach, voice traffic can be carried through the LTE eNB, and the Internet access traffic can be carried through the 5G gNB.

---

16 SU-MIMO stands for Single User-Multiple Input Multiple Output, and MU-MIMO stands for Multiuser-Multiple Input Multiple Output. MU-MIMO is also known as Space Division Multiple Access (MU-MIMO). SU-MIMO reuses the same radio resources while transmitting different layers or streams of data for a given user. MU-MIMO or SDMA reuses the same radio resources while transmitting different layers or streams of data for different users, and different users are distinguished through different beams.

3.4.2 5G NR Air Interface: Protocol Stack and Physical Layer Frame Structure

5G NR is a flexible and high-performance air interface.

3.4.2.1 5G-NR: Radio Protocol Stack

5G NR utilizes LTE radio protocol stacks as a baseline and modifies protocols to increase flexibility and enhance performance compared to LTE (Tripathi and Reed 2019). Like LTE, 5G NR utilizes RRC, PDCP, RLC, MAC, and PHY protocols in the control plane. In the user plane, 5G NR utilizes PDCP, RLC, MAC, and PHY protocols like LTE but adds a new protocol called Service Data Adaptation Protocol (SDAP) above PDCP. SDAP facilitates the implementation of 5G QoS by providing a mapping between the QoS flow and the DRB.

The RRC protocol in 5G NR supports LTE-like functions, including system information, paging, transport of NAS signaling, management of DRBs, and mobility signaling. It also adds new functions, such as support for Signaling Radio Bearer 3 (SRB3) to support dual connectivity between 5G NR and LTE and duplication of signaling to increase reliability.

The PDCP protocol in 5G NR supports LTE-like functions, including RoHC and radio interface security (e.g., ciphering and integrity protection of signaling messages). It also adds new functions, such as support for integrity protection for the user plane and duplication of signaling and user traffic for enhanced reliability. While the RLC protocol in LTE reorders PDUs, the PDCP in 5G NR performs such reordering.

The RLC protocol in 5G NR supports transparent, unacknowledged, and acknowledged modes like LTE. Additionally, 5G NR's RLC protocol supports variable length TTIs resulting from different OFDM numerologies that utilize different SCS values.

The MAC protocol in 5G NR supports multiplexing of RLC PDUs, prioritization among and within UEs, and scheduling like LTE. Additionally, 5G NR has a more flexible HARQ that supports Code Block Group-based HARQ.

The PHY layer in 5G NR supports OFDM and MIMO like LTE but adds significant flexibility to both OFDM and MIMO. 5G NR utilizes new coding techniques of polar coding and LDPC coding to increase spectral efficiency. 5G NR supports massive MIMO and mmW spectrum to significantly increase peak and user-experienced throughput. 5G NR is inherently a beamformed radio interface, where fewer beams per cell are used at low frequencies, and more beams per cell are used at high frequencies.

3.4.2.2 5G-NR: PHY Frame Structure

5G NR PHY layer supports OFDM and OFDMA like LTE but adds significant flexibility. For example, 5G NR utilizes the time units of frame, subframe, slot, and symbol shown in Figure 3.8. Additionally, 5G defines a 5 ms time unit of half-frame for additional flexibility supporting beamforming. However, 5G NR supports variable SCS instead of the typical SCS of 15 kHz in LTE. Example SCS values defined in the 3GPP Release 15 for 5G NR include 15, 30, 60, and 120 kHz for the PDSCH. As the OFDM symbol time is inversely proportional to the SCS, the symbol time is no longer fixed in 5G NR. The slot in 5G NR is defined to be 14 OFDM symbols for the normal Cyclic Prefix and 12 symbols for the extended Cyclic Prefix. The 5G slot has a variable length due to its dependence on the SCS. While LTE uses the subframe for scheduling, 5G NR uses the slot for scheduling.

5G NR supports a unified frame structure for FDD and TDD, unlike LTE, which has different frame formats for FDD and TDD. Different slot formats are defined in 5G NR to support FDD and various downlink/uplink allocations for TDD.

3.4.3 Summary of 5G NR Operations

Examples of key operations or procedures in 5G include initial registration, PDU session setup, random-access procedure, RRC connection setup, DL data transfer, UL data transfer, handover, and idle mode activities. Release 15 is used in the discussions in this section although future releases can also be operated using the procedures described below with additional enhancements. These 5G operations are briefly discussed in Sections 3.4.3.1 to 3.4.3.7 for the Standalone NR with the 5GC architecture. The Standalone NR with the 5GC architecture corresponds to the architecture where a 5G gNB can serve a 5G UE via the 5G NR air interface without any assistance from an LTE eNB, and this 5G gNB utilizes the 5GC as the core network that provides connectivity toward one or more data networks including the Internet.

3.4.3.1 Initial Registration in 5GS

Before a 5G NR UE can exchange traffic, prerequisites such as registration and PDU session setup must be met. Registration makes the network aware of the UE's reachability and involves mutual authentication and security setup. PDU session setup establishes a PDU session (which is associated with a Network Slice) between the UE and a data network to enable exchange of user traffic.

Figure 3.25 shows the major steps of initial registration in 5G NR (Tripathi and Reed 2019). Figure 3.26 shows the PDU session setup. While the LTE attach procedure includes the EPS bearer setup, 5G has separate registration and PDU session setup procedures.

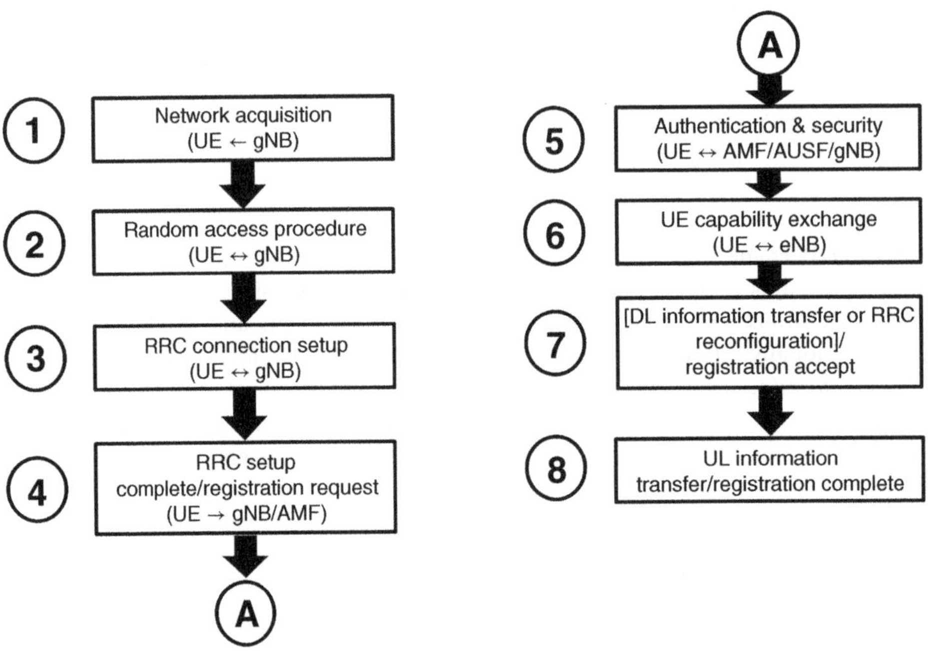

Figure 3.25 Initial Registration in 5GS.

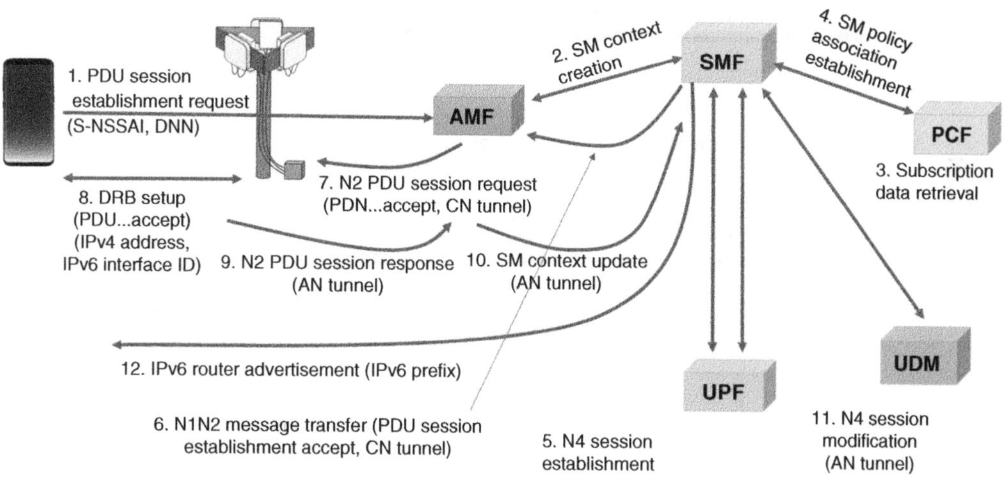

Figure 3.26 PDU Session Setup in 5GS. *Source:* nadir_insan / Adobe Stock Photo.

Step 1: Network Acquisition

Upon power-up, the 5G NR UE detects a 5G NR cell by observing PSS and SSS and determining the PCI of the 5G NR cell. The 5G NR UE then processes the PBCH to obtain the MIB containing critical information about the system. The MIB in NR contains information about the SFN, parameters for intra-frequency cell selection and reselection, and cell barring information (3GPP 2024e). Furthermore, the MIB also specifies the information, such as the SCS and the PDCCH configuration, to enable the UE to retrieve SIB1. The MIB is sent with a periodicity of 80 ms, and repetitions are made within such 80-ms period.

The 5G NR is a beamformed air interface. The Synchronization Signal/Physical Broadcast Channel (SS/PBCH) Block is sent in different beams in the cell at separate times, with all SS/PBCH beams transmitted in a 5-ms period. A UE at a given location should be able to detect the SSPBCH Block (SSB) at least once in a 20-ms period. The 5-ms SSB transmission period may be repeated by the gNB four times in the 20-ms period.

After processing the MIB, the 5G NR UE obtains SIB1, which provides scheduling information about other SIBs. SIB1 also supports unified access control and provides access category-barring information. SIB1 also provides random-access parameters. SIB1 is transmitted with a periodicity of 160 ms.

Step 2: Random-Access Procedure

After acquiring SIB1, the UE is ready to initiate the RA procedure. The UE sends an RA preamble to the 5G NR cell and receives an RAR. Using the UL grant specified by the RAR, the UE initiates establishing an RRC signaling connection. See Figure 3.27 for details on the random-access procedure (Tripathi and Reed 2019).

Step 3: RRC Connection Setup

The RRC connection is a dedicated signaling connection on the NR radio interface between the UE and the serving NR cell. The 5G NR UE sends an RRC Setup Request message toward the end of the RA procedure and obtains radio interface configuration information from the serving 5G NR cell. Furthermore, the UE initiates connectivity with the 5GC by sending a NAS message in the last RRC message of the RRC connection setup procedure. Figure 3.28 provides more details on the RRC connection setup procedure (Tripathi and Reed 2019).

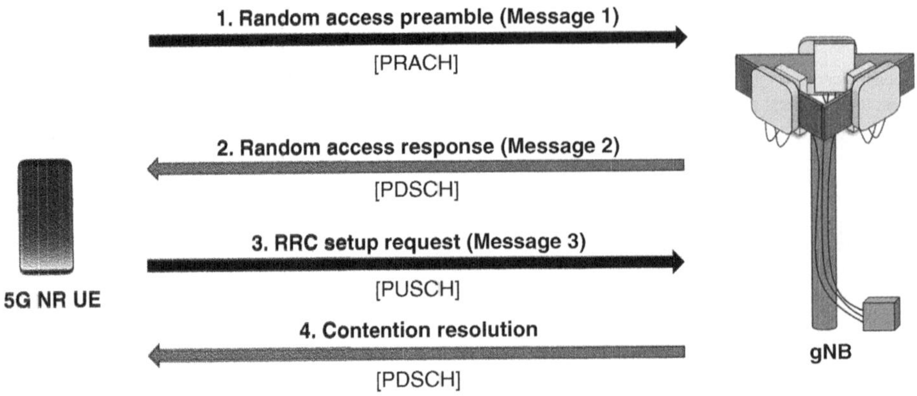

Figure 3.27 Random-Access Procedure in 5G NR. *Source:* nadir_insan / Adobe Stock Photo.

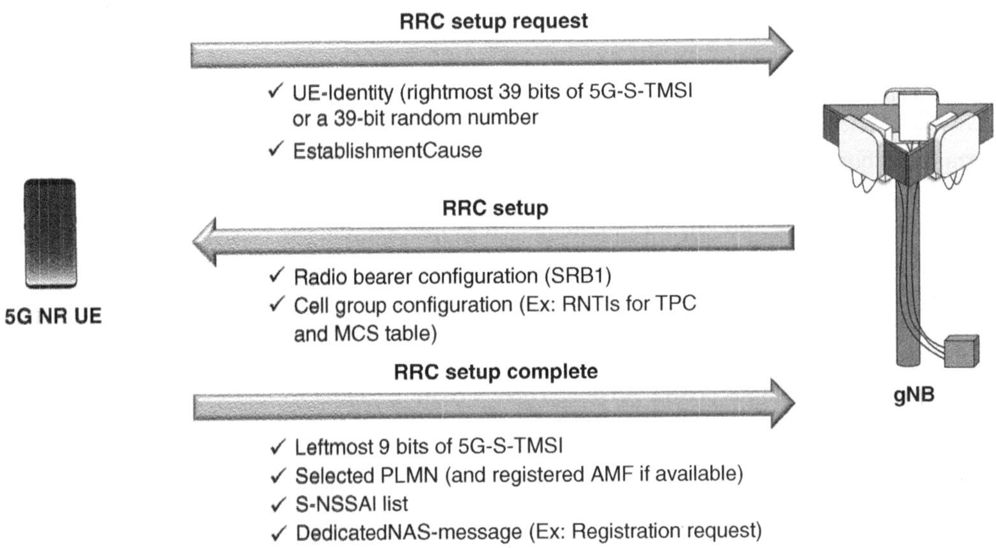

Figure 3.28 RRC Connection Setup in 5G NR. *Source:* nadir_insan / Adobe Stock Photo.

Step 4 Registration Request

The 5G NR UE sends the NAS Registration Request message to the AMF via the gNB. This NAS message is encapsulated by the 5G NR UE in the RRC Setup Complete message. The 5G NR eNB selects a suitable AMF for the UE. The Registration Request message contains parameters such as Registration Type (e.g., initial registration or mobility/periodic registration update), UE ID (e.g., SUCI[17]), security parameters, requested Network Slice Selection Assistance Information (NSSAI),[18] requested Discontinuous Reception (DRX) parameters, and selected UE capabilities.

17 SUCI is Subscription Concealed Identifier and can be used to hide the Subscription Permanent Identifier.

18 NSSAI contains the network slices requested by the UE. Examples of these network slices in Release 15 include eMBB, URLLC, and mIoT.

The UE also indicates a follow-on request if it intends to exchange additional signaling after registration (e.g., PDU session setup signaling). DRX enables the UE to periodically process the radio interface information instead of continuously processing such information.

Step 5: Authentication and Security Activation

Mutual authentication between the 5G NR UE and 5GC occurs. Specifically, the AUSF obtains authentication information from the UDM and exchanges authentication messages with the UE using the NAS signaling connection between the UE and the AMF. Upon successful mutual authentication, the NAS security between the UE and the AMF and the AS security between the UE and the gNB are set up. The RRC and NAS messages can both be ciphered and integrity protected. The AMF obtains relevant policy information, such as the allowed NSSAI from the PCF. The AMF provides the NAS Registration Accept message to the gNB.

Step 6: UE Capability Exchange

The 5G NR UE conveys its 5G NR capabilities to the gNB by sending the RRC UE Capability Enquiry message. For example, the gNB may ask the UE to provide capabilities related to one or more RATs (e.g., NR or E-UTRA-NR) in its UE Capability Enquiry messages to the 5G NR UE. The 5G NR UE replies with the UE Capability Information message and specifies its capabilities related to the RAT type specified by the gNB. Examples of capability parameters include capabilities of various protocols such as PHY, MAC, RLC, and PDCP (3GPP 2024f). The UE also provides RF parameters, measurements, and mobility parameters. For example, the UE specifies supported SCS values, modulation schemes (e.g., QPSK, 16-QAM, 64-QAM, and 256-QAM), bandwidth (e.g., support for 400 MHz in FR2), the number of MIMO layers (2, 4, or 8 for DL and 1, 2, or 4 for UL), power class (e.g., 23 dBm EIRP for power class 3 and 26 dBm for power class 2 in FR1 and 55 dBm for Fixed Wireless Access for power class 1 and 43 dBm for power class 2 to 4 in FR2), and support for different codebooks for MIMO. While LTE, LTE-M, and NB-IoT devices specify the UE Category in the UE Capability Information message, 5G NR does not define any UE categories.

Step 7: Registration Accept

The gNB sends an RRC message, such as RRC Reconfiguration or DL Information Transfer, to the 5G NR UE. This RRC message encapsulates the NAS Registration Accept message that the gNB has received from the AMF. The NAS Registration Accept message specifies parameters such as 5G-GUTI, Registration Area (equivalent to the TAI list containing one or more tracking areas), allowed NSSAI, periodic registration update timer, and accepted DRX parameters.

Step 8: Registration Complete

The 5G NR UE replies to the Registration Accept message with the Registration Complete message by encapsulating such NAS message in the RRC UL Information Transfer message.

Note that no PDU session is set up during registration in 5G. If the UE wishes to set up a PDU session, it can initiate such a procedure any time after the successful registration. Figure 3.26 shows the major steps of the PDU session setup (Tripathi and Reed 2019). A PDU session can be viewed as the association between the UE and a DN for a given network slice that enables the exchange of PDUs containing user traffic. A PDU session exists inside a Network Slice and cannot span across multiple Network Slices. Inside a Network Slice, identified by the S-NSSAI, there can be multiple PDU sessions. Each PDU session has a distinct identifier allocated by the UE and conveyed to the 5GC by the UE. The user plane connection for a PDU session consists of (i) the DRB on the NR air interface between the UE and the gNB in a given cell and (ii) the N3 connection (which is a GTP-U tunnel) between the gNB and the UPF.

Step 1: PDU Session Establishment Request

The 5G NR UE, any time after successful registration, sends a PDU Session Establishment Request to establish connectivity with a DN. The UE sends the NAS UL NAS Transport message encapsulated in the RRC UL Information Transfer message. The NAS UL NAS Transport message has the NAS payload container type of "N1 SM information" with the N1 SM Payload being the PDU Session Establishment Request message. The NAS UL NAS Transport includes S-NSSAI to identify the Network Slice and the Data Network Name (DNN) to identify the DN toward which the UE requests the PDU session. The PDU Session Establishment Request message contains the PDU Session ID determined by the UE, requested PDU Session Type (e.g., IPv6 or IPv4), requested SSC mode[19], and 5GSM Capability (e.g., support for reflective QoS[20]). The AMF selects a suitable SMF based on the SMF capabilities, S-NSSAI, and the DNN.

Step 2: SM Context Creation

The AMF sends Nsmf_PDUSession_CreateSMContext Request to the SMF to create the UE-specific SM context. Along with the N1 SM payload, this message includes parameters such as SUPI, S-NSSAI, DNN, PDU Session ID, and AMF ID (i.e., GUAMI[21]).

Step 3: Subscription Data Retrieval

The SMF obtains subscriber data from the UDM. Examples of subscription data retrieved by the SMF from the UDM include allowed PDU session types, allowed SSC modes, default 5G QoS Identifier (5QI) and Allocation and Retention Priority (ARP), and Session-AMBR[22]. The SMF can subscribe to the UDM for notifications. The SMF then replies to the AMF with Nsmf_PDUSession_CreateSMContext Response containing the SM Context ID and cause.

Step 4: SM Policy Association Establishment

The SMF exchanges signaling messages with the PCF to establish the association of the SM policy. The PCF provides the SMF with policy information such as authorized Session-AMBR and authorized default 5QI and ARP.

Step 5: N4 Session Establishment

The SMF selects a suitable UPF as a PDU Session Anchor (PSA) for the UE's PDU session toward the DN. The SMF establishes the UE-specific N4 session with the UPF. The SMF specifies the packet detection rule to the UPF. The Core Network (CN) Tunnel Info (e.g., Tunnel Endpoint ID (TEID)) for the UE-specific N3 Tunnel between the gNB and the UPF. More specifically, the CN Tunnel information specifies the UPF side of the GTP-U tunnel between the gNB and the UPF on the N3 interface. The CN Tunnel Info may be determined by the UPF or the SMF.

---

19 SSC stands for Session and Service Continuity. The SSC mode 1 preserves the IP address for the UE, while the SSC modes 2 and 3 involve a change in the UE's IP address. The SSC mode 2, the connection with a new UPF is established after the connection with the old UPF is released. In the SSC Mode 3, the UE has simultaneous connectivity with both the old UPF and the new UPF for a period.

20 Reflective QoS implies the use of the same QoS parameters for the uplink and the downlink.

21 GUAMI is globally unique AMF ID.

22 AMNR stands for aggregate maximum bit rate.

Step 6: N1N2 Message Transfer

The SMF sends the Namf_Communication_N1N2MessageTransfer message to the AMF to (i) transport the N1 SM container with the PDU Session Establishment Accept message and (ii) provide the PDU session related information such as CN Tunnel information, PDU Session ID, QoS flow indicator (QFI), QoS Profile, S-NSSAI, Session-AMBR, and PDU Session Type. The PDU Session Establishment Accept message itself, to be sent to the UE by the AMF, contains QoS Rules and QoS flow level parameters, selected SSC Mode, S-NSSAI, DNN, IPv4 address and/or IPv6 interface ID, Session-AMBR, Selected PDU Session Type, and P-CSCF address.

Step 7: N2 PDU Session Request

The AMF sends an NGAP[23] N2 PDU Session Resource Setup Request message to the gNB. This message contains (i) PDU Session Resource Setup Request Transfer with details of QoS parameters (which enables the gNB to perform admission control) and (ii) the NAS PDU. The NAS PDU carries the DL NAS Transport message, including the PDU Session ID and the N1 SM info containing the PDU Session Establishment Accept message.

Step 8: DRB Setup and PDU Session Establishment Accept

The gNB considers the QoS requirements for the DRB associated with the PDU session and sets up a DRB for the UE. The gNB also transports the NAS PDU Session Establishment Accept message in the RRC message. The gNB can utilize the RRC Reconfiguration message to set up the DRB and encapsulate the DL NAS Transport message containing the PDU Session Establishment Accept message.

Step 9: N2 PDU Session Response

After setting up the DRB and sending the PDU Session Establishment Accept message, the gNB sends N2 PDU Session Resource Setup Response to the AMF and specifies the gNB side of the GTP-U tunnel for the UE-specific N3 connection between the gNB and the UPF in the form of the Access Network (AN) Tunnel Information.

Step 10: SM Context Update

The AMF sends the SM context update to the SMF and provides the AN tunnel information as part of N2 SM information.

Step 11: N4 Session Modification

The SMF modifies the N4 session toward the UPF by providing the AN Tunnel Information. The GTP-U Tunnel between the gNB and the UPF is now complete. As the DRB has already been set up in Step 8, the UP connection for the PDU session between the UE and the UPF is fully established.

Step 12: IPv6 Router Advertisement

If the UE is allocated an IPv6 address, the SMF sends an IPv6 Router Advertisement message through the UPF and the gNB to the UE in the form of an IP layer message to specify the network prefix portion of the IPv6 address.

---

23 NGAP stands for Next-Generation Application Protocol. NGAP governs the operation of the N2 interface between the gNB and the AMF.

3.4.3.2 Random Access in 5G NR

As mentioned earlier, before initiating the RA procedure, the UE completes network acquisition by synchronizing with the 5G NR cell using the PSS and the SSS and acquiring the MIB and SIB1. Figure 3.26 summarizes the four-step RA procedure in NR. This procedure is similar to that in LTE and LTE-M. Similar to LTE-M and NB-IoT, random access in NR enables the 5G UE to be synchronized in the uplink with all other 5G UEs in the NR serving cell so that signals from all 5G UEs in the cell arrive at the gNB at approximately the same time. However, the RA configuration is more flexible in 5G NR compared to LTE and LTE-M.

Recall that the 5G NR is a beamformed radio interface. Hence, transmission and reception occur in beams. For RA, the NR UE finds the best SSB beam and utilizes the PRACH configuration associated with such an SSB beam. The UE sends an RA preamble (considered to be message 1) on the PRACH using the beam-specific PRACH configuration specified in SIB2. The RA preamble is based on Zadoff-Chu sequences like LTE. Both short and long sequences are supported. While LTE uses the fixed SCS of 1.25 kHz on the PRACH, 5G NR supports flexible subcarrier spacing such as 1.25 to 30 kHz in sub-7 GHz spectrum and 60 and 120 kHz in the mmW spectrum. Furthermore, in contrast to the fixed channel bandwidth to accommodate six PRBs (or 1.08 MHz) of the PRACH in LTE, 5G NR supports flexible channel bandwidths for the PRACH, that is, 1.25 to 5 MHz for sub-7 GHz spectrum and 10 and 20 MHz for mmW spectrum.

The gNB looks for the RA preambles from UEs during certain periods and in specific SSB beams based on the mapping between the SSB beam and the PRACH configuration. The 5G NR gNB responds with the RAR (considered to be message 2) that contains typical configuration parameters for the UE such as the Timing Advance Command to facilitate UL synchronization for the UE and C-RNTI.

Using the allocated UL grant, the 5G NR UE sends Message 3 (considered to be message 3 or Msg 3), for example, the RRC Setup Request message containing its identity, such as 5G-S-TMSI if it has previously been allocated a 5G-GUTI by an AMF. 5G-S-TMSI is a short form of 5G-GUTI. If the UE does not have 5G-S-TMSI, it generates a random number and includes it in the RRC Setup Request message. As more than one 5G NR UE might have used the same RA preamble ID, more than one 5G NR UE may send the RRC Setup Request message using the UL grant specified by the gNB in the RAR. The presence of the UE ID, 5G-S-TMSI, or a random number enables the gNB to resolve the conflict among the UEs that might have used the same RA preamble simultaneously. The 5G NR gNB sends the Contention Resolution message (containing the UE identity) to the UE that the gNB has successfully detected.

A successful RA procedure enables the RRC connection setup procedure to continue. If a 5G NR UE's RA procedure fails, the UE retries the RA procedure by sending another RA preamble after a random wait time.

3.4.3.3 RRC Connection Setup in 5G NR

Figure 3.28 summarizes key steps of the RRC connection setup for 5G NR (Tripathi and Reed 2019).

Like LTE, 5G NR utilizes a three-way handshake between the 5G NR UE and the gNB to set up the RRC connection. As mentioned earlier, the RRC Setup Request message includes the UE ID, i.e., such as 39 rightmost bits of 5G-S-TMSI or a 39-bit random number. The 5G NR UE indicates the purpose behind the message by specifying establishmentCause (e.g., mobile-originated signaling).

The gNB responds with the RRC Setup message. The RRC Setup message provides a configuration for UE-specific SRB1. This message also specifies 5G NR-specific cell group configuration, such as C-RNTIs for Transmit Power Control[24] (TPC) and Modulation and Coding Scheme (MCS) table[25].

Now that SRB1 is configured, the 5G NR UE uses SRB1 to send the RRC Setup Complete, which is the response to the RRC Setup message. The RRC Setup Complete message conveys the leftmost 9 bits of 5G-S-TMSI, selected PLMN (and registered AMF if available), S-NSSAI list, and dedicated NAS-message (e.g., Registration Request).

3.4.3.4 Example Downlink Data Transfer in 5G NR

After one or more PDU sessions are set up, the 5G NR UE can exchange traffic with the gNB.

5G provides a flexible framework for DL data transfer; various implementations are possible. The main steps of the DL data transfer in 5G NR in an example implementation are illustrated in Figure 3.29 (Tripathi and Reed 2019). These are similar to those for LTE, but there is significantly more flexibility in 5G NR in DL data transfer.

In *Step 1*, the gNB configures the UE with DRBs and measurements. The gNB may configure the UE to make measurements of SS/PBCH only, Channel State Information-Reference Signal (CSI-RS) only, and both SS/PBCH and CSI-RS. The gNB may also configure the UE to make measurements of Channel State Information-Interference Measurement (CSI-IM). 5G NR aims to minimize the overhead of these signals. While CRS in LTE is transmitted across the entire cell, a given CSI-RS may be transmitted in a specific portion of the cell in a CSI-RS beam. CSI-RS can be viewed as

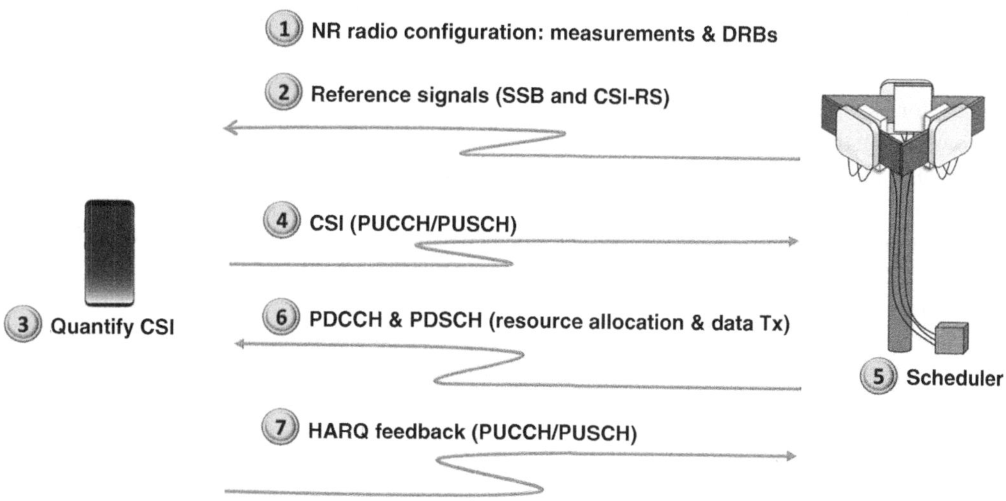

Figure 3.29 Downlink Data Transfer in 5G NR. *Source:* Tripathi and Reed (2019) / The Wireless University; nadir_insan / Adobe Stock Photo.

24 The gNB can send uplink power control commands to the UE to influence the transmit power of the UE using the TPC indications.

25 The gNB may configure the UE with more than one MCS table. A suitable MCS table can be used for the UE to help achieve a target error rate at the physical layer.

UE-specific instead of cell specific although multiple UEs may be configured to observe the same CSI-RS. The gNB can specify the frequency band, the number of PRBs, OFDM symbols, and subcarriers in a PRB where the UE can observe CSI-RS. The CSI-RS scrambling ID and periodicity of CSI-RS are also specified. As 5G NR is a beamformed air interface, the beam identity in the form of Transmission Configuration Indication (TCI) is also specified by the gNB.

The CSI in 5G NR includes LTE-like CSI components such as CQI, PMI, and RI and new CSI components such as Layer 1-RSRP (L1-RSRP)[26], Layer Indicator (LI)[27], CSI-RS Resource Indicator (CRI), and SSB Resource Indicator (SSBRI). CRI refers to the CSI-RS beam, and SSBRI refers to the CSI-RS beam based on the configured TCIs. The CSI reporting can be periodic, semi-persistent, or aperiodic (e.g., based on a DCI command).

5G NR supports multiple CQI tables. CQI Table 1 supports (i) BLER of 0.1, (ii) modulation schemes of QPSK, 16QAM, and 64QAM, (iii) the least spectral efficiency of 0.1523 bits per modulation symbol, and (iv) the highest spectral efficiency of 5.5547 bits per modulation symbol. CQI Table 2 supports (i) BLER of 0.1; (ii) modulation schemes of QPSK, 16QAM, 64QAM, and 256QAM; (iii) the least spectral efficiency of 0.1523 bits per modulation symbol; and (iv) the highest spectral efficiency of 7.4063 bits per modulation symbol. CQI Table 3 supports (i) BLER of 0.00001; (ii) modulation schemes of QPSK, 16QAM, and 64QAM; (iii) the least spectral efficiency of 0.0586 bits per modulation symbol; and (iv) the highest spectral efficiency of 4.5234 bits per modulation symbol.

5G NR also supports much more comprehensive PMIs. Type-I codebooks suitable for SU-MIMO and Type-II codebooks suitable for MU-MIMO are supported. Single-panel and multi-panel antenna configurations are supported.

The gNB may configure CA for a UE. Up to four bandwidth parts (BWPs) may be defined on a given carrier frequency or component carrier. One BWP on a carrier frequency is active at a time, and the gNB allocates radio resources for the active BWP.

In *Step 2*, the 5G NR UE, based on the configuration by the serving gNB, observes the reference signals, for example, the CSI-RS of the serving 5G NR cell. The CSI-RS may be transmitted in a narrower beam than an SSB beam. The CSI-RS beams may cover the same region as the SSB beams.

In *Step 3*, the UE quantifies the prevailing DL radio channel conditions as CSI.

In *Step 4*, the 5G NR UE sends the channel condition report to the gNB to enable the gNB to determine optimal PDSCH transmission parameters. This report at a given instant contains one or more CSI components mentioned in Step 1 (i.e., CQI, PMI, RI, L1-RSRP, LI, CRI, and SSBRI). The channel condition report is sent on the PUCCH or the PUSCH.

In *Step 5*, the gNB executes an implementation-specific scheduling algorithm that considers the channel condition report along with numerous factors such as the UE capabilities, the number of PRBs in the UE's active BWP per carrier, the available resources on different carrier frequencies for CA, QoS needs of 5G QoS flows of NR UEs, and gNB capabilities to allocate DL radio resources.

The gNB can convey allocated DL radio resources to the UE using DCI Format 1_0 or 1_1 on the PDCCH in *Step 6*. DCI Format 1_0 is simpler and compact, while DCI Format 1_1 is more comprehensive.

Both DCI formats can convey parameters such as (i) frequency domain assignment (e.g., PRBs), (ii) time domain assignment (i.e., starting OFDM symbol number and length of the allocation), (iii) MCS from the configured CQI table, (iv) NDI[28], (v) Redundancy Version (RV)[29], (vi) HARQ

26 L1-RSRP is the RSRP measured on the CSI-RS or the SSS.
27 LI indicates the strongest layer among the spatial multiplexing layers.
28 NDI indicates whether it is new transmission or retransmission.
29 RV indicates which set of LDPC code symbols are transmitted by the gNB.

process number (i.e., one of sixteen processes), (vii) PDSCH-to-HARQ feedback timing indicator, and (viii) PUCCH Resource Indicator for UL feedback to the DL transmission. 5G NR utilizes LDPC coding on the PDSCH to increase the effectiveness of channel coding and reduce the error rate while reducing energy consumption. Additionally, DCI Format 1_1 can indicate TCI for future communications and Code Block Group (CBG)[30] transmission information.

The gNB may utilize dynamic resource allocation by specifying radio resources dynamically using a DCI. The gNB may also utilize semi-persistent scheduling, where the resources are specified through RRC signaling and activated or deactivated based on DCI signaling. The time domain allocation may use the slot (other than the OFDM symbols occupied by PDCCHs), a mini slot (i.e., a few symbols of the slot), or multiple slots. The frequency domain allocation may use PRBs or Resource Block Groups (RBGs), where an RBG consists of multiple PRBs.

The 5G NR UE retrieves the packet from the PDSCH transmitted by the gNB based on the PDCCH-specified parameters in *Step 6*.

The 5G NR UE attempts to decode the packet on the PDSCH. In *Step 7*, the UE conveys the HARQ ACK or NACK feedback to the gNB depending on whether it has successfully decoded the packet or not based on "PDSCH_to_HARQ_feedback_timing_indicator" specified by the DCI. The UE sends this HARQ feedback on the PUCCH or the PUSCH.

The gNB sends a new packet if it has received HARQ ACK or retransmits a previously sent packet if it has received a HARQ NACK if the maximum number of HARQ retransmissions has not been reached yet.

3.4.3.5 Example Uplink Data Transfer in 5G NR

5G provides a flexible framework for UL data transfer; various implementations are possible. The main steps of the UL data transfer in 5G NR in an example implementation are illustrated in Figure 3.30 (Tripathi and Reed 2019). These steps are similar to those for LTE, but there is significantly more flexibility in 5G NR in UL data transfer.

The gNB configures the 5G NR UE to facilitate UL data transfer in *Step 1* by providing configuration related to SR, SRS, BSR, and Power Headroom Report (PHR). The CSI-RS configuration is helpful for both DL data transfer and UL data transfer. The gNB provides configurations for different PUCCH formats that the UE can use depending on the content transmitted on the UL. As part of the SR configuration, a set of PUCCH format 0/1 resources are specified. The periodicity, offset, and the number of symbols for the SR-related PUCCH are specified. In the absence of SR configuration or SR failure, the UE can use the RA procedure to make a scheduling request to the gNB.

The gNB specifies SRS configuration to the 5G NR UE, such as the number of OFDM symbols for the SRS transmission (e.g., Last 1, 2, or 4 OFDM symbols), the SRS transmission bandwidth with two transmission combs (e.g., SRS on every 2nd or 4th subcarrier), and the number of ports (e.g., 1, 2, or 4). The SRS transmission may be aperiodic, semi-persistent, or periodic.

The gNB also specifies BSR configurations, such as periodic BSR timers. A short BSR provides UL buffer information (e.g., amount of data) about one Logical Channel Group (LCG), while a long BSR provides UL buffer information about multiple LCGs. The gNB specifies PHR-related parameters, for example, a periodic timer for periodic PHR. Type 1 PHR reports the difference between the UE's maximum transmit power and the estimated PUSCH power. Type 2 PHR reports the

30 A large packet may be transmitted in different CBGs so that an error in a CNG requires retransmission of a specific CBG only. If CBGs are not used, a single error in a large packet ("transport block") requires retransmission of the entire large packet.

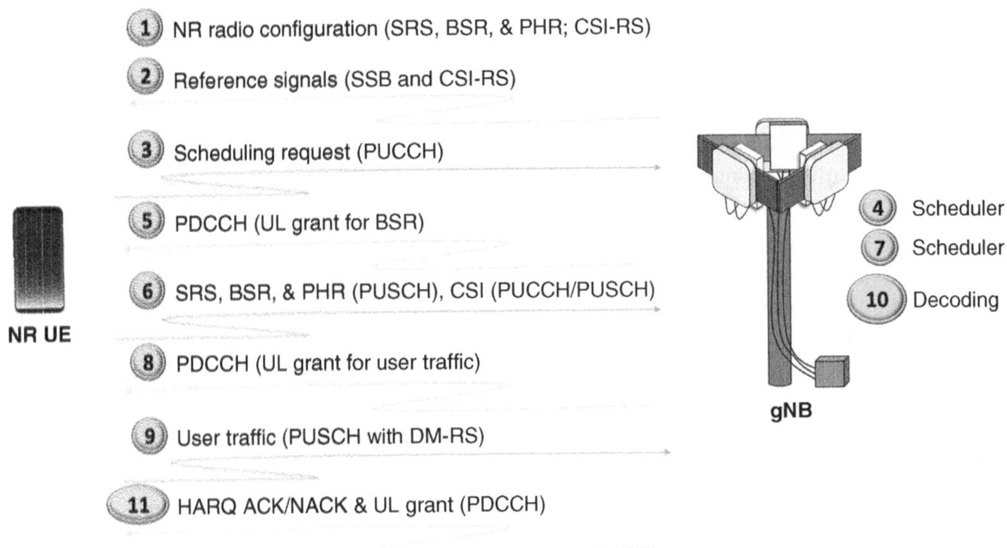

Figure 3.30 Uplink Data Transfer in 5G NR. *Source:* nadir_insan / Adobe Stock Photo.

difference between the UE's maximum transmit power and the E-UTRA power. Type 3 PHR reports the difference between the UE's maximum transmit power and the estimated SRS power.

In *Step 2*, the 5G NR UE observes the SSB and the CSI-RS transmitted by the gNB to quantify the radio channel conditions in the CSI. While the CSI is directly helpful for the DL data transfer, it is also helpful for the UP data transfer, especially for beam selection by the gNB to receive signals from the 5G NR UE.

In *Step 3*, when the 5G NR UE has data, such as a signaling message or traffic, to send, it sends a SR on the configured PUCCH.

The gNB executes a scheduling algorithm in *Step 4* to allocate adequate UL resources to the 5G NR UE so that the 5G NR UE can send the BSR in *Step 6*.

The gNB specifies the UL grant on the PDCCH for the 5G NR UE's BSR in *Step 5*.

In *Step 6*, the 5G NR UE sends the SRS, BSR, and PHR on the PUSCH per the configuration specified by the gNB. The UE also sends the CSI on the PUCCH or the PUSCH.

The gNB executes a scheduling algorithm in *Step 7*. The gNB learns about the UL resource requirements of the UL data waiting at the 5G NR UE by processing the BSR. The gNB evaluates the UL channel conditions by observing the SRS and determines the power availability at the 5G NR UE by considering the PHR. The gNB considers a variety of factors including the 5G NR UE capabilities (e.g., half-duplex FDD), UL channel conditions, available transmit power at the UE, available PRBs in the active BWP in the uplink, the number of carrier frequencies, the type and the amount of data in the 5G NR UE's UL buffer, and QoS needs of all active 5G NR UEs to allocate UL radio resources.

The gNB, in *Step 8*, conveys allocated resources to the 5G NR UE using the DCI format 0_0 or 0_1 on the PDCCH. Both DCI formats can convey parameters such as (i) frequency domain assignment (e.g., PRBs) with or without hopping, (ii) time domain assignment (i.e., starting OFDM symbol number and length of the allocation), (iii) MCS from the configured CQI table up to 64QAM or 256QAM, (iv) NDI, (v) RV, (vi) HARQ process number (i.e., one of sixteen processes), (vii) TPC command for scheduled PUSCH, and (viii) UL/Supplementary Uplink (SUL_ Indicator for UEs configured with SUL). 5G NR utilizes LDPC coding on the PUSCH to increase the effectiveness of channel coding and reduce the error rate while reducing energy consumption. Additionally, DCI Format 0_1 can indicate SRS Resource Indicator (SRI) and precoding information.

The 5G NR UE in *Step 9* transmits a packet on the PUSCH based on the received UL grant in Step 8. The PUSCH can be transmitted with or without transform precoding. Up to four transmission layers can be used on the PUSCH. The PUSCH in 5G NR can use QPSK, 16QAM, 64QAM, and optionally 256QAM. Frequency hopping may or may not be used on the PUSCH; both intra-slot and inter-slot frequency hopping are supported.

In *Step 10*, the gNB attempts to decode the packet on the PUSCH in a suitable beam through receive beamforming in an implementation-specific manner and conveys the HARQ ACK or NACK feedback to the 5G NR UE depending on whether it has successfully decoded the packet or not. The 5G NR gNB sends this HARQ feedback on the PDCCH. The 5G NR gNB also allocates UL resources for new packet transmission or retransmission on the PUSCH depending on the HARQ feedback that it is sending to the 5G NR UE.

3.4.3.6 Handover in 5G NR

Handover in Release 15 5G NR is similar to handover in LTE and LTE-M. Figure 3.31 summarizes key steps of 5G NR handover (Tripathi and Reed 2019). Like LTE, 5G NR supports UE-assisted network-controlled handover, where the 5G NR UE in the RRC_CONNECTED state provides measurements to the serving the gNB. The gNB makes a handover decision and instructs the UE to switch the RRC connection to a target cell by sending a handover command to the UE. The serving gNB communicates with the target (new) gNB before issuing a handover command to the 5G NR UE so that the target gNB can reserve resources for the 5G NR UE.

The gNB configures the 5G NR UE with suitable measurements using the RRC Reconfiguration message in *Step 1*. The gNB specifies the carrier frequency or frequencies that the 5G NR UE should measure. On a given carrier frequency, the 5G NR UE autonomously detects cells (i.e., PCIs). The gNB can also specify whitelisted cells that the 5G NR UE considers for higher-priority measurements. The gNB also specifies what measurements to make, such as RSRP, RSRQ, and signal-to-interference plus noise ratio (SINR). Furthermore, the gNB configures the measurement

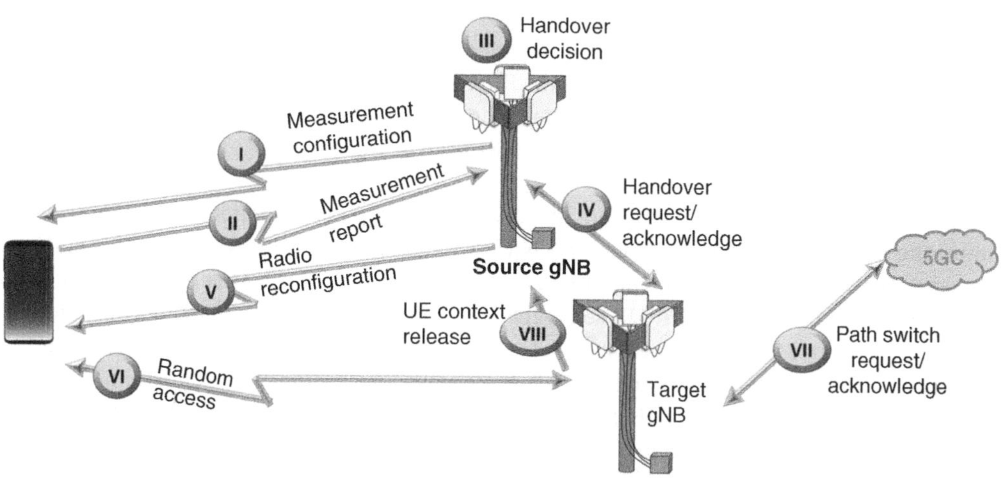

Figure 3.31 Handover in 5G NR. *Source:* nadir_insan / Adobe Stock Photo.

reporting or handover events for which the UE needs to send a Measurement Report message. The gNB may specify the periodicity of measurement reports in case of periodic reporting.

In the first release of 5G (i.e., the 3GPP Release 15), six intra-RAT handover events and two inter-RAT handover events are defined. A1 to A6 are intra-RAT handover events, while B1 and B2 are inter-RAT handover events. See Section 3.2.2 for brief descriptions of these events; details can be found in 3GPP (2024e).

In *Step 2*, the 5G NR UE follows the measurement configuration specified by the gNB and makes measurements such as RSRP, RSRQ, and SINR of the currently serving cell and autonomously detects neighboring cells. The specific time periods during which the UE can make measurements are influenced by the gNB-specified SSB Measurement Timing Configuration (SMTC). On a given carrier frequency, the received signal contains signals from the serving cell and the neighboring cells, and the 5G NR UE can use the PCIs of cells to determine the RSRP and RSRQ of cells. When the reporting criterion for the measurement reporting is met (e.g., the occurrence of an event such as A3), the 5G NR UE sends the Measurement Report message containing the cell identity (i.e., PCI) and RSRP and RSRQ of the serving and relevant neighbor cell(s). As 5G NR is a beamformed air interface, beam-level measurements per cell and cell-level measurements are made and reported. Cell-level measurements involve averaging the beam-level measurements of the beams that are sufficiently strong (i.e., exceeding a given measurement threshold).

In *Step 3*, the serving gNB processes the 5G NR UE's measurement report and executes an implementation-specific handover algorithm. If the serving gNB makes a handover decision, it exchanges handover signaling messages with the target gNB.

In *Step 4*, the serving gNB sends a Handover Request message to the target gNB. The Handover Request message contains the target PCI, and information related to the UE-specific N2 and N3 connections (e.g., AMF's IP address for the N2 connection and the UPF Tunnel Information). The 5G QoS (e.g., QoS flows and UE-AMBR) parameters are also specified. The target gNB executes an implementation-specific admission control algorithm and evaluates the QoS needs of the current 5G QoS flows of the UE against the availability of radio resources. If the target gNB decides to accept the handover request, it responds to the serving gNB with the Handover Request Acknowledge message, which contains parameters such as C-RNTI allocated to the UE in the new cell, a dedicated random-access preamble (to avoid contention with other UEs using the PRACH in the new cell), and system information. This Handover Request Acknowledge message includes the list of 5G QoS flows admitted by the target gNB. If packet forwarding is intended for the 5G NR UE, the Xn-U tunnel information is also included in this message.

In Step 5, the currently serving gNB sends the RRC Reconfiguration to the 5G NR UE. This message contains the handover command and the PCI of the target cell for handover. It also conveys random-access-related information, such as the dedicated random-access preamble for the target cell.

After receiving the handover command, the UE in *Step 6* initiates the random-access procedure toward the target cell of the new serving 5G NR gNB and typically utilizes a dedicated random-access preamble as part of the contention-free random-access procedure. After the successful random-access procedure, the 5G NR UE sends the RRC Reconfiguration Complete message to the new serving gNB in response to the RRC Reconfiguration message.

In *Step 7*, the gNB-core network connectivity is modified. For example, the UE-specific N2 connection is changed so that the AMF can exchange UE-specific signaling messages with the new serving gNB. Furthermore, the N3 connection is modified to enable the UPF to exchange user traffic with the new serving gNB. To modify N2 and N3 connections, the target gNB (i.e., new serving gNB) sends the Path Switch Request to the AMF. The AMF exchanges PDU session update messages with the SMF (i) to provide the SMF with information about the target gNB side of the N3 tunnel and (ii) to obtain information from the SMF about the UPF side of the N3 tunnel. Next, the AMF responds to the new serving gNB with the Path Switch Request Acknowledge message to inform the gNB about the change in the network's UE-specific N2 and N3 connections.

The 3GPP has made enhancements to the overall handover process in later releases by introducing features such as Conditional Handover (CHO), Dual-Active Protocol Stack (DAPS) handover, and L1/L2 Triggered Mobility (LTM). CHO involves preparing more than once candidate handover target cells to reduce the overall handover interruption. DAPS handover involves simultaneous use of two protocol stacks with the currently serving cell and the target neighbor cell to reduce handover interruption. While typical handover relies on Layer 3 RRC signaling, LTM exploits lower-layer signaling to reduce signaling overhead and interruption time.

3.4.3.7 Activities of a 5G NR UE in the RRC_IDLE and RRC_INACTIVE States

The RRC_CONNECTED state enables the 5G NR UE to actively exchange traffic with the network. In a typical scenario, in the absence of any data traffic for a period of a few seconds (e.g., five seconds), the UE-specific inactivity timer expires. The gNB asks the UE to enter inactive or idle mode. The UE context is not preserved in the NG-RAN when the UE is in idle mode. However, when the UE is in the inactive mode, the radio connection is released, but the UE context is preserved in the last serving gNB. Hence, when the UE needs to enter the connected mode, it can make a faster transition due to faster restoration of the radio connection made possible by the preserved UE context in the NG-RAN. The idle and connected modes reduce the amount of signaling and processing for both the UE and the network.

Figure 3.32 summarizes the main operations related to the RRC_IDLE UE in 5G (Tripathi and Reed 2019). The gNB sends the NGAP UE Context Release Request to the AMF, and the MME responds to the eNB with the UE Context Release Command. The 5G NR gNB then asks the 5G

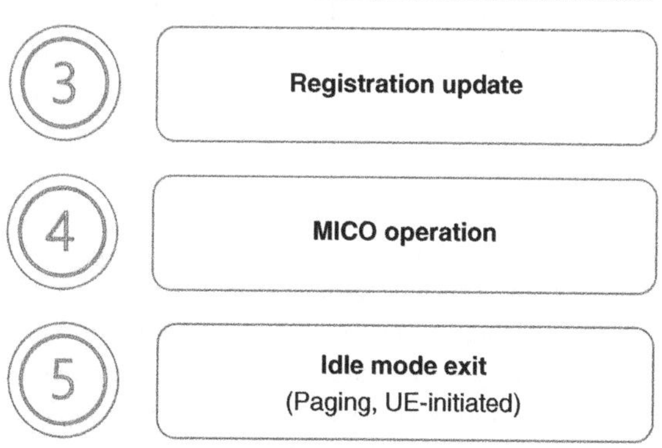

Figure 3.32 Cell Reselection and Tracking Area Update in 5G NR.

NR UE to enter the RRC_IDLE state by sending the RRC Release command in *Step 1*. The gNB sends the NGAP UE Context Release Complete to the AMF. The AMF preserves the UE context with the information such as the 5G-GUTI allocated to the UE and the location of the UE (e.g., the Registration Area where the UE is registered by the AMF; this location information is used to support paging). The IP address allocated to the 5G NR UE also remains valid in the RRC_IDLE state of the UE.

In *Step 2*, the RRC_IDLE UE typically camps on the strongest cell to ensure a high probability of success in receiving any incoming page message and establishes a radio connection by initiating the random-access procedure for any UE-initiated data transfer. Hence, in an implementation-specific manner, the 5G NR UE searches for strong cells and performs cell reselection to a neighbor cell if such a neighbor cell can provide a better signal compared to the currently monitored cell.

In *Step 3*, the 5G NR UE carries out a Registration Update when necessary. For example, if the periodic registration timer expires, the UE executes the registration update procedure. Furthermore, when the 5G NR UE performs cell reselection, the 5G NR UE checks if the new cell belongs to the tracking area that is part of the Registration Area. The AMF determines the UE-specific Registration Area that consists of one or more tracking areas. The Registration Area in 5G is equivalent to the TAI list used in 4G LTE. If the new cell's TA is not in the 5G NR UE's Registration Area, the UE executes the mobility-based registration update procedure.

5G supports a new type of connectivity in the idle mode called Mobile Initiated Connection Only (MICO) as mentioned in *Step 4*. When the 5G NR UE is in the MICO mode, only the 5G NR UE can initiate the connection with the network; the network cannot send a page message to the 5G NR UE in the case of any incoming traffic. The 5G NR UE is unreachable by the network, but the UE can contact the network anytime. The MICO mode enables the 5G NR UE to save power by minimizing network monitoring for paging.

There may be a few reasons for the 5G NR UE to exit the idle mode. For example, to execute the registration update procedure, the UE exits the idle mode, carries out the random-access procedure, establishes an RRC signaling connection, and exchanges NAS signaling with the AMF. Specifically, the 5G NR UE sends the Registration Request message to the AMF, and the AMF replies with the Registration Accept message. The UE sends the Registration Complete message to complete the registration update procedure. The AMF (especially if it is an AMF that has not allocated the current 5G-GUTI to the 5G NR UE) allocates a new 5G-GUTI to the UE and provides an updated Registration Area to the UE. In the future, when the AMF needs to contact the RRC_IDLE 5G NR UE, it sends a paging message to the gNBs that are part of the TAs of the UE-specific Registration Area. Upon completion of the registration update procedure, the 5G NR UE usually re-enters the idle mode.

The 5G NR UE also exits the RRC_IDLE mode when it needs to send user traffic. Furthermore, the 5G NR UE exits the idle mode to monitor for an incoming paging message. To exit the idle mode, the UE carries out the random-access procedure and establishes an RRC signaling connection. The gNB establishes suitable UE-specific radio bearers, such as SRBs and DRBs, to facilitate the transferring of signaling messages and user traffic.

In support of paging, the 5G NR UE monitors a specific paging occasion (a specific subframe) in a paging frame. There is one paging occasion per DRX cycle, and the 5G NR UE monitors for PDCCHs in multiple slots. The specific paging occasion and the paging frame that the 5G NR UE monitors for paging are a function of the UE's ID (e.g., IMSI). The 5G NR UE monitors the paging occasion during the ON period of the DRX paging cycle. The overall DRX paging cycle that includes ON and OFF periods is configurable by the gNB and typically lasts for about two seconds. The idle mode 5G NR UE wakes up every two seconds or so and looks for a page message during a specific paging occasion based on its permanent identity (i.e., IMSI). The gNB sends the RRC Paging message during the paging occasion that the 5G NR UE is expected to monitor. The paging message typically includes 5G-S-TMSI as the identity of the UE.

The RRC_INACTIVE state is a new state introduced by 5G beyond the typical states of RRC_CONNECTED and RRC_IDLE inherited from LTE. In support of the RRC_INACTIVE state, a tracking area is divided into multiple RAN Notification Areas (RNAs), and the user mobility is managed from the perspective of RNAs. For example, as long as the user stays in the same RNA, no RNA-specific signaling needs to be exchanged between the UE and the RAN. The 5G NR UE is still considered accessible by the 5GC as if the 5G NR UE were in the RRC_IDLE state. When a signaling message or user traffic needs to be exchanged with the UE in the RRC_INACTIVE state, the RAN manages paging and the resumption of the UE's RRC connection. The transition from the RRC_INACTIVE state to the RRC_CONNECTED state is fast due to the preservation of the UE's AS context in the last serving gNB, which is the gNB controlling the cell where the UE entered the RRC_INACTIVE state.

In case of user inactivity, the serving gNB may ask the 5G NR UE to enter the RRC_INACTIVE state by sending the RRC Release message with SuspendConfig containing parameters that influence the inactive state operations. Example parameters of SuspendConfig, include RAN paging cycle (e.g., 32, 64, 128, or 256 frames), periodic RNA update timer (e.g., 5 to 720 min), a 40-bit full Inactive-Radio Network Temporary Identifier (I-RNTI), and one 24-bit short I-RNTI, RNA information (i.e., the cell list constituting the current RNA and the RAN area specified as the combination of the Tracking Area Code and the RAN area code). The RRC_INACTIVE UE looks for a RAN paging message every RAN paging cycle. Periodic RNA updates can be viewed as a keep-alive mechanism, where the 5G NR UE exchanges signaling with the RAN to make the RAN aware of its presence in an RNA. I-RNTI is the identity assigned to the RRC_INACTIVE UE by the serving gNB.

To exit the RRC_INACTIVE state, the 5G NR UE sends an RRC Resume Request containing its Short I-RNTI and Resume cause (e.g., mobile-originated signaling, mobile-originated data, and RNA update). The gNB retrieves the UE context from the last serving gNB and replies to the UE with the RRC Resume message. Finally, the 5G NR UE sends RRC Resume Complete to complete the process of the resumption of the RRC connection. The serving gNB switches the UE-specific N2 connection toward the AMF and the N3 connection toward the UPF by exchanging path switch-related signaling messages with the AMF.

3.5 LTE-M and NB-IoT Enhancements Beyond Release 13

The LTE-M and NB-IoT descriptions in Sections 3.1 to 3.3 are based on Release 13 specifications. The 3GPP has enhanced LTE-M and NB-IoT in later releases. Figure 3.33 lists such enhancements for LTE-M and NB-IoT (Tripathi and Reed 2019). These enhancements are briefly discussed later (Qualcomm 2018).

3.5.1 Enhancements Common to LTE-M and NB-IoT

Single-Cell Multicast. The LTE feature of Single Cell-Point-to-Multipoint (SC-PTM) is extended to LTE IoT in Release 14. A given piece of information, for example, a message applicable to a group of IoT devices, can be sent in a multicast manner in a cell to reduce resource utilization and increase efficiency. This mechanism facilitates efficient firmware upgrades for many devices (e.g., environmental sensors), improves battery life by reducing the amount of time that a device needs to stay awake to receive information from the network and increases the network capacity due to the use of the multicast transmission instead of unicast transmissions.

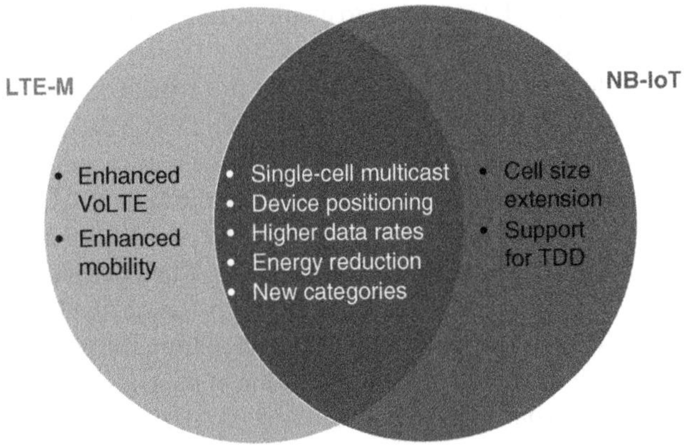

Figure 3.33 Post-Release 13 LTE-M and NB-IoT Enhancements.

Device Positioning. Support for the device positioning is added in Release 14 to enable new use cases such as asset tracking and eCall[31] Services. The Observed Time Difference of Arrival (OTDOA) method can complement satellite-based positioning in dense urban areas and in-building scenarios. Positioning Reference Signals with narrower bandwidths are introduced to facilitate positioning in IoT devices.

Higher Data Rates. To increase data rates, support for larger TBSs and more HARQ processes have been added to LTE-M and NB-IoT.

Energy Reduction. New wake-up signals (WUSs) are introduced to enable low-power channel monitoring. Furthermore, a lower transmit power class corresponding to the maximum EIRP of 14 dBm is defined.

Higher Density. Access class barring is enhanced to support a higher density of cells and users.

New Device Categories. While Release 13 defines Category M1 devices for LTE-M and Category NB1 devices for NB-IoT, Release 14 defines additional categories: Category M2 (Cat-M2) for LTE-M and Category NB2 (Cat-NB2) for NB-IoT. Cat-M2 devices support the channel bandwidth of 5 MHz and peak data rates of 2.4 Mbps in the DL and 2.6 Mbps in the uplink based on the TBS of 4,008 bits. Cat-M2 devices support half-duplex voice, positioning, and single-cell multicast. Beginning Release 14, some Cat-M1 devices may also support half-duplex voice, positioning, and single-cell multicast. Cat-NB2 supports peak data rates of 120 kbps in the downlink and 160 kbps in the uplink. Cat-NB2 devices may support positioning, single-cell multicast, and energy reduction. Beginning Release 14, some Cat-NB1 devices may also support positioning, single-cell multicast, and energy reduction.

3.5.2 LTE-M Enhancements

VoLTE Enhancements. Support for VoLTE is enhanced for smart wearables so that half-duplex voice can be supported more efficiently.

Mobility Enhancements. Support for inter-frequency measurements and high-mobility measurements in extended coverage scenarios is added, enhancing overall mobility performance.

Higher Data Rates. LTE-M is enhanced to support a 5 MHz channel bandwidth instead of the typical 1.4 MHz channel bandwidth.

31 eCall is an emergency call.

NB-IoT Enhancements

Cell Size Extension. Additional cyclic prefixes are supported to support a cell radius of at least 100 km.

Support for TDD. Support for TDD facilitates deployment in higher frequency bands. The TDD operations are optimized for small cells.

3.6 Major Takeaways

The 3GPP has two parallel work streams related to the NTN: NR-NTN and IoT-NTN. NR-NTN utilizes an NR-based air interface and the 5GC. The IoT-NTN supports LTE-M and NB-IoT air interfaces and utilizes the 4G EPC. Here are the major takeaways from this chapter.

- Two main wireless IoT technologies supported by the 3GPP are LTE-M and NB-IoT. The 5G use case of mIoT is addressed by LTE-M and NB-IoT.
- The 3GPP introduced LTE-M and NB-IoT in Release 13 and enhanced these technologies in subsequent releases.
- LTE, LTE-M, NB-IoT, and 5G NR utilize an OFDM-based radio interface.
- Release 13 LTE-M devices support the maximum transmission bandwidth of six PRBs corresponding to 1.08 MHz. Release 13 NB-IoT devices support the maximum transmission bandwidth of one PRB corresponding to 180 kHz.
- LTE-M and NB-IoT devices are much simpler and significantly less expensive than LTE devices.
- To increase battery life, LTE-M and NB-IoT use eDRX and PSM.
- LTE and LTE-M support FDD, TDD, and half-duplex FDD, while Release 13 NB-IoT devices support half-duplex FDD.
- LTE-M and NB-IoT devices have one receive antenna and a single associated receive chain to reduce complexity and cost.
- LTE-M and NB-IoT heavily rely on repetitions to enhance link budget and coverage.
- LTE-M is suitable for in-band deployments with LTE, while NB-IoT supports in-band, guard-band, and standalone deployments.
- While LTE-M and NB-IoT devices may use the maximum EIRP of 23 dBm, some devices may also use lower transmit power levels such as 20 and 14 dBm.
- The attach and EPS bearer setup procedures for LTE-M and NB-IoT are similar.
- 5G uses separate procedures for registration and PDU session setup.
- Key NR features include scalable and flexible OFDM, unified and flexible frame structure, support for various bands, massive MIMO and advanced beamforming, advanced channel coding, and MR-DC.
- The DL and UL data transfer procedures are similar for LTE, LTE-M, and NR, with LTE-M using simpler physical layer features and NR using advanced physical layer features. NB-IoT uses simpler physical layer features compared to LTE-M.
- LTE-M and NR support LTE-like mobility management, where the UE provides measurements based on the occurrence of measurement events, the serving eNB/gNB makes a handover decision, the target eNB/gNB performs admission control for handover, and the UE performs handover toward the target eNB/gNB upon receiving the handover command.
- NB-IoT relies on cell reselection instead of handover for mobility management.
- Examples of post-Release 13 enhancements in LTE-M and NB-IoT include single-cell multicast, device positioning, higher data rates, energy reduction, and new device categories (e.g., Cat-M2 and Cat-NB2).

References

3GPP (2024a). *General Packet Radio Service (GPRS) enhancements for Evolved Universal Terrestrial Radio Access Network (E-UTRAN) access.* TS23.401. https://portal.3gpp.org/desktopmodules/Specifications/SpecificationDetails.aspx?specificationId=849 (accessed 29 August 2024).

3GPP (2024b). *System architecture for the 5G System (5GS).* TS23.501. https://portal.3gpp.org/desktopmodules/Specifications/SpecificationDetails.aspx?specificationId=3144 (accessed 29 August 2024).

3GPP (2024c). *Evolved Universal Terrestrial Radio Access (E-UTRA) and Evolved Universal Terrestrial Radio Access Network (E-UTRAN): Overall description.* TS36.300. https://portal.3gpp.org/desktopmodules/Specifications/SpecificationDetails.aspx?specificationId=2430 (accessed 29 August 2024).

3GPP (2024d). *Evolved Universal Terrestrial Radio Access (E-UTRA): User Equipment (UE) radio access capabilities.* TS36.306. https://portal.3gpp.org/desktopmodules/Specifications/SpecificationDetails.aspx?specificationId=2434 (accessed 29 August 2024).

3GPP (2024e). *NR and NG-RAN overall description.* TS38.300. https://portal.3gpp.org/desktopmodules/Specifications/SpecificationDetails.aspx?specificationId=3191 (accessed 29 August 2024).

3GPP (2024f). *NR: User Equipment (UE) radio access capabilities.* TS38.306. https://portal.3gpp.org/desktopmodules/Specifications/SpecificationDetails.aspx?specificationId=3193 (accessed 29 August 2024).

Qualcomm (2018). Leading the LTE IoT evolution to connect the massive Internet of Things. July 2018, https://www.qualcomm.com/content/dam/qcomm-martech/dm-assets/documents/powerpoint_messaging_-_leading_the_lte_iot_evolution_to_connect_the_massive_iot_july18.pdf ().

Tripathi, N.D. and Reed, J.H. (2019). *5G Cellular Communications: Journey and Destination.* The Wireless University. https://thewirelessuniversity.com/ (accessed 29 August 2024).

4

Challenges of an NTN

A Non-Terrestrial Network (NTN) poses distinct radio environment-related challenges compared to a terrestrial network (TN). This chapter describes such NTN challenges and their implications on the design and operations of the NTN. For example, the NTN typically has a long propagation delay and often a significantly varying propagation delay, requiring timing synchronization and timer enhancements. A non-geostationary NTN platform leads to large and varying Doppler shifts, and suitable frequency synchronization adjustments are needed. Compared to a typical TN, a next-generation Node B (gNB) in an NTN covers a much larger geographic area, reducing the amount of radio resources per user, increasing the differential propagation delays among the users, and potentially increasing handover signaling load due to the need to move numerous users from one NTN platform to another in a short period of time. While a TN cell is fixed, an NTN cell can move, requiring enhanced cell reselection, handover, and registration area management. An NTN may use Earth-fixed beams, quasi-Earth-fixed beams, or Earth-moving beams, requiring different mobility management approaches. The type of NTN payload, such as the transparent payload or the regenerative payload, affects the architecture, design, and operations of the NTN. The propagation path loss can be quite large for NTN platforms such as the geostationary earth orbit (GEO) satellites, influencing receiver design, link budget, and achievable user and network performance. Long-term signal strength characteristics differ between the TN and the NTN, requiring a different mobility management approach in the NTN. Special atmospheric effects such as the Faraday effect and scintillations may need to be considered for the NTN. Certain NTN-specific operations, such as the feeder link switch and noncontiguous coverage, must be managed. The chapter also provides a high-level overview of the NTN solutions that the 3GPP has developed to address these NTN challenges.

4.1 An Overview of NTN-Specific Challenges

The NTN possesses quite distinct characteristics compared to the TN (3GPP 2023) (Stuhlfauth 2022). Hence, the challenges unique to the NTN emerge. Figure 4.1 lists the NTN-specific challenges. These challenges are briefly introduced here and discussed in more detail in Sections 4.2 to 4.11.

3GPP-based Non-Terrestrial Networks in 5G and 6G: Expanding the Frontiers of Wireless Communications, First Edition.
Nishith D. Tripathi and Jeffrey H. Reed.

Companion website: www.wiley.com/go/tripathi5g

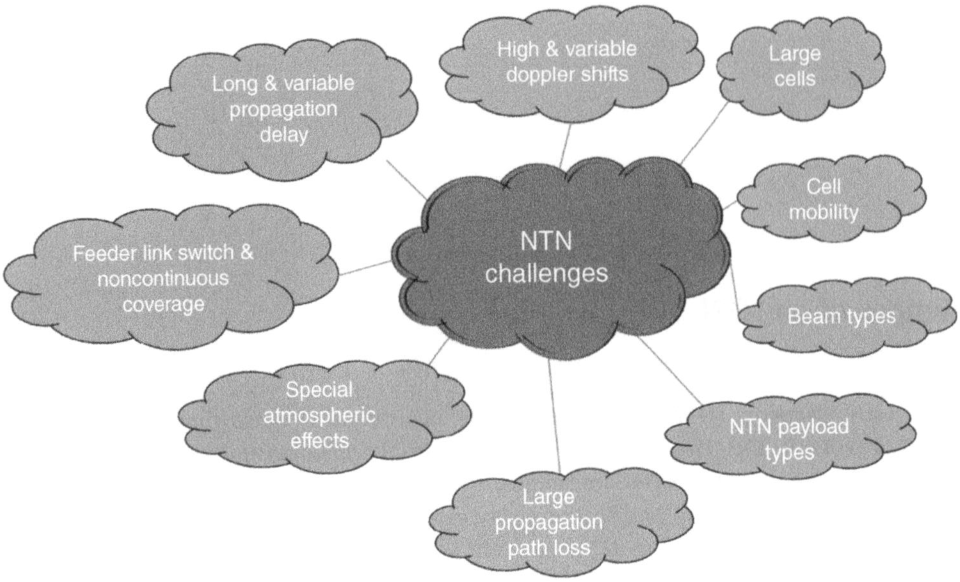

Figure 4.1 Major Challenges of the NTN.

Propagation Delay. The distance between the user equipment (UE) and the satellite can be significantly longer than between the UE and a TN gNB. Such long distances lead to long propagation delays. Furthermore, in the case of a non-geostationary orbit (NGSO) satellite, the UE experiences a variable propagation delay as the satellite moves along its orbit.

Doppler Shifts. A Low Earth orbit (LEO) satellite may move at a speed of about 7 km/s, causing large Doppler shifts. Furthermore, the angle between the UE and the satellite keeps changing, resulting in varying Doppler shifts.

Cell Size. Compared to a typical TN, a cell in an NTN covers a much larger geographic area. For a given amount of channel bandwidth, the available amount of radio resources per user may be reduced. Furthermore, differences in propagation delays among the users in the cell may be large, requiring timing adjustments. A large cell size, in conjunction with cell mobility, implies increased signaling load.

Cell Mobility. An NGSO satellite leads to moving cells on the Earth's surface. Such NTN cell mobility requires even stationary UEs to experience cell reselection and handover. An NTN UE would need to carry out registration updates or tracking area updates when the tracking area changes due to cell change.

Types of Beams. An NTN may use Earth-fixed beams, quasi-Earth-fixed beams, or Earth-moving beams. These beams have different effects on the UEs. Hence, different mobility management approaches are needed for different types of beams.

NTN Payload Types. While Release 17 and Release 18 support only a transparent payload, future releases, such as Release 19, may support a regenerative payload. Different types of payloads result in different NTN network architectures, requiring modifications in the operations of the NTN.

Propagation Path Loss. The propagation path loss can be quite large for NTN platforms such as GEO satellites. Such large path loss leads to significant attenuation of the signal, weakening the received signal and preventing the use of aggressive modulation and coding schemes (MCSs). A high-gain antenna at the NTN platform would be essential to ensure adequate radio link performance.

Signal Strength Characteristics. In a TN, the path loss near the gNB is relatively small, and the path loss near the cell edge is large. There is a significant difference in the received signal strength at the UE (i) near the gNB and (ii) near the cell edge. However, in the case of the NTN, the difference in the signal strength near the cell center and the cell edge may not be significant. Hence, reference signal received power (RSRP) as the sole handover criterion may be inadequate, requiring a different mobility management approach in the NTN.

Atmospheric Effects. Signals in a typical TN are relatively close to the Earth's surface. However, the signal between the UE on the Earth's surface and a high-altitude satellite passes through the ionosphere and hence may experience the Faraday effect (where the polarization of the signal changes) and scintillations (which are rapid variations in the signal strength). Such special atmospheric effects need to be considered for the NTN system design.

Feeder Link Switch. When an NTN platform goes out of sight of the NTN Gateway (NTN-GW), the NTN platform needs to be connected to another suitable NTN Gateway. Such a change in the NTN Gateway is called a feeder link switch because of the change in the feeder link connectivity. Such NTN-specific operations need to be managed effectively.

Noncontiguous/Noncontinuous Coverage. An NGSO satellite covers different geographic areas at different instants or during different periods. Hence, when all the planned NGSO satellites to provide continuous coverage in a target service area (e.g., global or regional coverage) are not operational, some geographic areas may not be served by any NGSO satellite, leading to noncontiguous coverage on the Earth's surface. Such noncontiguous coverage requires enhanced mobility management.

4.2 Long and Variable Propagation Delays

The propagation delay between the UE and the gNB depends on the type of the NTN platform (e.g., a high altitude platform station (HAPS), an LEO satellite, or a GEO satellite) and the type of the NTN payload (i.e., transparent or regenerative). Figure 4.2 compares the UE-NTN platform delays for different types of NTN platforms. In general, propagation delays in a typical NTN can be expected to be much longer than those in a typical TN.

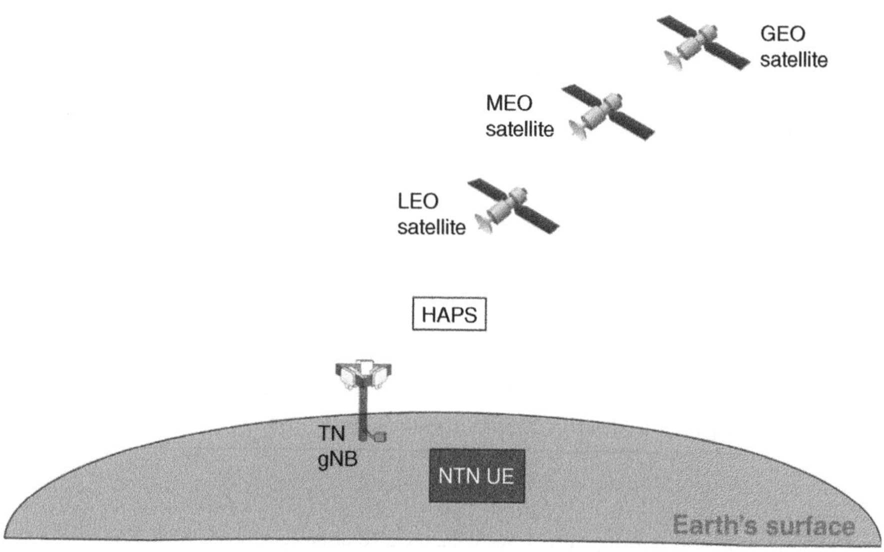

Figure 4.2 Propagation Delays for Different NTN Platforms.

The maximum distance between a TN UE and a typical TN gNB is usually a few kilometers in a suburban environment, a few hundred meters in an urban environment, and tens of kilometers in a rural environment[1]. In contrast, as mentioned in Section 2.1 in Chapter 2, a HAPS may be placed at an altitude of about 20 km, an LEO satellite may have an altitude of 300 to 1,500 km, and a GEO satellite has an altitude of about 35,786 km. Hence, the shortest distance between the NTN UE and the NTN platform is significantly longer than the typical longest distance between the TN UE and the TN gNB. These long UE-NTN platform distances translate into long propagation delays between the NTN UE and the NTN gNB. For a given NTN UE location, the actual propagation delay between the NTN UE and the NTN gNB depends on the position of the NTN platform (i.e., the elevation angle of the path between the NTN UE and the NTN platform), the type of the NTN payload (i.e., transparent or regenerative), and the position of the NTN Gateway relative to the NTN platform (i.e., the elevation angle of the path between the NTN Gateway and the NTN platform) for the transparent payload (3GPP 2020). The NTN Gateway is assumed to be collocated with the other functions of the NTN gNB.

The radio frequency (RF) signal in a wireless communication system (approximately) travels at the speed of light, which is 3×10^8 m/s. This speed can be used along with the distance traveled to determine the propagation delay. Table 4.1 summarizes the one-way propagation delay and round-trip time (RTT) for several NTN configurations. The RTT is between the NTN UE and the NTN gNB. These delays mentioned in Table 4.1 are based on the 3GPP estimates (3GPP 2020). Round-trip

Table 4.1 Example Propagation Delays in the NTN.

Scenario	NTN Configuration	1-Way Distance (NTN UE-NTN gNB) (km)	1-way Propagation Delay (ms)	Round-Trip Time (RTT) (ms)
1	GEO satellites with a transparent payload	81,712	272	544
2	GEO satellites with a regenerative payload	40,586	135	270
3	LEO satellites with a transparent payload (altitude of 600 km)	4,261	14	28
4	LEO satellites with a regenerative payload (altitude of 600 km)	1,932	6.4	12.8
5	LEO satellites with a transparent payload (altitude of 1,500 km)	7,749	26	52
6	LEO satellites with a regenerative payload (altitude of 1,500 km)	3,648	12	24
7	MEO satellites with a transparent payload (altitude of 10,000 km)	28,558	95	190
8	MEO satellites with a regenerative payload (altitude of 10,000 km)	14,018	47	94
9	HAPSs with a regenerative payload	229	0.75	1.5
10	HAPSs with a transparent payload	$229 * 2 = 458$	1.5	3.0

1 In special cases such as boomer cells in vast plains of Australia, the gNB-distance may exceed 100 km.

delays (RTDs) or RTTs that consider the two-way propagation delays are important in various radio interface operations such as the random-access procedure, timing advance, and downlink (DL) and uplink (UL) data transfer latency.

For the GEO satellite-based NTN operating in the transparent payload configuration, the one-way propagation delay between the UE and the NTN Gateway/gNB is estimated to be about 272 ms for a total distance of about 81,712 km for the NTN UE-GEO satellite-NTN Gateway path (3GPP 2020). The two-way propagation delay or the RTT for this configuration is ($2 * 272 = 544$ ms). For the GEO satellite-based NTN operating in the regenerative payload configuration with the integrated gNB on the NTN platform, the one-way propagation delay between the UE and the gNB is estimated to be about 135 ms, and the RTT is estimated to be 270 ms. Note that the regenerative payload configuration has a shorter RTT than the transparent payload configuration due to the shorter distance between the NTN UE and the gNB in the case of the regenerative payload.

The 3GPP has evaluated propagation characteristics for two altitudes of LEO satellites, 600 and 1,500 km. For the LEO satellite-based NTN operating at an altitude of 600 km in the transparent payload configuration, the one-way propagation delay between the UE and the NTN Gateway/gNB is estimated to be about 14 ms for a total distance of about 4,261 km (3GPP 2020). The two-way propagation delay or the RTT for this configuration is 28 ms. For the LEO satellite-based NTN operating at an altitude of 600 km in the regenerative payload configuration, with the integrated gNB on the NTN platform, the one-way propagation delay between the UE and the gNB is estimated to be about 6.4 ms, and the RTT is estimated to be 12.8 ms.

For the LEO satellite-based NTN operating at an altitude of 1,500 km in the transparent payload configuration, the one-way propagation delay between the UE and the NTN Gateway/gNB is estimated to be about 26 ms for the total distance of about 7,749 km (3GPP 2020). The two-way propagation delay, or the RTT for this configuration, is 52 ms. For the LEO satellite-based NTN operating at an altitude of 1,500 km in the regenerative payload configuration with the integrated gNB on the NTN platform, the one-way propagation delay between the UE and the gNB is estimated to be about 12 ms, and the RTT is estimated to be 24 ms.

As MEO satellites operate between LEO satellite altitudes and GEO satellite altitudes, propagation delays for the MEO satellite-based NTNs are between the two extremes of LEO satellite-based NTNs and GEO satellite-based NTNs. For the Medium Earth Orbit (MEO) satellite-based NTN operating at an altitude of 10,000 km in the transparent payload configuration, the one-way propagation delay between the UE and the NTN Gateway/gNB is estimated to be about 95 ms for the total distance of about 28,558 km (3GPP 2020). The two-way propagation delay or the RTT for this configuration is 190 ms. For the MEO satellite-based NTN operating at an altitude of 10,000 km in the regenerative payload configuration with the integrated gNB on the NTN platform, the one-way propagation delay between the UE and the gNB is estimated to be about 47 ms, and the RTT is estimated to be 94 ms.

HAPSs operate at much lower altitudes than even the lowest altitude LEO satellites. Hence, propagation delays for the HAPS satellite-based NTNs are less than the propagation delays of LEO satellite-based NTNs. For the HAPS-based NTN operating at an altitude of 20 km in the transparent payload configuration and the elevation angle 5°, the one-way propagation delay between the UE and the NTN Gateway/gNB is estimated to be about 1.5 ms for the total distance of about 458 km (3GPP 2020). The distance between the NTN UE and the HAPS and between the HAPS and the NTN Gateway is assumed to be the same (i.e., 229 km). The two-way propagation delay or the RTT for this configuration is 3 ms. For the HAPS-based NTN operating at an altitude of 20 km in the regenerative payload configuration with the integrated gNB on the NTN platform, the

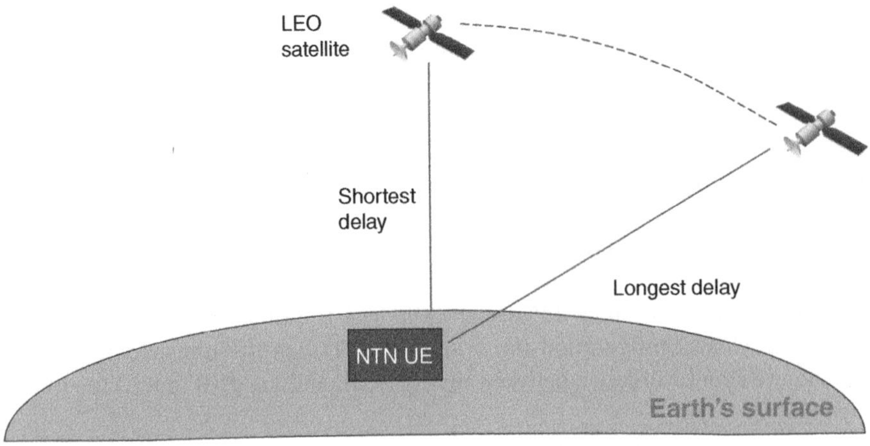

Figure 4.3 Propagation Delay Variations in an NTN.

one-way propagation delay between the UE and the gNB is estimated to be about 0.75 ms, and the RTT is estimated to be 1.5 ms.

The UE-TN gNB distances are much shorter in typical TN deployments, leading to propagation delays of only a few microseconds (or even less in the case of small cells), which are much shorter than the NTN delays specified in Table 4.1. Furthermore, in the case of a NGSO satellite, the UE experiences variable propagation delay as the satellite moves along its orbit, as shown in Figure 4.3.

When an LEO satellite is directly above the NTN UE (i.e., at the nadir position), the distance between the NTN UE and the NTN platform is minimal, leading to the shortest propagation delay. In contrast, when the NTN UE is communicating with an LEO satellite that is just above the horizon (e.g., at a certain elevation angle associated with the edge of coverage (EOC) situation), the distance between the NTN UE and the NTN platform is maximum, leading to the longest propagation delay. For example, the differential one-way propagation delay between the nadir and the EOC paths is 4.44 ms for the differential distance of 1,332.2 km for an LEO satellite operating at an altitude of 600 km (3GPP 2020). The differential one-way propagation delay between the nadir and the EOC paths is (i) 7.2 ms for the differential distance of 2,147.5 km for an LEO satellite operating at an altitude of 1,500 km and (ii) 13.4 ms for the differential distance of 4,018.16 km for an MEO satellite operating at an altitude of 10,000 km (3GPP 2020).

4.3 High and Variable Doppler Shifts

Relative mobility between the transmitter and receiver causes a Doppler shift, which is a change in the frequency of the received signal compared to the transmitted signal's frequency (Tripathi and Reed 2014). GEO satellites and HAPSs can essentially be considered stationary for the purpose of calculating Doppler shifts. Hence, for GEO satellites and HAPSs, the Doppler shift is primarily caused by the UE movement. In the case of NGSO satellites, the Doppler shift due to satellite movement is much higher than the Doppler shift caused by a typical UE. Hence, the impact of the UE movement for NGSO satellites can generally be ignored. Figure 4.4 illustrates the geometry influencing the Doppler shift.

In Figure 4.4, the NTN UE is located at point M, and the NTN platform such as an LEO satellite is located at point S. Point O is the Earth's center, R is the Earth's radius, and h is the height or altitude

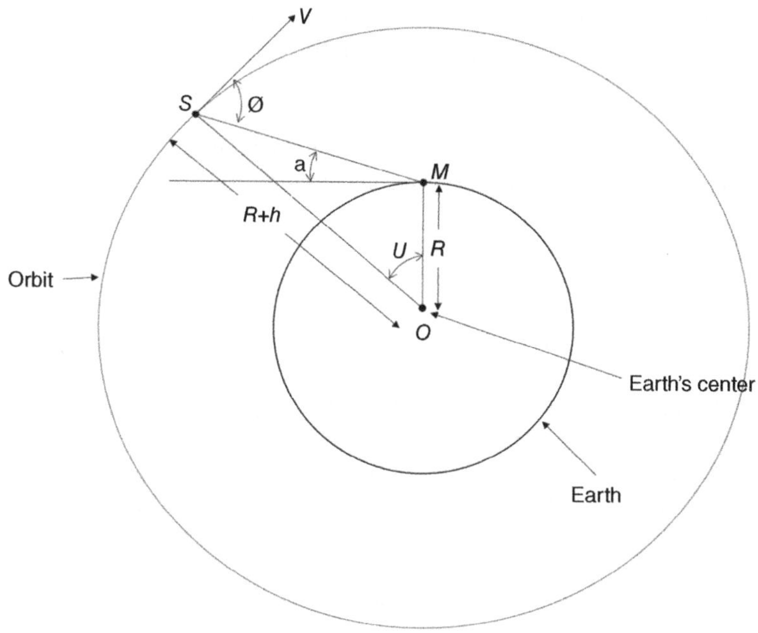

Figure 4.4 Doppler Shifts in an NTN.

of the NTN platform. The NTN platform's velocity at M is represented by the vector V. The angle alpha is the elevation of the NTN platform for the UE at point M. The carrier frequency of the transmitted signal, the magnitude of the vector V, and the angle theta between the vector SM and the vector V are vital parameters that determine the Doppler shift. The NGSO satellite speed V is 7.5622 km/s at an altitude of 600 km, 7.1172 km/s at an altitude of 1,500 km, and 4.9301 km/s at an altitude of 10,000 km (3GPP 2020). The angle theta can be determined based on the geometry of Figure 4.4 using R, h, and the angle u; see 3GPP (2020) for details.

Table 4.2 summarizes the conclusions of the 3GPP evaluation of the maximum Doppler shifts for the carrier frequencies of 2 and 30 GHz (3GPP 2020).

As the altitude increases, the maximum Doppler shift and the Doppler shift variations decrease at a given carrier frequency. As the Doppler shift is directly proportional to the carrier frequency, the

Table 4.2 Doppler Shifts in the NTN.

Scenario Number	Scenario Description	Carrier Frequency (GHz)	Maximum Doppler Shift (kHz)
1	LEO satellite at a 600 km altitude	2	±48
2	LEO satellite at a 600 km altitude	30	±720
3	LEO satellite at a 1,500 km altitude	2	±40
4	LEO satellite at a 1,500 km altitude	30	±600
5	MEO satellite at a 10,000 km altitude	2	±15
6	MEO satellite at a 10,000 km altitude	30	±225

Doppler shift at 30 GHz is 15 times the Doppler shift at 2 GHz. For a TN, the subcarrier spacing of 15 kHz in sub-7 GHz is adequate to overcome the Doppler shifts at the maximum target speeds of about 500 km/h. The Doppler shift in such a case is a fraction of the subcarrier spacing. However, in the NGSO satellite-based NTN, the Doppler shifts are quite large compared to the subcarrier spacing, and such Doppler shift-induced inter-subcarrier interference cannot be addressed by the subcarrier spacing. Hence, a mechanism such as frequency precompensation becomes important to ensure proper functioning NTN.

4.4 NTN Cell Size

Compared to a typical TN cell, a cell in an NTN covers a much larger geographic area. Figure 4.5 shows relative cell sizes of an NT cell for different NTN platforms).

A HAPS has a lower altitude than satellites and tends to cover a relatively small geographic area, although the HAPS cell covers an area larger than a typical TN cell. A LEO satellite's beam covers a larger area than a HAPS beam, and a GEO satellite's beam covers a much larger area than the beams of LEO and MEO satellites.

As an NTN cell is much larger than a typical TN cell, the amount of radio resources per user is reduced in an NTN cell for a given amount of channel bandwidth and user density.

Furthermore, differences in propagation delays among the users in the NTN cell may be large, requiring a wider range of timing adjustments compared to a TN cell. The 3GPP has carried out an analysis of delays in 3GPP (2020). For example, in the case of a TN cell with a 10 km cell radius, the maximum delay difference between the cell center and the cell edge is 0.00333 ms. In contrast, the HAPS cell at an altitude of 20 km has a delay difference between the EOC and nadir is 0.697 ms. The delay difference between the EOC and nadir is 4.44 ms for an LEO satellite at an altitude of 600 km and 16 ms for a GEO satellite at an altitude of 35,786 km. A large cell size, in conjunction with cell mobility, implies increased signaling load.

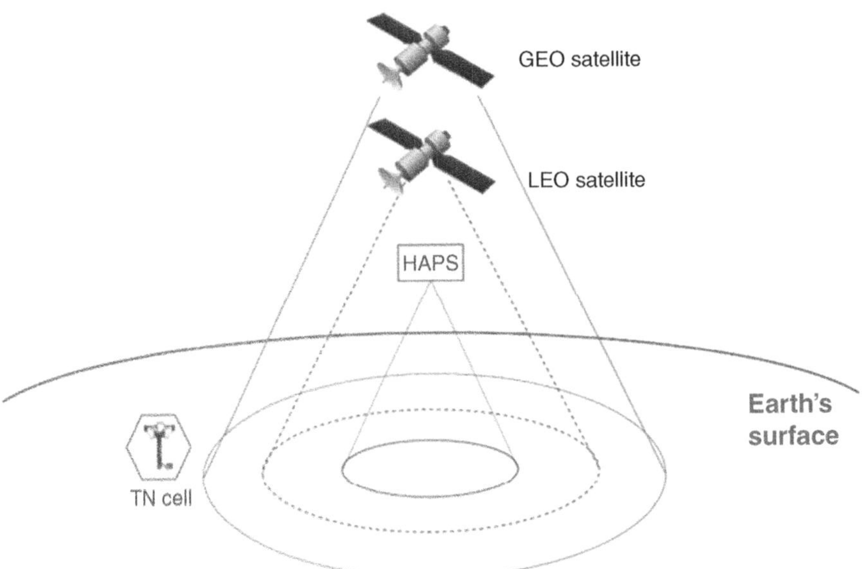

Figure 4.5 Cell Sizes in an NTN.

4.5 NTN Cell Mobility

A typical TN cell covers a fixed geographic area. In contrast, an NGSO satellite leads to cells moving on Earth's surface. Hence, even if an NTN UE is stationary, it experiences a change in the best cell, requiring cell reselection for a UE in the RRC_IDLE state and handover for a UE in the RRC_CONNECTED state.

Prior to Release 17, a TN cell can only be associated with a single tracking area (TA). The Access and Mobility Management Function (AMF) registers the UE in a registration area consisting of one or more Tas (Tripathi and Reed 2019). Consider Figure 4.6, where an NTN UE is stationary.

At time t_1, the NTN UE is served by an NTN cell with a physical cell identifier (PCI)=A. The AMF has registered the UE in the TA identified by TAI=X. The NTN cell with PCI=A broadcasts TAI=X. At time t_2, the NTN cell with PCI=A no longer serves the geographic area of the NTN UE. A new NTN cell with PCI=B provides service to the geographic area, where the NTN UE is located. As PCI=B is associated with TAI=Y and as the UE's registration area only contains TAI=X, the NTN UE performs a registration area update. An NTN cell can be very large as discussed in Section 4.4, and there can be numerous UEs in a large NTN cell. As one NTN cell moves out of a given geographic area and another NTN cell moves in, numerous UEs experience a change in their serving cell. If the cell broadcasts a single TAI, the NTN UE needs to perform a registration update. Registration updates from many UEs result in excessive access stratum (AS) signaling and non-access stratum (NAS) signaling. Furthermore, radio resource utilization also increases significantly (e.g., random-access resources and dedicated radio channel resources). The processing load on the NG-RAN and the 5GC (especially the AMF) also increases significantly. Hence, an enhancement to the TA management is needed in an NTN.

4.6 Type of Beams

An NTN may use Earth-fixed beams, quasi-Earth-fixed beams, or Earth-moving beams. These beams have different effects on the UEs. Hence, different mobility management approaches are needed for different types of beams.

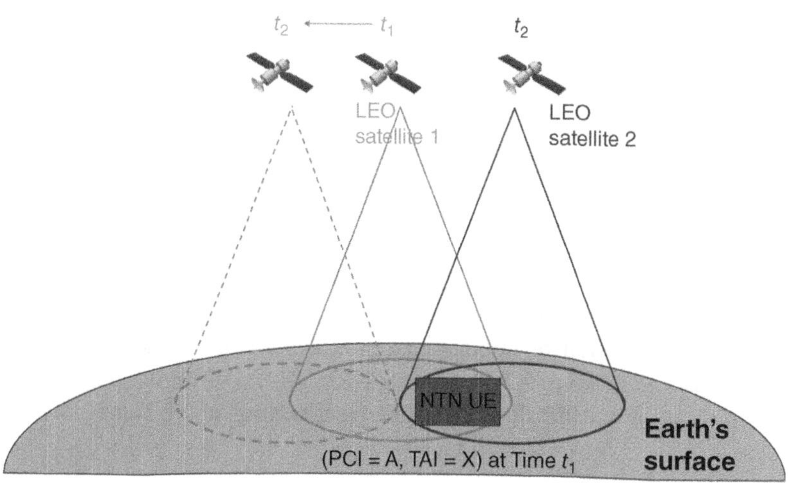

Figure 4.6 Impact of NTN Cell Mobility on Registration Updates.

An NTN may use one of three types of beams to provide radio coverage: Earth-fixed beams, Earth-moving beams, or quasi-Earth-fixed beams (3GPP 2024b). Section 2.4 in Chapter 2 explains these beams in detail. A given NTN payload utilizes one of such beams based on its capabilities and configuration.

An Earth-fixed beam always covers the same geographic area. GEO satellites and HAPSs use Earth-fixed beams. GEO satellites and HAPSs cannot be absolutely fixed in their positions. Due to the slight movement of GEO satellites and HAPSs, the actual coverage on the Earth's surface changes a little from time to time but is considered to be fixed for planning and operations. Mobility management mechanisms such as cell reselection and handover primarily occur due to the movement of the NTN UE and the movement of the NTN cells because the NTN cells are essentially fixed from the perspective of the NTN UEs.

An Earth-moving beam covers a different geographic area on the Earth's surface from one instant to the very next instant. When NGSO satellites do not steer their beams at different angles but use the same angle to point their beams, the radio coverage provided by such fixed beams changes from one instant to the next. As the radio coverage area illuminated by Earth-moving beams continuously changes and as different NTN cells serve a given geographic area, mobility management mechanisms such as cell reselection and handover occur on an ongoing basis. Enhancements to TN-oriented mobility management are needed to ensure effective mobility management for Earth-moving beams.

A quasi-Earth beam continues to cover one geographic area during a finite period even as the NGSO satellite moves along its orbit. Then, such a beam covers a different geographic area during another finite period. For example, a given service area is served by one NGSO satellite during one period and another NGSO satellite during another period. Two NGSO satellites need to overlap over time in these service periods to enable reliable and high-performance transfer of UEs from one NGSO satellite to another. Enhancements to TN-oriented mobility management are needed to ensure effective mobility management for quasi-Earth-moving beams.

4.7 Types of NTN Payloads

As shown in Figure 4.7, there are two main types of NTN payloads: a transparent payload and a regenerative payload. Section 1.2 in Chapter 1 provides a brief introduction to the transparent payload.

In the case of the transparent NTN payload, the NTN payload does not carry out technology-specific baseband processing. The NTN payload performs RF filtering, frequency conversion, and amplification functions. The NTN payload essentially carries out the conversion of RF signals between the service link and the feeder link. For the downlink data transfer, user traffic arrives at the gNB/eNB from the 5GC/EPC. The gNB/eNB carries out 5G/LTE-specific baseband processing and RF processing and provides the 5G/LTE signal to the NTN Gateway (referred to as "NTN GW" in Figure 4.7). The NTN Gateway transmits the 5G/LTE RF signal to the NTN payload via the feeder link using an implementation-specific approach (e.g., the use of implementation-specific frequency band and transmission of implementation-specific control signaling). The transparent NTN payload maps the received RF signal into NR/LTE Uu RF signal for the UE and performs suitable amplification and RF filtering before transmitting an RF signal toward UEs of the NTN cell.

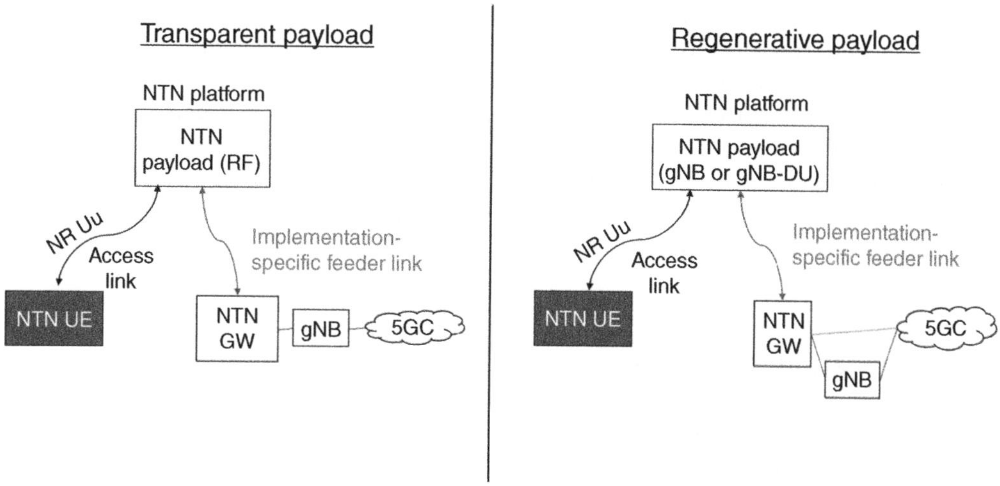

Figure 4.7 Transparent and Regenerative Payloads in an NTN.

Consider uplink data transfer for the transparent payload. User traffic is transmitted by the NTN UE on the access/service link using the NR/LTE Uu RF signal. The transparent NTN payload maps the received uplink RF signal into an RF signal that NTN Gateway can process (and it may also include any implementation-specific control signaling). The transparent payload performs suitable amplification and RF filtering before transmitting an RF signal toward the NTN Gateway. The NTN Gateway provides the NR/LTE signal to the gNB/eNB, and the gNB/eNB carries out 5G/LTE-specific baseband processing to retrieve user traffic. The gNB/eNB sends user traffic to the 5GC/EPC.

The main difference between the transparent payload and the regenerative payload is that the regenerative payload carries out at least some technology-specific baseband processing. Figure 4.7 shows two examples of a regenerative payload for 5G; an equivalent scenario also exists for LTE. In one example, the NTN payload implements the integrated gNB. In another example, the NTN payload implements the gNB-distributed unit (gNB-DU). In the case of 5G, when the integrated gNB is on the NTN platform, the regenerative NTN payload carries out complete NR Uu baseband processing and RF processing and exchanges NR Uu RF signals with the NTN UE on the access or service link. NR-Uu refers to the NR based radio interface between the UE and the gNB. The regenerative payload communicates with the NTN Gateway via the feeder link in an implementation-specific manner to transfer NG interface information, such as N2 signaling toward the AMF and N3 user traffic toward the UPF. The NTN Gateway carries out suitable processing for the information transfer to or from the 5GC.

When the gNB-DU is on the NTN platform, the regenerative NTN payload carries out selected NR Uu baseband processing (i.e., implementation of PHY, MAC, and RLC layers) and RF processing and exchanges NR Uu RF signals with the NTN UE on the access or service link. The regenerative payload communicates with the NTN Gateway via the feeder link in an implementation-specific manner to transfer F1 interface information, such as F1-C signaling and F1-U traffic. The NTN Gateway interfaces with the gNB-central unit (gNB-CU), which implements radio resource control (RRC), service data adaptation control (SDAP), and packet data convergence protocol (PDCP). The NTN Gateway performs suitable processing to exchange F1-C signaling and F1-U traffic with the gNB-CU.

As mentioned in Chapter 1, Release 17 and Release 18 support only a transparent payload. Release 19 supports a regenerative payload. As shown in Figure 4.7, different types of payloads result in different NTN network architectures and hence require suitable modifications in the operations of the NTN. For example, when the regenerative payload implements the integrated gNB, an inter-satellite link (ISL) can be supported to enable a series of satellites to be connected via the ISL so that connectivity with a faraway NTN Gateway can be established. The ISL interface would need to be designed. See Chapter 5 for a detailed description of the NTN architecture with transparent and regenerative payloads.

4.8 Propagation Path Loss

The propagation path between the NTN UE and the NTN platform can often be line-of-sight (LOS). In contrast, the propagation path between the TN UE and the TN gNB is often non-line-of-sight (NLOS). The propagation path loss exponent can thus be two for the LOS propagation in an NTN, while it is much larger, such as 3.5 in a TN. However, that is the only good news for the NTN path loss! As the distances between the NTN UE and the NTN platform are usually significantly larger than the TN UE-gNB distances, the overall path loss is much larger in an NTN than in a TN. Such large path loss leads to significant attenuation of the signal, weakening the received signal and preventing the use of aggressive MCSs. Hence, data rates in an NTN are expected to be smaller than those in a TN. A high-gain antenna at the NTN platform is important to ensure adequate radio link performance.

4.9 Long-Term Signal Strength Characteristics

In a TN, the path loss near the gNB is relatively small, and the path loss near the cell edge is quite large. There is a significant difference in the received signal strength at the UE (i) near the gNB and (ii) near the cell edge. However, in the case of the NTN, the difference in the signal strength near the cell center and the cell edge may not be significant. Hence, RSRP as the sole handover criterion is often inadequate, requiring a different mobility management approach in the NTN.

In a typical TN, the distance between the UE and the gNB is much shorter than the distance between the NTN UE and the NTN platform. Hence, propagation path losses are usually larger in an NTN than in a TN, leading to weaker RSRP values in an NTN compared to a TN. Figure 4.8 shows the trend of RSRP measured at the UE for a TN and an NTN.

In both cases, as expected, the RSRP decreases as the UE moves away from the cell center to the cell edge because of the increased distance-based propagation path loss. However, in the case of the TN, there is a significant difference between the RSRP near the cell center and the RSRP near the cell edge. Additionally, as the TN UE moves away from its current serving cell and toward a neighbor cell, the RSRP differences are significant enough such that the RSRP of the neighbor cell can be compared with the RSRP of the serving cell to make a handover or cell reselection decision.

In contrast, there is not a large RSRP difference between the cell center and cell edge in a typical NTN, because the propagation path loss is essentially dictated by the more significant distance between the Earth's surface and the NTN platform; the distance between (i) a cell center UE and the NTN platform and (ii) a cell edge UE and the NTN platform is almost the same. Furthermore, there is insignificant RSRP difference near the cell edge between the serving and neighboring

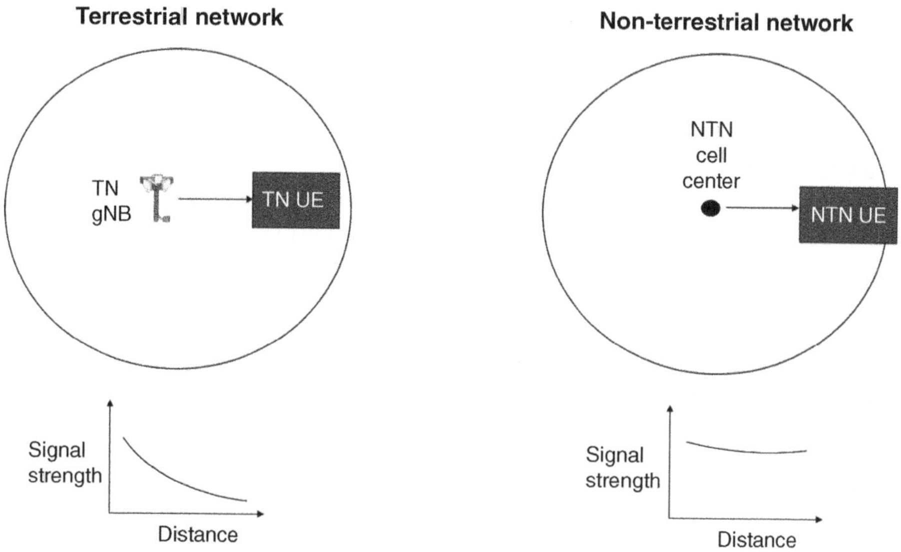

Figure 4.8 Long-Term Signal Strength Characteristics in an NTN.

cells. Hence, the long-term signal strength or RSRP characteristics imply that RSRP alone cannot be a reliable criterion for handover. New handover and cell reselection criteria are required because of these RSRP characteristics.

4.10 Special Atmospheric Effects

The RF signals between the gNB/eNB and the UE in a typical TN travel close to the Earth's surface. However, the RF signals between the gNB/eNB and the UE in an NTN pass through the upper portion of troposphere and/or ionosphere, depending on the altitude of the NTN platform. The RF signals are affected by special atmospheric effects such as scintillations and Faraday rotation as they travel through the troposphere and the ionosphere.

The troposphere is the lowest layer of Earth's atmosphere with the bottom of the troposphere at Earth's surface. The top of the troposphere is about 10 km above sea level (UCAR 2024). The height of the top of the troposphere is a function of latitude (i.e., lowest over the poles and highest at the equator) and season (i.e., lower in winter and higher in summer). The height of the top of the troposphere is as much as 20 km near the equator and as little as 7 km over the poles in winter. Air temperature, air pressure, and air density decrease with altitude in the troposphere.

The ionosphere is a series of regions in parts of the mesosphere and thermosphere where the Sun radiation has caused electrons to be loose from their atoms and molecules (UCAR 2024). These electrically charged atoms and molecules are referred to as ions, providing the ionosphere with certain unique properties[2]. The ionosphere extends from about 80 (or 50) to 600 (or 1,000) km above the Earth's surface (National Oceanic and Atmospheric Administration (NOAA) 2024).

2 By the way, the aurora, often called Northern Lights or Southern Lights, occurs in the layers of the ionosphere.

4.10.1 Faraday Rotation

Cross-pol antennas are commonly used in a TN to transmit RF signals from a TN gNB/eNB. The cross-polarized RF signals in the downlink are received by a suitable antenna system at the UE. Similarly, the RF signal transmitted by the UE's antenna system is received by the cross-pol antenna systems of the gNB/eNB in the uplink.

In an NTN, the polarization of the RF signal changes because of Faraday rotation. The Faraday rotation refers to the rotation of the polarization due to the interaction of the electromagnetic wave with the ionized medium in the ionosphere. Hence, when an NTN platform is above the ionosphere, the Faraday rotation comes into the picture. In other words, if an NTN is using LEO, MEO, and/or GEO satellites, the Faraday rotation is applicable. If an NTN uses HAPSs at an altitude of about 20 km, the Faraday rotation is not applicable. For an improved signal reception in the presence of Faraday rotation, the NTN antenna system may be modified.

4.10.2 Scintillation

Scintillation refers to rapid fluctuations of the received signal amplitude and phase. The scintillation effect exists in the ionosphere and the troposphere. See 3GPP (2020) for instructions on modeling the effects of ionospheric and tropospheric scintillation.

The 3GPP considers scintillation in the ionosphere propagation to be significant for frequencies below 6 GHz. Scintillation is one of the most severe disruptions for a propagation path traversing the ionosphere for signals below 3 GHz. Scintillation may occur occasionally up to 10 GHz (3GPP 2020). Scintillations are a function of location, time of day, season, solar, and geomagnetic activity. Strong levels of scintillation rarely occur in midlatitude regions (except during strong geomagnetic storms) but may occur daily during postsunset hours in low-latitude regions. At high latitudes, such as auroral and polar latitudes, moderate to strong levels of scintillations can occur. Scintillations are caused by convective plasma processes. Although amplitude scintillations are dominant at low latitudes and phase scintillations are dominant at high latitudes, both scintillations are not exclusive to each other.

The 3GPP considers scintillation in the troposphere for frequencies above 6 GHz. The effect of tropospheric scintillation increases with the carrier frequency, especially above 10 GHz. Furthermore, the effects of scintillation increase with low elevation angles due to the longer propagation path and wide beam widths. Tropospheric scintillation is caused by sudden changes in the refractive index due to temperature variation, water vapor content, and barometric pressure.

4.11 Feeder Link Switch

The feeder link between the NTN platform and the NTN Gateway is a unique interface in an NTN. The feeder link enables the NTN UE to access data networks via the core network. Maintenance of the feeder link requires a clear propagation path between the NTN platform and the NTN Gateway. If an NTN platform loses the LOS path toward an NTN Gateway, it needs to find an LOS path toward another NTN Gateway. When the NTN platform changes its anchor NTN Gateway, feeder link switch occurs because of the change in the endpoints of the feeder link. Consider Figure 4.9 as a feeder link switch example.

As shown in Figure 4.9, the satellite has visibility of NTN Gateway1 at time t_1, and hence the feeder link between the satellite and NTN Gateway1 is used for all UEs served by the NTN cells

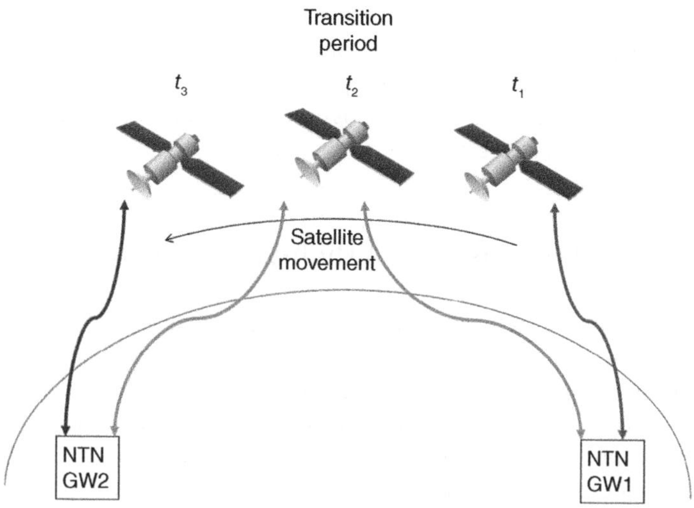

Figure 4.9 Feeder Link Switch in an NTN.

of the satellite. Note that an NTN platform can create multiple beams, with one beam comprising an NTN cell per beam or multiple beams comprising an NTN cell. At time t_1, the satellite does not have the visibility of NTN Gateway2. At time t_2, the satellite has visibility of both NTN Gateway1 and NTN Gateway2. As the satellite is moving away from NTN Gateway1 and toward NTN Gateway2, the feeder link switch can be executed by moving the UEs from one feeder link to another (or equivalently, one NTN Gateway to another) during the transition period when the satellite has visibility of both outgoing and incoming gateways. At time t_3, the satellite has visibility of only NTN Gateway2, and all UEs associated with the satellite are using the new feeder link between the satellite and NTN Gateway2.

While Figure 4.9 shows the satellite movement scenario, there may be other reasons for a feeder link switch, such as maintenance and traffic offloading. A suitable feeder link switch mechanism needs to be designed to ensure a seamless service experience for NTN users. For example, in the transparent payload case, a feeder link switch can be executed by a network-initiated procedure by using two distinct radio resources simultaneously. In another example, a feeder link switch can use a network-initiated make-before-break strategy in the regenerative payload case.

4.12 Noncontiguous/Noncontinuous Coverage

As mentioned in Section 2.2 in Chapter 2, three GEO satellites, six MEO satellites, or hundreds of LEO satellites can provide global radio coverage. For enhanced performance, such as higher capacity or larger throughput, more than one satellite can simultaneously cover the same geographic area. Some operators may aim to provide global coverage, while others may only cover specific regions.

An NGSO satellite serves one geographic area at one instant or during one period and another geographic area at another instant or during another period. NGSO satellites orbit the Earth with a velocity of a few kilometers per second, illuminating different geographic areas at different times. Hence, an adequate number of NGSO satellites must be operational to provide continuous coverage in a given geographic area. However, the required number of NGSO satellites to provide continuous coverage in the target service area may not be operational when commercial NTN services are first offered. This implies noncontiguous coverage, where some geographic area does not get illuminated by any of the operational NGSO satellites. The noncontiguous coverage introduces challenges to seamless service experience and mobility management.

4.13 3GPP Solutions to the NTN Challenges: A High-Level Overview

Examples of 3GPP-defined NTN solutions that address the NTN challenges include enhanced timing and frequency synchronization, protocol timer adjustments, enhanced system information, enhanced mobility management, enhanced registration area management, hybrid automatic repeat request (HARQ) enhancements, random-access procedure enhancements, and a new 5G QoS class for the NTN (3GPP 2018) (3GPP 2021) (3GPP 2024a). *These solutions are described in detail in various chapters of the book.* A brief summary of example NTN solutions is given later.

Enhanced timing and frequency synchronization. Allow larger timing adjustments for NTNs to address long and variable propagation delays. The NTN UE estimates the Doppler shift using its location and the location of the NTN platform advertised in system information. It also performs precompensation for frequency synchronization.

Protocol timer adjustments. To accommodate long propagation delays, timers that are a function of the RTT (e.g., RLC timers) are extended in an NTN.

Enhanced System Information. The NTN UE obtains NTN-specific information (e.g., satellite ephemeris information) using enhanced system information. Through system information, the NTN UE can learn about the type of network it is observing (e.g., a TN or an NTN).

Enhanced mobility management. As the RSRP criterion alone is inadequate in numerous NTN scenarios, additional mobility management criteria such as location-based and time-based criteria (e.g., the remaining serving time of an NTN cell in case of quasi-Earth-fixed beams) are needed for mobility management.

Enhanced Registration Area management. Earth-fixed tracking areas are defined to avoid frequent registration updates resulting from moving NTN cells. An NTN cell broadcasts different tracking area identifiers (TAIs) as it moves across different geographic areas associated with such TAIs.

HARQ enhancements. More HARQ processes are supported in an NTN to accommodate longer RTTs. Furthermore, HARQ can be disabled to just make use of RLC retransmissions.

Random-access procedure enhancements. Various random-access parameters related to timing are extended to accommodate long propagation delays.

5G QoS enhancement. A new 5G QoS indicator (5QI) is introduced for the GEO-based NTN to reflect long propagation delays for GEO satellites.

4.14 Major Takeaways

An NTN poses distinct challenges compared to a TN, requiring NTN-specific solutions. This chapter describes NTN challenges and their implications for the design and operations of NTN. Here are the major takeaways from this chapter.

- The NTN typically has a long propagation delay and often a significantly varying propagation delay, requiring timing synchronization and timer enhancements.
- A non-geostationary NTN platform leads to large and varying Doppler shifts, and suitable frequency synchronization adjustments are needed.
- Compared to a typical TN, a gNB in an NTN covers a much larger geographic area, reducing the amount of radio resources per user, increasing the differential propagation delays among the users, and potentially increasing handover signaling load due to the need to move numerous users from one NTN platform to another in a short period of time.
- While a TN cell is fixed, an NTN cell can move, requiring enhanced cell reselection, handover, and registration area management.
- An NTN may use Earth-fixed beams, quasi-Earth-fixed beams, or Earth-moving beams, requiring different mobility management approaches.
- The type of NTN payload, such as the transparent payload or the regenerative payload, affects the NTN's architecture, design, and operations.
- The propagation path loss can be very large for NTN platforms such as the GEO satellites, influencing the receiver design, the link budget, and achievable user and network performance.
- Long-term signal strength characteristics differ between the TN and the NTN, requiring a different mobility management approach in the NTN.
- Special atmospheric effects such as the Faraday effect and scintillations may need to be considered for the NTN.
- Certain NTN-specific operations, such as the feeder link switch, must be managed.
- Examples of 3GPP-defined NTN solutions that address the NTN challenges include enhanced timing and frequency synchronization, protocol timer adjustments, enhanced system information, enhanced mobility management, enhanced registration area management, HARQ enhancements, random-access procedure enhancements, and a new 5G QoS class for the NTN.

References

3GPP (2018). *Study on using satellite access in 5G.* TR 22.822. https://portal.3gpp.org/desktopmodules/Specifications/SpecificationDetails.aspx?specificationId=3372 (accessed 29 August 2024).

3GPP (2020). *Study on New Radio (NR) to support non-terrestrial networks.* TR 38.811. https://portal.3gpp.org/desktopmodules/Specifications/SpecificationDetails.aspx?specificationId=3234 (accessed 29 August 2024).

3GPP (2021). *Solutions for NR to support non-terrestrial networks (NTN).* RP-211557. 3GPP Work Item Description. 3GPP TSG RAN meeting #91-e, e-meeting, March 22-26, 2021, https://www.3gpp.org/ftp/tsg_ran/TSG_RAN/TSGR_92e/Docs (accessed 29 August 2024).

3GPP (2023). *Solutions for NR to support non-terrestrial networks (NTN).* TR 38.821. https://portal.3gpp.org/desktopmodules/Specifications/SpecificationDetails.aspx?specificationId=3525 (accessed 29 August 2024).

3GPP (2024a). *System architecture for the 5G System (5GS).* TS23.501. https://portal.3gpp.org/desktopmodules/Specifications/SpecificationDetails.aspx?specificationId=3144 (accessed 29 August 2024).

3GPP (2024b). *NR and NG-RAN overall description.* TS38.300. https://portal.3gpp.org/desktopmodules/Specifications/SpecificationDetails.aspx?specificationId=3191 (accessed 29 August 2024).

National Oceanic and Atmospheric Administration (NOAA) (2024). IONOSPHERE. https://www.swpc.noaa.gov/phenomena/ionosphere (accessed 29 August 2024).

Stuhlfauth, R. (2022). 5G NTN Takes Flight: Technical Overview of 5G Non-Terrestrial Networks. Rohde & Schwarz White Paper. Version 1.0. July 2022. https://www.rohde-schwarz.com/us/solutions/satellite-testing/landingpages/white-paper-5g-ntn-takes-flight-technical-overview-of-5g-non-terrestrial-networks_255919.html (accessed 8 August 2024).

Tripathi, N.D. and Reed, J.H. (2014). *Cellular Communications: A Comprehensive and Practical Guide*. Wiley. https://www.wiley.com/en-kr/Cellular+Communications%3A+A+Comprehensive+and+Practical+Guide-p-9780470472071 (accessed 29 August 2024).

Tripathi, N.D. and Reed, J.H. (2019). *5G Cellular Communications: Journey and Destination*. The Wireless University. https://thewirelessuniversity.com/ (accessed 29 August 2024).

UCAR (2024). Layers of Earth's Atmosphere. https://scied.ucar.edu/learning-zone/atmosphere/troposphere (accessed 29 August 2024).

5

NTN Architectures

The 3GPP introduced the NTN as a 5G feature in Release 17. The 3GPP carried out various NTN studies, including potential architectures for the NTN. This chapter illustrates various candidate architectures of the NTN. While the 3GPP has defined an NTN architecture with a transparent payload in Release 17 and Release 18 NTN specifications, the 3GPP has also investigated other NTN architectures, such as an NTN with a regenerative payload and multi-connectivity NTNs. An NTN payload can be viewed as an NTN platform's subsystem that performs processing to support communications. The regenerative payload is within the scope of the 3GPP Release 19 specifications. In the NTN architecture with a transparent payload, the on-the-ground base station carries out baseband technology-specific processing such as NR and LTE processing (e.g., LTE-M processing or NB-IoT processing), and the NTN payload in the NTN platform performs RF functions such as frequency translation, filtering, and amplification. In the case of the NTN architecture with a regenerative payload, the NTN platform contains an NTN payload that performs gNB or gNB-DU functions. Furthermore, in the case of a multi-connectivity NTN, an NTN UE simultaneously communicates with (i) two NTNs (e.g., one GEO satellite-based NTN and another LEO satellite-based NTN) or (ii) one TN and one NTN. Non-terrestrial core and services networks can be quite beneficial in several scenarios described in this chapter. An example NT architecture with non-terrestrial core and services networks is also illustrated. Enhanced tracking area management and enhanced QoS for an NTN are also described. Optical communication may play a key role in NTN deployments. A brief overview of optical communication in the context of an NTN is also given. The set of NTN architectures discussed in the chapter provides a glimpse of potential NTN architectures that may be deployed to achieve the goals of anywhere mobile broadband connectivity.

5.1 NTN Network Architectures in a Nutshell

The 3GPP has studied various architectures for the NTN. Figure 5.1 lists the main types of NTN architectures. Details of these NTN architectures are discussed in detail in Sections 5.2–5.5.

Figure 5.1 classifies an NTN architecture into two groups, single-connectivity NTN and multi-connectivity NTN. In the single-connectivity NTN, the NTN UE communicates with the NTN only and utilizes on-the-ground 5GC and IMS. In the multi-connectivity NTN, the NTN UE simultaneously communicates with multiple radio networks or core networks.

In the single-connectivity scenario, an NTN UE may communicate with the NTN gNB (and the 5GC and potentially IMS) via a transparent NTN payload or a regenerative NTN payload. An NTN

3GPP-based Non-Terrestrial Networks in 5G and 6G: Expanding the Frontiers of Wireless Communications, First Edition.
Nishith D. Tripathi and Jeffrey H. Reed.

Companion website: www.wiley.com/go/tripathi5g

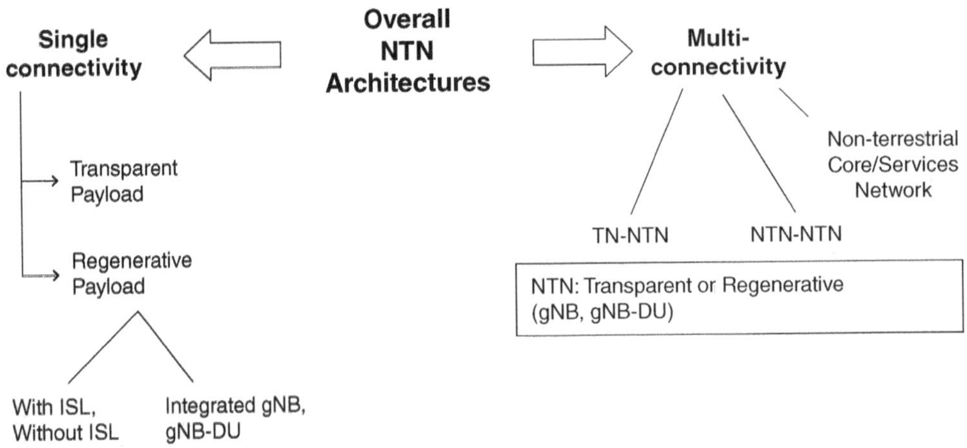

Figure 5.1 NTN Architecture Options.

payload can be viewed as an NTN platform's subsystem that performs processing to support communications. See Section 5.5 for more details on the NTN payload. An NTN platform may also be referred to as a spaceborne or airborne vehicle. In the case of the regenerative payload, the regenerative payload may house the full gNB functions or only the gNB-DU functions. 5G supports a disaggregated gNB that consists of a gNB-central unit and a gNB-distributed unit. A gNB-central unit implements upper layers of the NR radio protocol stack, including Radio Resource Control (RRC), Service Data Adaptation Protocol (SDAP), and Packet Data Convergence Protocol (PDCP). A gNB-distributed unit implements lower layers of the NR radio protocol stack, including Radio Link Control (RLC), Medium Access Control (MAC), and Physical (PHY) layer. When the regenerative NTN payload houses the gNB functions, two NTN payloads may have an inter-satellite link (ISL) between them. An ISL enables a series of NTN payloads to connect to each other in a chain to allow connectivity with a distant NTN Gateway on the ground. For example, an NTN payload providing middle-of-the-ocean coverage at a given instant may not have visibility of any NTN Gateway. An ISL enables such an NTN payload to connect to an NTN Gateway.

In the multi-connectivity scenario, an NTN UE may simultaneously communicate with a TN and an NTN; the NTN may use a transparent or regenerative NTN payload. An NTN UE may communicate with two NTNs, such as two LEO satellite-based NTNs or one GEO satellite-based NTN and one LEO satellite-based NTN. These NTNs may use regenerative or transparent payloads. In another example of multi-connectivity, an NTN UE may connect to both (i) on-the-ground 5GC network functions and (ii) non-terrestrial 5GC network function(s). Furthermore, one or more IMS components may also be located on an NTN satellite.

Note that the 3GPP-based NTN supports only the transparent payload in Release 17 and Release 18. 3GPP Release 19 extends support to an NTN with a regenerative payload. See Chapter 10 for a glimpse of Release 19 NTN enhancements targeted by the 3GPP.

5.2 An NTN with a Transparent Payload

Figure 5.2 shows a simplified NTN architecture with a transparent NTN payload (3GPP 2024b). Both 5G NR and 4G IoT technologies are shown. Note that only a transparent payload is supported in Release 17 and Release 18.

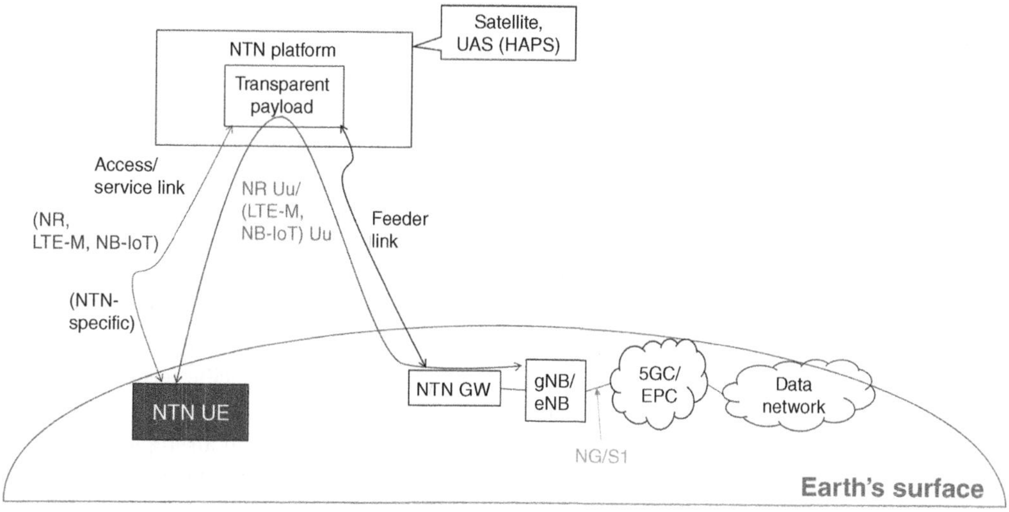

Figure 5.2 An NTN with a Transparent NTN Payload.

5.2.1 NR-NTN

The NTN that supports the NR air interface is referred to as an NR-NTN. In the 5G NTN architecture, the 5G NR-NTN UE communicates with on-the-ground gNB using a 5G NR Uu interface that passes through a transparent NTN payload and an NTN Gateway. The service link or the access link between the NR-NTN UE and the transparent NTN payload is based on the NR Uu interface defined by the 3GPP. The feeder link between the transparent NTN payload and the NTN Gateway is implementation specific and beyond the scope of 3GPP specifications. In one example implementation, the feeder link may use a proprietary waveform and possibly a non-3GPP frequency band. In another example implementation, the feeder link can use the 5G NR waveform used on the service/access link.

The baseline (i.e., pre-Release 17) NR-based air interface is enhanced by the 3GPP to address the NTN-specific challenges described in Chapter 4. The NR-NTN UE utilizes such NTN-centric enhanced 5G NR air interface to transmit and receive information from an NR-NTN gNB. A suitable NR frequency band is used on the enhanced NR-NTN air interface between the 5G NTN UE and the transparent NTN payload. In support of the information transfer from the NR-NTN UE to the network, the NR-NTN UE carries out suitable NR baseband processing to send information such as user traffic, an access stratum message, a non-access stratum message, or physical layer information (e.g., a random-access preamble and a scheduling request). The NR-NTN UE's physical layer also performs suitable RF processing so the transparent NTN payload can process the 3GPP-compliant NTN-specific uplink RF signal. The transparent payload observes the target channel bandwidth of the uplink radio channel in the NR-NTN cell to capture uplink RF signals of all NR-NTN UEs in the cell. The transparent payload performs functions such as frequency translation to map the received composite NR Uu uplink signal to the carrier frequency targeted for the feeder link. In an implementation-specific manner, the transparent payload also performs RF filtering and amplification. The NTN Gateway receives the RF signal from the transparent NTN payload via an implementation-specific feeder link. The NTN Gateway provides the received signal to the gNB. The logical interface between the NTN Gateway and the gNB is undefined. Indeed, the NTN Gateway function can be considered an internal function of the NR-NTN gNB. The NR-NTN gNB carries out typical TN gNB-like baseband processing to retrieve the uplink information received from NTN UEs. The NR-NTN gNB interacts with the 5GC network functions, such as the AMF and the UPF, as needed per NTN UE. For example, in the case of user traffic, the NR-NTN gNB forwards the NR-NTN UE's IP packets toward a specific UPF that forwards the packet toward a suitable data network based on the network slice associated with an established PDU session.

In support of the information transfer from the 5G network to the NR-NTN UE, the UPF receives the NR-NTN UE's IP packets from a data network associated with an established PDU session. The UPF forwards these IP packets to the NR-NTN gNB using a specific GTP-based N3 tunnel. The NR-NTN gNB carries out typical TN gNB-like baseband processing to send packets of different NR-NTN UEs in the downlink. Specifically, the NR-NTN gNB scheduler allocates radio resources to UEs and determines transmission parameters (e.g., packet size and MCS) for the selected NR-NTN UEs. The NR-NTN gNB sends a 5G NR DL signal to the NTN Gateway. The transparent NTN payload receives the RF signal from the NTN Gateway via an implementation-specific feeder link. The transparent payload performs functions such as frequency translation to map the received NR Uu downlink signal to the carrier frequency targeted for the service link. In an implementation-specific manner, the transparent payload also performs RF filtering and amplification. The transparent payload transmits the downlink RF signal in the NR-NTN cell. The NR-NTN UEs observe the PDCCHs to check their resource allocation. The NR-NTN UE that has been allocated radio resources carries out suitable NR baseband processing to retrieve the information sent by the NR-NTN gNB.

5.2.2 IoT-NTN

The NTN that supports LTE-based IoT technologies such as LTE-M and NB-IoT is referred to as an IoT-NTN. In the IoT-NTN architecture, the IoT-NTN UE communicates with on-the-ground IoT-NTN eNB using the air interface that passes through a transparent NTN payload and an NTN Gateway. The service link or the access link between the IoT-NTN UE and the transparent NTN payload is based on the enhanced LTE-M or NB-IoT interface defined by the 3GPP. As mentioned earlier, in the case of NR-NTN, the feeder link between the transparent NTN payload and the NTN Gateway is implementation specific and beyond the scope of 3GPP specifications.

The 3GPP has enhanced the pre-Release 17 LTE-M/NB-IoT-based air interface to address the NTN-specific challenges described in Chapter 4. The IoT-NTN UE utilizes such NTN-centric enhanced LTE-M/NB-IoT air interface to transmit information to and receive information from an IoT-NTN eNB. A suitable frequency band is used on the enhanced IoT air interface between the IoT-NTN UE and the transparent NTN payload. In support of the information transfer from the IoT-NTN UE to the network, the IoT-NTN UE carries out suitable LTE-M/NB-IoT baseband processing to send information such as user traffic, an access stratum message, a non-access stratum message, or physical layer information. The IoT-NTN UE's physical layer also performs suitable RF processing so that the transparent NTN payload can process the 3GPP-compliant LTE-M/NB-IoT-specific uplink RF signal. The transparent payload observes the target channel bandwidth of the uplink radio channel in the IoT-NTN cell to capture uplink RF signals of all IoT-NTN UEs in the cell. The transparent payload performs functions such as frequency translation to map the received composite LTE-M/NB-IoT uplink signal to the carrier frequency targeted for the feeder link. Similar to the NR-NTN case, the transparent payload may also perform RF filtering and amplification. The NTN Gateway receives the RF signal from the transparent NTN payload via an implementation-specific feeder link and provides the received signal to the IoT-NTN eNB. The IoT-NTN eNB carries out typical TN eNB-like baseband processing to retrieve the uplink information received from IoT-NTN UEs. The IoT-NTN eNB interacts with the EPC network elements, such as the MME and the S-GW, as needed per IoT-NTN UE. For example, in the case of IP-based user traffic, the IoT-NTN eNB forwards the IoT-NTN UE's IP packets toward a specific S-GW that forwards the packet toward a suitable P-GW. The P-GW, in turn, forwards the packet to a suitable data network identified by an access point name (APN).

In support of the IP-based information transfer from the LTE network to the IoT-NTN UE, the P-GW receives the IoT-NTN UE's IP packets from the data network associated with an APN. The P-GW forwards these IP packets to the relevant S-GW via a specific GTP-based S5/S8 bearer, and the S-GW forwards the IP packets to the appropriate IoT-NTN eNB using a specific GTP-based S1-U bearer. The IoT-NTN eNB performs typical TN eNB-like baseband processing to send packets of different UEs in the downlink. Specifically, the IoT-NTN eNB scheduler allocates radio resources to UEs and determines transmission parameters (e.g., packet size and MCS) for the selected IoT-NTN UEs. The IoT-NTN eNB sends an LTE-M/NB-IoT DL signal to the NTN Gateway. The transparent NTN payload receives the RF signal from the NTN Gateway via an implementation-specific feeder link. The transparent payload performs functions such as frequency translation to map the received LTE-M/NB-IoT downlink signal to the carrier frequency targeted for the service link. In an implementation-specific manner, the transparent payload may also perform RF filtering and amplification. The transparent payload transmits the downlink RF signal in the IoT-NTN cell. The IoT-NTN UEs observe the PDCCHs to check their resource allocation. The IoT-NTN UE that has been allocated radio resources carries out suitable LTE-M/NB-IoT baseband processing to retrieve the information sent by the IoT-NTN eNB.

For NR-NTN and IoT-NTN with a transparent payload, the NTN gNB/eNB is on the ground. Hence, the NTN gNB/eNB can communicate with the core network using an on-the-ground transport network just like a TN gNB/eNB. Any hardware changes or software updates for the NTN gNB/eNB can be carried out like a TN gNB/eNB. Furthermore, as the NTN gNB/eNB is on the ground, troubleshooting becomes relatively easier. In summary, the transparent payload simplifies the design and maintenance of the NTN compared to an NTN with a regenerative payload. The drawback of the NTN with a transparent payload is that the NTN platform must have direct visibility of the NTN Gateway.

5.3 An NTN Network Architecture with Regenerative Payloads

An NTN with a regenerative payload enables an NTN operator to address more use cases, such as enhanced coverage, including middle-of-the-ocean coverage. Figure 5.3 illustrates an NTN with a regenerative payload, where the regenerative payload houses the NTN gNB or the IoT-NTN eNB.

A regenerative NTN payload opens the possibility of using an ISL. An optional ISL can use a chain of NTN satellites (or gNBs/eNBs) NTN Gateway to reach an on-the-ground NTN Gateway that does have the visibility of such NTN Gateway.

5.3.1 NR-NTN

In Figure 5.3, the 5G NR-NTN UE communicates with the NR-NTN gNB residing on the regenerative NTN payload using 5G NR Uu interface on the service link (also known as the access link). The feeder link carries the NG interface information, such as N2 or NGAP signaling and N3 user plane traffic. The NTN Gateway interfaces with (i) the AMF via the N2 interface and (ii) the UPF via the N3 interface (which is implemented as a GTP tunnel).

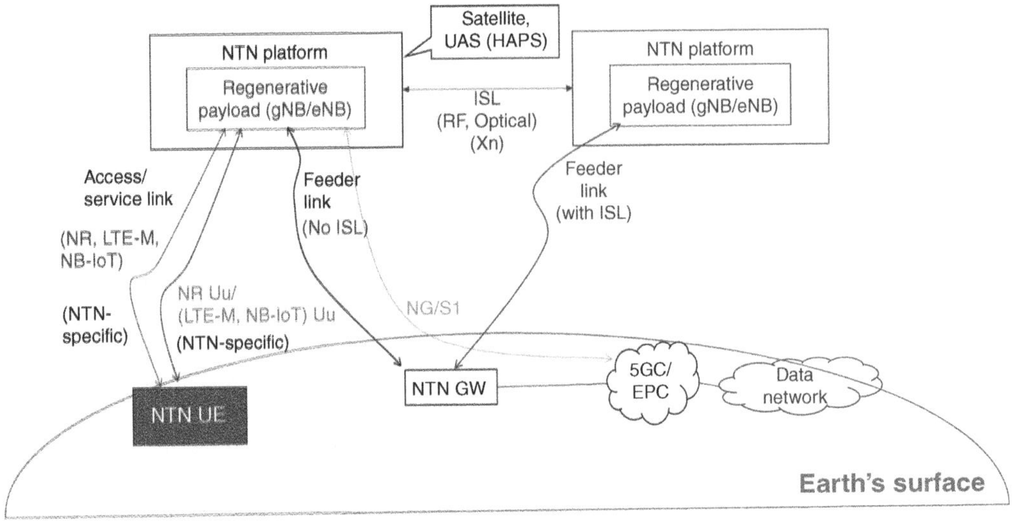

Figure 5.3 An NTN with a Regenerative NTN Payload Containing a gNB/eNB.

The feeder link between the regenerative NTN payload and the NTN Gateway is implementation specific and beyond the scope of 3GPP specifications. For example, the feeder link may use a proprietary waveform and possibly a non-3GPP frequency band or the 5G NR waveform.

5.3.2 IoT-NTN

In Figure 5.3, the IoT-NTN UE communicates with the IoT-NTN eNB residing on the regenerative NTN payload using LTE-M or NB-IoT interface on the service link or the access link. The feeder link carries the S1 interface information, such as S1-MME or S1AP signaling and S1-U or access bearer traffic. The NTN Gateway interfaces with (i) the MME via the S1-MME interface and (ii) the S-GW via the S1-U interface (which is implemented as a GTP tunnel).

The feeder link between the regenerative NTN payload and the NTN Gateway is implementation specific and beyond the scope of 3GPP specifications. The feeder link may use a proprietary waveform or the LTE-M/NB-IoT waveform utilized on the service/access link. A 3GPP-defined or a non-3GPP-defined frequency band may be utilized on the feeder link.

For NR-NTN and IoT-NTN with a regenerative payload housing a gNB/eNB, the NTN gNB/eNB is non-terrestrial, making hardware changes impractical and software updates more complex compared to a transparent NTN payload. Additionally, as the NTN gNB/eNB is on the NTN payload, troubleshooting becomes more complicated. In summary, the regenerative payload with a gNB/eNB opens up the possibility of supporting more use cases at the expense of increased complexity.

There is another architectural option to create a regenerative NTN payload for an NR-NTN: the use of a disaggregated gNB. As shown in Figure 5.4, the regenerative payload houses the gNB-DU instead of the full gNB.

In Figure 5.4, the 5G NR-NTN UE communicates with the NR-NTN gNB-DU residing on the regenerative NTN payload using the 5G NR Uu interface on the service link or the access link. The feeder link carries the F1 interface information, such as F1-C signaling and F1-U user plane traffic between the NTN Gateway and the gNB-DU. The feeder link between the regenerative NTN payload and the NTN Gateway is implementation specific and beyond the scope of 3GPP

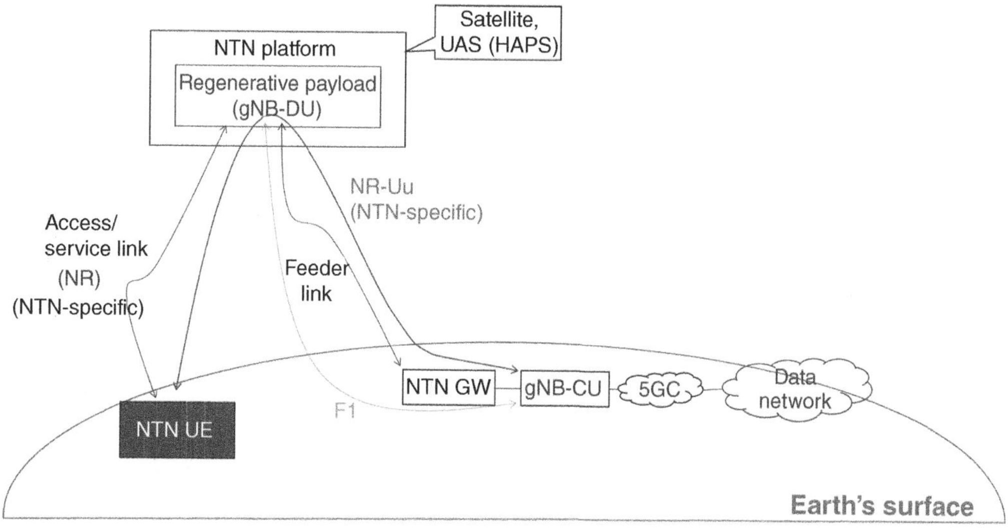

Figure 5.4 An NTN with a Regenerative NTN Payload Containing a gNB-DU.

specifications. The NTN Gateway is implemented on the ground and the on-the-ground gNB-CU exchange F1-C signaling and F1-U user plane traffic in an implementation-specific manner. In one approach, the NTN Gateway and the gNB-CU can be designed as a single entity that performs the combined functions. Such an entity interfaces with the NTN payload and the 5GC.

For NR-NTN with a regenerative payload housing the gNB-DU, the NTN gNB-CU is on the ground. Hence, the NTN gNB-CU can communicate with the core network using an on-the-ground transport network just like a TN gNB. Any hardware changes or software updates for the NTN gNB-CU can be carried out like a TN gNB, but hardware changes in the gNB-DU are impractical. Troubleshooting of the terrestrial gNB-CU is easier than troubleshooting the non-terrestrial gNB-DU. Mobility management for the gNB-DU-based regenerative payload is simpler than the gNB-based regenerative payload due to on-the-ground gNB-CU. However, ISL is not useful because two gNB-DUs do not communicate directly with each other. In summary, the regenerative payload with a gNB-DU makes the design and maintenance of the NTN more complex than an NTN with a transparent payload.

5.4 Multi-connectivity NTNs

3GPP Release 17 and Release 18 support mobility between an NTN and a TN. *However, simultaneous connectivity of a UE with (i) multiple NTNs or (ii) a TN and an NTN is not yet supported at the time of this writing.* The 3GPP has studied various multi-connectivity scenarios (3GPP 2020). Such multi-connectivity scenarios are discussed later.

Figure 5.5 illustrates a multi-connectivity scenario where a UE is simultaneously connected to a TN and an NTN with a transparent NTN payload.

In Figure 5.5, the NTN UE is in a geographic area with radio coverage from both a TN and an NTN. The NR NTN UE communicates with the TN gNB using the 5G NR Uu air interface. The TN gNB is connected to the 5GC via the NG interface. Specifically, the TN gNB is connected to the AMF via the N2 interface, and the UPF connection via the N3 interface is implemented as a GTP tunnel.

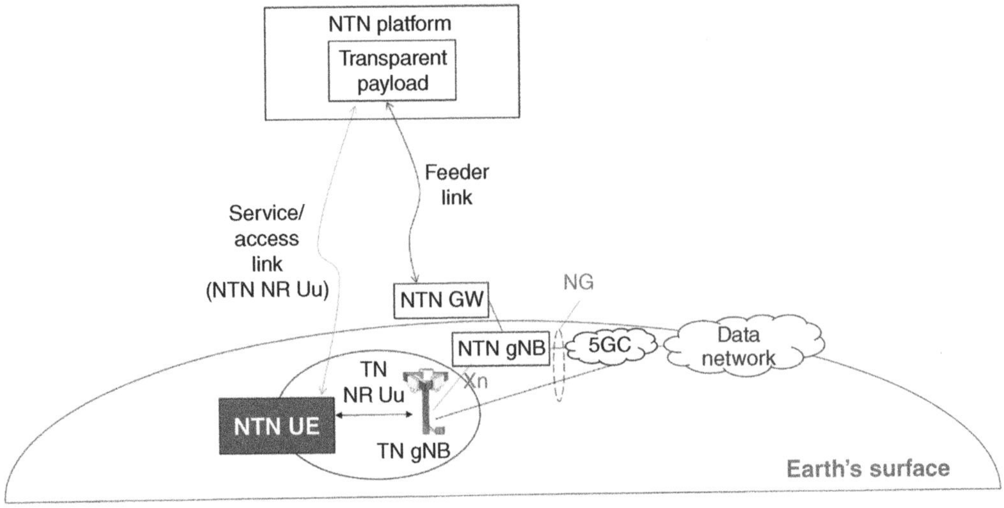

Figure 5.5 A Multi-connectivity NTN with a Transparent NTN and a TN.

The NTN in Figure 5.5 is using a transparent NTN payload. Hence, the main NTN gNB functions are performed by on-the-ground NTN gNB, and the transparent NTN payload essentially acts as a repeater. The 5G NR-NTN UE communicates with the on-the-ground gNB using the enhanced NTN NR air interface that passes through a transparent NTN payload and an NTN Gateway. The service link or the access link between the NR-NTN UE and the transparent NTN payload is based on the NTN NR Uu interface. In contrast, the feeder link between the transparent NTN payload and the NTN Gateway is implementation specific. The NTN gNB is connected to the 5GC via the NG interface. More specifically, the NTN-gNB connects to the AMF via the N2 interface and the UPF via the N3 interface. The TN gNB and the NTN gNB can communicate via the Xn interface, which can make use of a traditional optical transport network.

As a TN usually provides much lower latency than an NTN, latency-sensitive applications can be supported through the TN. In contrast, delay-tolerant services of an NTN UE can be served through the NTN. The NTN can also be used to send data (e.g., buffered videos or file offloads) in a large geographic area using efficient broadcast and multicast mechanisms instead of a unicast transport mechanism of a TN. Load balancing between a TN and an NTN can also be carried out, especially for best-effort data services such as Internet access.

The multi-connectivity NTN architecture shown in Figure 5.5 makes mobility management more complex than a single-connectivity scenario. As the NTN cell is much larger than a TN cell, the UE mobility will often result in the serving TN cell changing while the serving NTN cell remains the same. This is especially likely when the NTN uses a GEO satellite or a HAPS as the NTN platform. Another possibility is that a stationary or slowly moving NTN UE may be connected to the same TN cell for a long time, but the NGSO satellite coverage changes, causing a change in the serving NTN cell.

Figure 5.6 illustrates a multi-connectivity scenario, where a UE is simultaneously connected to a TN and an NTN with a regenerative NTN payload housing a gNB.

In Figure 5.6, the NTN UE is in a geographic area with visibility of both a TN and an NTN with a regenerative payload containing the NTN gNB functions. The NR NTN UE communicates with the TN gNB using the 5G NR Uu air interface. The TN gNB is connected to the AMF via the N2 interface and the UPF via the N3 interface (implemented as a GTP tunnel).

The NTN in Figure 5.6 uses a regenerative NTN payload housing the full gNB. The 5G NR-NTN UE communicates with the NTN gNB residing on a regenerative NTN payload using an enhanced

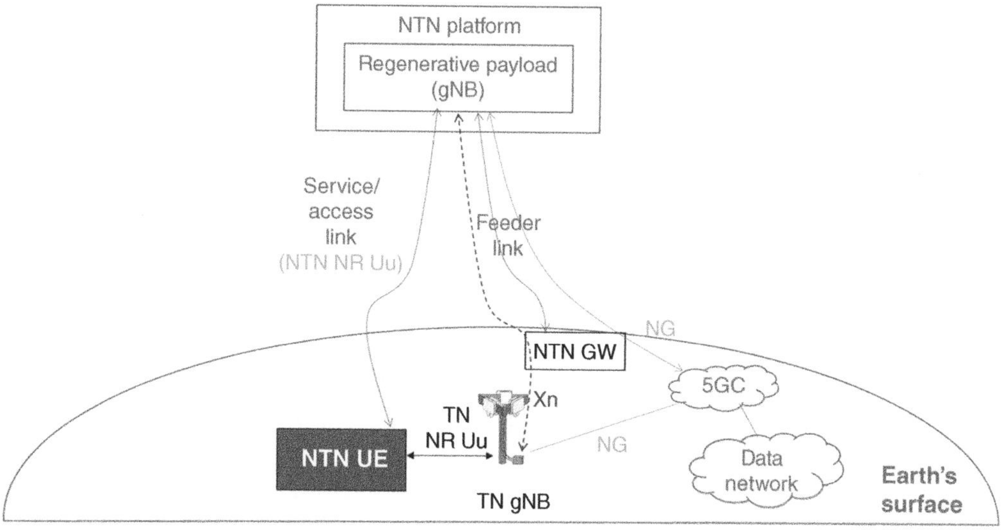

Figure 5.6 A Multi-connectivity NTN with a gNB-Based Regenerative NTN and a TN.

NTN NR air interface. The feeder link connects such non-terrestrial gNB with the NTN Gateway in an implementation-specific manner. The NTN Gateway connects to the AMF via the N2 interface and the UPF via the N3 interface. If the Xn interface between the TN gNB and the NTN gNB is supported, XnAP signaling and Xn user plane need to be transported on the feeder link. If the Xn interface between the TN gNB and the NTN gNB is not supported in Figure 5.6, certain features such as Xn handover are not available.

Note that the 3GPP has not (3GPP 2023c) yet addressed this architecture of Figure 5.6 in 3GPP (2023c) because the need to transport the Xn protocol on the feeder link makes this architecture very complex. The 3GPP has designated such architecture for further study. This architecture may become viable for GEO or HAPS-based NTNs if the 5GC (along with potentially IP multimedia subsystem [IMS]) also resides on the NTN platform.

Figure 5.7 illustrates another multi-connectivity scenario with a regenerative NTN payload. However, Figure 5.7 places the gNB-DU on the regenerative NT payload instead of the gNB (which is done in Figure 5.6). The NTN UE is simultaneously connected to a TN and an NTN with a regenerative NTN payload housing a gNB-DU.

As shown in Figure 5.7, the NTN UE has radio coverage from both a TN and an NTN with a regenerative payload containing the NTN gNB-DU functions. The NR NTN UE communicates with the TN gNB using the 5G NR Uu air interface. The TN gNB is connected to the AMF via the N2 interface and the UPF via the N3 interface. The 5G NR-NTN UE communicates with the NTN gNB-DU residing on a regenerative NTN payload using an enhanced NTN NR air interface. The feeder link connects such non-terrestrial gNB-DU with the NTN Gateway in an implementation-specific manner. The NTN Gateway connects to the NTN gNB-CU via the F1 interface (i.e., F1-C for F1Ap signaling and F1-U for user plane traffic).

The NTN gNB-CU and the TN gNB communicate via the Xn interface. This Xn interface supports XnAP signaling and Xn-U for user traffic (e.g., traffic forwarding during handover if needed). The NTN gNB-CU and the TN gNB connect to the AMF via the N2 interface for NGAP signaling and the UPF via the N3 interface to transport user traffic.

While Figure 5.7 shows an integrated TN gNB, the TN gNB may also be implemented using a disaggregated gNB with TN gNB-DU and the TN gNB-CU. In such a case, the TN gNB-CU and the NTN gNB-CU connect to each other via the F1 interface.

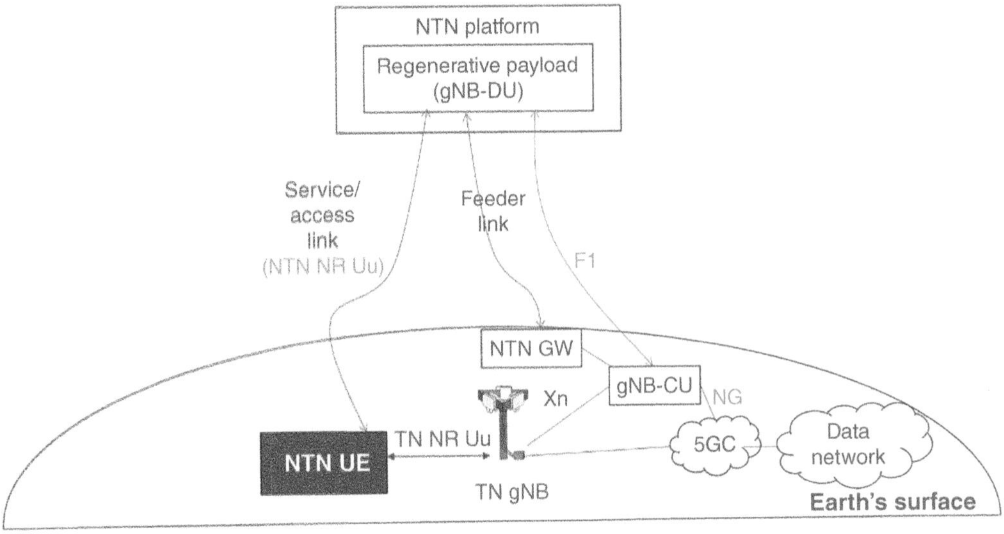

Figure 5.7 A Multi-connectivity NTN with a gNB-DU-Based Regenerative NTN and a TN.

The NTN architecture shown in Figure 5.7 is much easier to design and implement compared to the NTN architecture shown in Figure 5.6, because the Xn interface in Figure 5.7 is on the ground and can use on-the-ground transport network infrastructure.

The 3GPP has also studied an NTN architecture that involves simultaneous connectivity of an NTN UE with multiple NTNs, as shown in Figure 5.8.

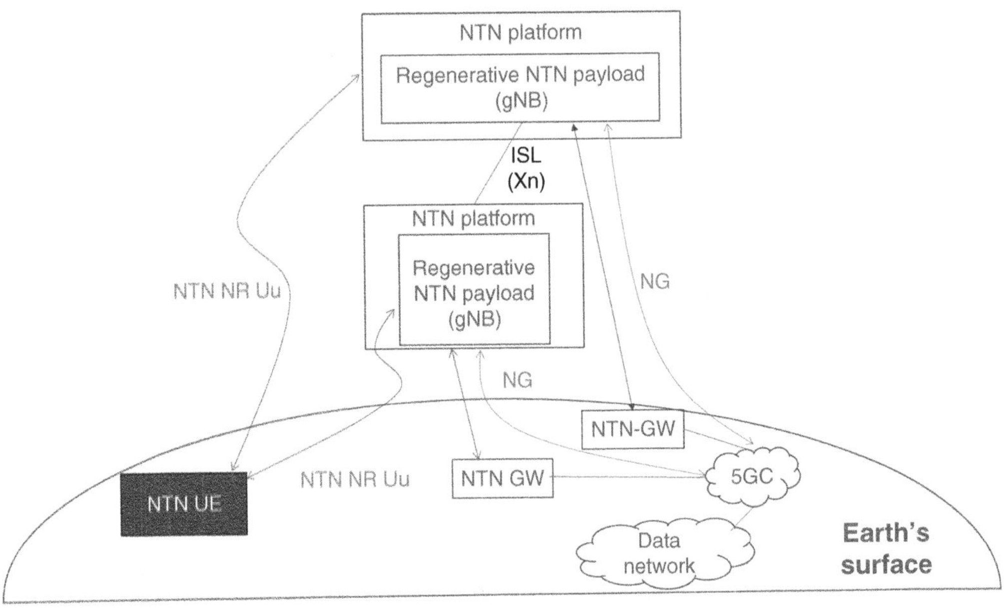

Figure 5.8 A Multi-connectivity NTN with Two Regenerative NTNs.

In Figure 5.8, the NTN UE has radio coverage from two NTNs with regenerative NTN payloads containing the NTN gNB functions. The NR NTN UE communicates with two NTN gNBs using the 5G NR Uu air interface that has been enhanced to support the NTN. The NTN gNBs are connected to the NTN Gateway via an implementation-specific feeder link. The NTN Gateway connects to the AMF to transport the N2 interface information (e.g., NGAP signaling) and the UPF to transport the N3 interface information (e.g., user traffic) between the NTN-gNB and the 5GC. The connectivity between the NTN Gateway and the AMF/UPF is implementation specific. Two NTN gNBs can connect to each other using an ISL. The Xn interface can be implemented using the ISL so that XnAP signaling and Xn user plane can be transported between the NTN gNBs. In a typical TN, two gNBs are usually connected via an IP network, and an optical fiber is frequently used as a physical medium (i.e., as layer 1) to transport information in the IP network. In the NTN, wired connectivity between two NTN gNBs residing on two different NTN platforms is not possible. Hence, wireless connectivity is needed between the NTN gNBs. The ISL needs to be wireless. The ISL can be realized in an implementation-specific manner and may use RF or optical signals.

In a typical 5G deployment, both the 5GC and the IMS are stationary and on the ground. 3GPP studies the feasibility of non-terrestrial 5GC network functions and the IMS network as part of Release 19 enhancements (3GPP 2023a, 2023b). Figure 5.9 shows an example configuration of an NTN with non-terrestrial 5GC and IMS.

In an example scenario, consider the exchange of user traffic between two NTN UEs that are relatively far from each other so that direct or sidelink communication is impossible. The user traffic from NTN UE 1 reaches the NTN gNB residing on an NTN payload via the NR NTN air interface. The NTN-gNB forwards the user traffic to an on-board UPF via NTN UE 1-specific GTP-U tunnel on the N3 interface. The UPF forwards the user traffic from NT UE 1 to the NTN-gNB via NTN UE 2-specific GTP-U tunnel on the N3 interface. The NTN-gNB sends the user traffic received from NTN UE 1 to NTN UE 2 via the NR NTN air interface. As the user traffic between the UEs no longer needs to use on-the-ground infrastructure, the round-trip delay on the feeder link is eliminated, reducing the end-to-end delay for the user traffic between the NTN UEs. Of course, for certain

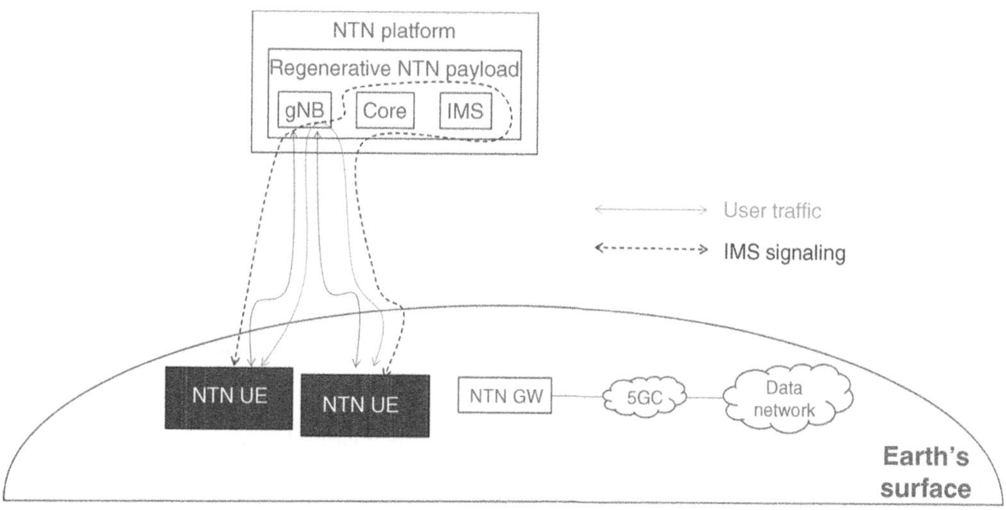

Figure 5.9 An NTN with Non-terrestrial Core/Service Networks.

traffic, such as the Internet traffic, the typical NTN UE-NTN gNB-(NTN Gateway)-UPF-data network path is followed.

It may also be advantageous to place one or more IMS network elements on the NTN payload. For example, if a GEO satellite or HAPS-based NTN platform also houses the IMS, IMS-centric services between the NTN devices would also obviate the need for the feeder link, reducing the delay for localized IMS-centric services. Note that a user plane function (UPF) may also be placed on the NTN platform along with the IMS to avoid the feeder link for data transfer for localized IMS-based services. As the 5GC components, including the AMF and the SMF, could still be on the ground, the feeder link is needed for initial NAS signaling and radio-core signaling.

5.5 An NTN with a Transparent Payload: A Closer Look

A logical architecture of an NTN with a transparent payload is illustrated in Figure 5.2 in Section 5.2. However, in a commercial NTN deployment, several additional functions are needed (3GPP 2024b). Annex B.4 of 3GPP (2024b) shows an example architecture implementation. Figure 5.10 illustrates such an example NTN implementation (3GPP 2024b).

Figure 5.10 depicts a transparent NTN payload, described in 3GPP Release 17 and Release 18. The NTN infrastructure consists of (i) the NTN payload residing on an NTN platform, (ii) the NTN Gateway, and (iii) the NTN control function. The NTN serving link provisioning system includes one or multiple NTN payloads residing on NTN platforms and one or multiple NTN Gateways. The NTN gNB consists of the NTN serving link provisioning system and the non-NTN infrastructure gNB functions.

The NTN payload resides on an NTN platform, such as a spaceborne vehicle (e.g., a satellite) or an airborne vehicle, such as a UAS (e.g., a HAPS). The NTN payload provides a physical structure, power, commanding capabilities, telemetry, attitude control[1] for the NTN platform, thermal environment, and radiation shielding.

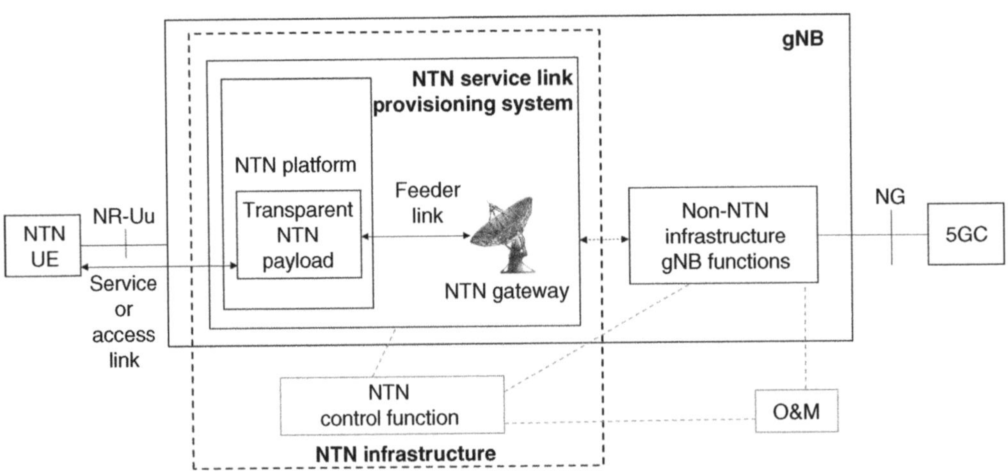

Figure 5.10 Example Implementation of an NTN with a Transparent Payload.

1 The attitude control of the platform involves control of the orientation of the platform.

The NTN service link provisioning system performs mapping between the 5G NR-Uu radio protocol and radio resources of the NTN infrastructure, such as beams, radio channels, and transmit power.

The NTN control function controls (i) NTN platforms such as spaceborne and airborne vehicles and (ii) radio resources of the NTN infrastructure, such as NTN payloads and NTN Gateways. The NTN control function also provides control data, such as ephemeris information, to the non-NTN infrastructure gNB functions of the gNB. Note that provisioning any required NTN control data in the NTN gNB is beyond the scope of 3GPP specifications. Furthermore, the transport of NR-Uu protocol between the NTN service link provisioning system and the non-NTN infrastructure gNB functions is also beyond the scope of 3GPP specifications.

Operations and maintenance (O&M) may provide the following NTN parameters per beam based on the beam type.

- **Earth-Fixed Beams.** (A) Mapping between (i) the NTN cell identifier for NG and Uu interfaces and (ii) the beam and (B) the cell's reference location (e.g., the cell's center and range).
- **Quasi-Earth-Fixed Beams.** (A) Mapping between (i) the NTN cell identifier for NG and Uu interfaces and the time window and (ii) the beam, (B) the cell's reference location (e.g., cell's center and range), (C) the time window of the successive feeder link and service link switchovers, and (D) the identifier and the time window of all serving NYN platforms (e.g., satellites and HAPSs) and NTN Gateways.
- **Earth-Moving Beams.** (A) Mapping between (i) the NTN cell identifier for Uu interface and (ii) the beam, (B) mapping between (i) the NTN cell identifier for the Uu interface and (ii) the fixed geographical areas reported on the NG interface, (C) information about the beam's direction and motion of the beam's footprint on the Earth's surface, (D) elevation relative to the NTN payload, (E) schedule of successive serving NTN Gateways and gNBs, and (F) schedule of successive feeder link and service link switchovers.

5.6 Enhanced Tracking Area Management for an NTN

Chapter 4 discussed the challenge of frequent registration updates in an NTN due to the mobility of NTN cells. Prior to Release 17, a TN cell can only be associated with a single tracking area identifier (TAI). However, different NGSO satellites cover the same geographic area at different times; the change in the NTN cell automatically leads to a TAI change when an NTN cell transmits the same TAI at all times. Suppose no change is made in an NTN. In that case, such rapid TAI changes cause numerous NTN UEs to carry out mobility-based registration updates, causing a tsunami of signaling and significantly increasing the radio resource utilization and processing load in the radio and core networks.

To address the challenge of a massive scale of registration updates resulting from moving NTN cells and the constraint of one TA/TAI per NTN cell, the 3GPP has defined Earth-fixed tracking areas (TAs). Figure 5.11 shows how Earth-fixed TAs can solve the challenge of registration updates associated with the mobility of NTN cells.

The Earth's surface area is divided into Earth-fixed tracking areas (TAs), with each TA associated with a tracking area code (TAC). Let us consider an example of how NGSO satellites provide radio coverage and enhance TA management in an NTN.

First, consider the operation of the NTN cell with physical cell identifier (PCI) = A. At time t1, the NTN cell with PCI = A is illuminating two TAs, TA1 and TAC = TA2. One TAC identifies TA1, and

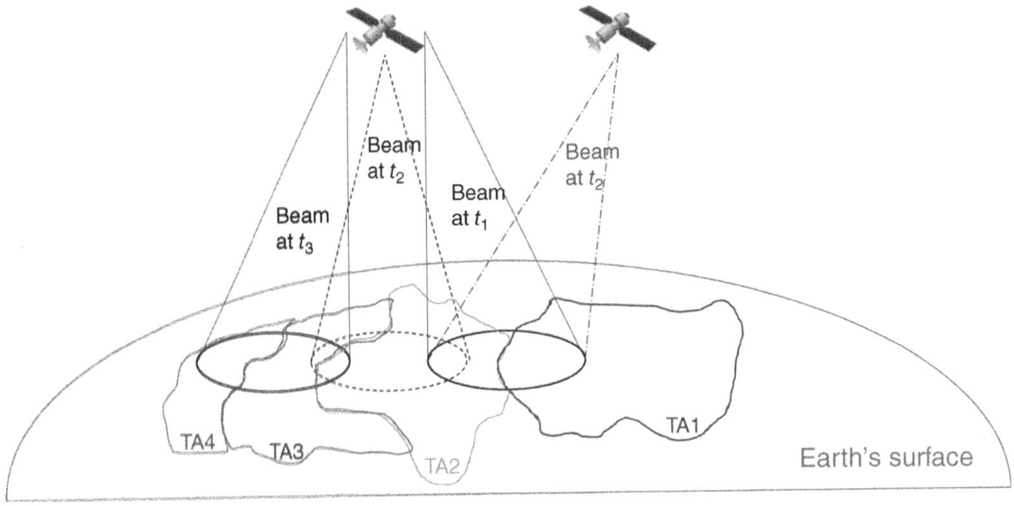

Figure 5.11 Enhanced Tracking Area Management in an NTN.

TA2 is identified by another TAC. Hence, it broadcasts TA1 and TA2 in the system information block (SIB 1) (3GPP 2024b). At time t2, the NTN cell with PCI = A illuminates TA2 and TA3. Hence, it broadcasts TA2 and TA3 in SIB 1. Finally, at time t3, the NTN cell with PCI = A is illuminating TA3 and TA4. Hence, it broadcasts TA3 and TA4 in SIB 1. At time t2, another NGSO satellite 2 starts covering the geographic area that NGSO satellite 1 was covering at t1. The NGSO satellite 2 with the NTN cell PCI=B illuminates TA1 and TA2 at time t2 and broadcasts TA1 and TA2 in SIB1 at time t2. While Figure 5.11 illustrates the example of quasi-Earth-fixed beams, a similar approach is feasible for Earth-moving beams.

Let's assume that the NTN UE registers around time t1 (and well before time t2). The AMF registers the NTN UE in both TA1 and TA2. Due to the movement of NGSO satellites 1 and 2, when the new NTN cell with PCI=B starts covering the NTN UE's geographic area at t2, the NTN UE still sees TA1 and TA2 being broadcast in SIB1. As the NTN UE's registration area consists of the TAI list {TA1, TA2}, the NTN UE does not perform (re) registration. In summary, Earth-fixed tracking areas, association, and broadcast of multiple TAs in an NTN cell and the update of the TA list in SIB1 as the NTN cell goes from one geographic area to another avoid the excessive registration update signaling even when NTN cells change for a given NTN UE.

As can be seen from Figure 5.11, this enhanced TA management approach requires extensive TA planning to define Earth-fixed TAs. Furthermore, the NTN-gNB must know what TAs to broadcast and during what periods. A change in the TAs being illuminated requires the NTN-gNB to change SIB1. Hence, the NTN-gNB needs to follow the rules for changing an SIB so that all the NTN UEs in the NTN cell can obtain the correct TAs in their geographic area.

As mentioned earlier, an NTN cell can now belong to multiple TAs to support the NTN. The RRC parameter "maxTAC-r17" defines the maximum number of tracking area codes to which a cell belongs to. This parameter is set to 12 (3GPP 2024c).

The information element (IE) "trackingAreaList-r17" defines a list of up to maxTAC-r17 (i.e., 12) TACs. The IE "trackingAreaList-r17" is part of the IE "PLMN-IdentityInfoList." The IE "PLMN-IdentityInfoList" is part of the IE "CellAccessRelatedInfo," which is an IE of SIB1.

5.7 Enhanced QoS for an NTN

5G defines a comprehensive QoS framework to support applications with diverse service requirements. These applications are classified into three usage categories: enhanced mobile broadband (eMBB), ultra-reliable low-latency communication (URLLC), and massive Machine-Type Communications (mMTC) in Release 15 (also known as 5G Phase 1). The 3GPP utilizes a scaler called 5G QoS indicator (5QI) as a reference to 5G QoS characteristics. The gNB uses the 5QI value to specify suitable QoS to a 5G QoS flow (e.g., a flow corresponding to video streaming or web browsing) through packet forwarding treatment, target error rate, and target latency. A given standardized 5QI value maps to a standardized combination of 5G QoS characteristics (3GPP 2024a).

One of the QoS characteristics associated with the 5QI is Packet Delay Budget, which is the one-way delay between the UE and the UPF. Prior to the introduction of the NTN in Release 17, the largest value of the PDB was 300 ms for 5QI = 8 and 9 (suitable for Internet applications such as email and web browsing) and 500 ms for 5QI = 74 and 76 (suitable for streaming). However, the delay in a GEO satellite-based NTN can be much longer than 500 ms. Hence, the 3GPP defined a new 5QI = 10 to meet the delay requirements of a GEO satellite-based NTN. As mentioned in 3GPP (2024a), a GEO satellite's worst-case one-way propagation delay is expected to be about 270 ms. The UL scheduling delay, corresponding to the two-way propagation delay, is 2×270 ms = 540 ms. Hence, the 3GPP has defined the PDB of 1100 ms that will accommodate the worst-case propagation delay for a GEO satellite-based NTN. The 3GPP introduces a new 5QI = 10 in Release 17 to address GEO satellites. The 5QI = 10 has the resource type = non-guaranteed bit rate (non-GBR), the default priority of 90, and the packet error rate of 10^{-6} (3GPP 2024a).

5.8 Optical Communication for an NTN

Optical communication may be utilized in an NTN in multiple ways. For example, the ISL and/or the feeder link can be realized using an optical communication link. In general, two major candidates for the implementation of the ISL and the feeder link include a radio frequency (RF) signal and an optical signal. Note that the specification of a transport network to realize the ISL is beyond the scope of the 3GPP. Similarly, the feeder link is beyond the scope of 3GPP specifications. Recall that the ISL between two NTN platforms can carry the Xn interface signaling and user traffic between two gNBs in an NTN with a regenerative payload. The ISL can also be used in a multi-hop network to enable a gNB without direct visibility of an on-the-ground NTN Gateway to connect to such an NTN Gateway by using the help of one or more other gNBs in the NTN. The feeder link helps connect the NTN platform to the on-the-ground wireless network infrastructure.

Lasers of free space optical (FSO) communication[2] make use of light propagating in free space for communications. An FSO communication link offers several advantages but also has some drawbacks (Kaushal and Georges Kaddoum 2017; Stuhlfauth 2022).

An FSO communication link may operate in a wide range of frequencies or wavelengths. The wavelength of an optical signal for FSO communication may range from about 510 to 1550 nm. Equivalently, the frequency of an optical signal for FSO communication may range from about 190 to 590 THz. Commercially available FSO systems typically use (i) near-infrared waves with the wavelength ranging from 750 to 1450 nm and (ii) short-infrared waves with the wavelength

2 FSO also stands for free space optics.

ranging from 1400 to 3000 nm (1530–1560 nm being a dominant spectral range for long distance communication) (Kaushal and Georges Kaddoum 2017).

Advantages of an FSO communication link at a frequency greater than 190 THz are (i) very high data rates (e.g., up to Tbps), (ii) support for distances up to 1000 km, and (iii) less prone to interference. Optical signals often have large bandwidths. Hence, they can support 100+ Gbps. Lasers of an optical communication system consume less power and lead to less energy dispersion. As lasers have narrow beams, they are relatively more difficult to intercept, increasing the security. A key disadvantage of an optical communication link is that the signals at these optical frequencies do not propagate through clouds. Another disadvantage of an optical communication link is the requirement of a tight acquisition, tracking, and pointing (ATP) system. Initial connection setup may take about 50 seconds, and reconnection (e.g., due to changes in the neighboring satellites in case of mega constellations) may require about 20 seconds. Furthermore, an optical link may be adversely affected by atmospheric and solar conditions.

SpaceX has 9000+ laser communication terminals (LCTs) or space lasers in orbit, with three LCTs per satellite (Kawade 2024). The space lasers operate at up to 200 Gbps (Starlink 2024).

5.9 Major Takeaways

The 3GPP carried out various NTN studies, including potential architectures for the NTN. This chapter illustrates various candidate architectures of the NTN. Here are the major takeaways of this chapter.

- The 3GPP has formally specified an NTN architecture with a transparent NTN payload in Release 17 and Release 18.
- Beyond an NTN with a transparent NTN payload, the 3GPP has investigated other NTN architectures, such as an NTN with a regenerative payload and multi-connectivity NTNs.
- An NTN with a regenerative payload is addressed in 3GPP Release 19 specifications.
- In the NTN architecture with a transparent payload, on-the-ground gNB/eNB carries out baseband technology-specific processing such as NR, LTE-M, or NB-IoT processing for NR-NTN and IoT-NTN. The NTN payload performs RF functions such as frequency translation, filtering, and amplification.
- The NTN with a regenerative payload implements gNB or gNB-DU functions in the NTN payload.
- In the NTN architecture with a regenerative payload housing the gNB, the feeder link transports the NG interface information, including NGAP signaling and user traffic. The NTN-UE communicates with the NTN-gNB via an enhanced NR-NTN air interface. The NTN Gateway exchanges the NG information, such as NGAP signaling on the N2 interface with the AMF and the user traffic on the N3 interface with the UPF.
- In the NTN architecture with a regenerative payload housing the gNB-DU, the feeder link transports the F1 interface information. The NTN-UE communicates with the NTN-gNB-DU via an enhanced NR-NTN air interface. The NTN Gateway exchanges the F1 information, such as F1AP signaling and F1-U user traffic, with an on-the-ground gNB-CU.
- In a multi-connectivity scenario, an NTN UE simultaneously communicates with two NTNs. These NTNs can both be LEO satellite-based NTNs, or one NTN can be a GEO satellite-based NTN, and another NTN can be an LEO satellite-based NTN.

- An NTN UE may simultaneously communicate with a TN and an NTN. The NTN may use a transparent payload or a regenerative payload.
- Non-terrestrial core and services networks can be quite beneficial in several scenarios. For example, the end-to-end latency can be reduced significantly for user traffic exchange between two NTN UEs using on-board gNB and UPF. Such architecture can avoid the feeder link for UE-to-UE data transfer.
- A wide variety of NTN architectures may be deployed to achieve the goal of "anywhere connectivity."

References

3GPP (2020). *Study on New Radio (NR) to support non-terrestrial networks*. TR 38.811. https://portal.3gpp.org/desktopmodules/Specifications/SpecificationDetails.aspx?specificationId=3234 (accessed 30 August 2024).

3GPP (2023a). *New SID on study on integration of satellite components in the 5G architecture Phase III*. S2-2309631. https://www.3gpp.org/ftp/tsg_sa/WG2_Arch/TSGS2_158_Goteborg_2023-08/Docs (accessed 30 August 2024).

3GPP (2023b). *Study on integration of satellite components in the 5G architecture*. TR23.700. https://portal.3gpp.org/desktopmodules/Specifications/SpecificationDetails.aspx?specificationId=4212 (accessed 30 August 2024).

3GPP (2023c). *Solutions for NR to support non-terrestrial networks (NTN)*. TR 38.821. https://portal.3gpp.org/desktopmodules/Specifications/SpecificationDetails.aspx?specificationId=3525(accessed 30 August 2024).

3GPP (2024a). *System architecture for the 5G System (5GS)*. TS23.501. https://portal.3gpp.org/desktopmodules/Specifications/SpecificationDetails.aspx?specificationId=3144 (accessed 30 August 2024).

3GPP (2024b). *NR and NG-RAN Overall Description*. TS38.300. https://portal.3gpp.org/desktopmodules/Specifications/SpecificationDetails.aspx?specificationId=3191 (accessed 30 August 2024).

3GPP (2024c). *System architecture for the 5G System (5GS)*. TS38.331. https://portal.3gpp.org/desktopmodules/Specifications/SpecificationDetails.aspx?specificationId=3197 (accessed 30 August 2024).

Kaushal, H. and Georges Kaddoum, G. (2017). Optical communication in space: challenges and mitigation techniques. *IEEE Communications Surveys & Tutorials* **19** (1): 57–96. https://ieeexplore.ieee.org/abstract/document/7553489 (accessed 30 August 2024).

Kawade, P. (2024). SpaceX's entry into the optical satcom terminal market could be a game-changer. 30 May 2024. https://www.analysysmason.com/research/content/articles/spacex-optical-terminal-nsi017/ (accessed 30 August 2024).

Starlink (2024). *Starlink Satellite Technology*. https://www.starlink.com/technology (accessed 30 August 2024).

Stuhlfauth, R. (2022). 5G NTN Takes Flight: Technical Overview of 5G Non-terrestrial Networks. Rohde & Schwarz White Paper. Version 1.0. July 2022. https://www.rohde-schwarz.com/us/solutions/satellite-testing/landingpages/white-paper-5g-ntn-takes-flight-technical-overview-of-5g-non-terrestrial-networks_255919.html (accessed 30 August 2024).

6

RF Planning and Design Considerations for an NTN

Radio frequency (RF) planning and design (RPD) aims to meet coverage, capacity, throughput, and quality of service (QoS) requirements in a wireless network. The goal of this chapter is to provide insights into NTN-specific aspects of RPD and RF propagation. An overall RPD framework is summarized. Potential NTN spectrum is discussed. Relevant propagation path loss models are specified, and NTN-specific propagation effects, such as atmospheric effects including scintillations, the Faraday effect, and weather effects, are discussed. Various aspects of the NTN link budget are described, including the implications of different NTN platforms, types of beams, and device types. Capacity planning for an NTN is also discussed. NTN configuration guidelines based on the NTN RPD are outlined. Finally, 3GPP-estimated link budgets and UE throughput are summarized.

6.1 RF Planning and Design for an NTN: An Overview

As the NTN possesses distinct characteristics compared to the terrestrial network (TN), RPD for an NTN has some different characteristics from RPD in a TN. For example, the link budget and a suitable propagation path loss model can be used to determine the coverage distance (i.e., the distance between the gNB and the UE) for a TN. The link budget procedure for a TN involves the determination of the maximum allowable path loss (MAPL), which is the maximum amount of attenuation or path loss that can be tolerated between the transmitter and the receiver while meeting a target quality of service (QoS). The MAPL in a TN, when used in conjunction with a suitable propagation path loss model (e.g., Okumura-Hata model), yields the cell radius for the TN (Tripathi and Reed 2014). In contrast, there are specific orbits for satellites in an NTN. Furthermore, UASs (e.g., high-altitude platform stations [HAPS]) operate at a certain altitude. Hence, the coverage distance between the device and the NTN platform is essentially determined by the maximum expected distance between the device and the NTN platform in an NTN. Such maximum expected distance is a function of the factors such as the orbit of the NTN platform and the minimum elevation angle between the NTN platform in case of a satellite-based NTN and the altitude of the NTN platform in case of a HAPS-based NTN. Additionally, propagation characteristics of TNs tend to create wide variations in signal strength for a TN at different device-gNB distances. In contrast, long-term signal strengths in an NTN cell depend significantly on the deployment scenarios. In case of an NTN that is aiming to provide line-of-sight coverage in an open environment, such as a rural environment,

3GPP-based Non-Terrestrial Networks in 5G and 6G: Expanding the Frontiers of Wireless Communications, First Edition.
Nishith D. Tripathi and Jeffrey H. Reed.

Companion website: www.wiley.com/go/tripathi5g

it does not have significant signal strength variations due to the absence of objects in the device-to-NTN platform path. In contrast, when an NTN is aiming to provide coverage in suburban and urban environments, blockages and shadowing effects occur, especially at low elevation angles. Additionally, signals are expected to be weak throughout an NTN cell due to the long distance traveled by the RF signal and the limited output power levels of a satellite and a device. Rain fades are likely to be a big issue in an NTN, especially at higher frequencies. Note that signals are much stronger near the cell centers and much weaker near the cell edge in a TN cell. Signal strengths are not significantly different at the cell center and at the cell edge in an NTN compared to a TN. Such different signal strength characteristics have implications for capacity planning in an NTN.

Figure 6.1 outlines key inputs and outputs of the RPD framework for an NTN.

The RPD framework in a TN helps determine the locations of cells and configurations of cells by performing a link budget and capacity planning and analysis. Cell configuration involves the determination of parameters such as cell identity, cell selection and reselection parameters, and random-access parameters. As shown in Figure 6.1, the RPD framework for an NTN can be used to determine link budgets, estimate capacity and throughput, plan for target capacity and throughput, and develop NTN configuration guidelines. To produce such outputs, the RPD framework needs to consider a variety of inputs, including the radio access technology, target service requirements, available NTN configurations, NTN spectrum, and types of devices.

6.1.1 RPD Inputs

6.1.1.1 Radio Access Technology

Different Radio Access Technologies (RATs) have different capabilities. For example, the 5G NR is a high-performance radio interface and supports advanced capabilities such as advanced antenna techniques. In contrast, LTE-M and NB-IoT are low-complexity RATs suitable for low-data-rate IoT devices. The RAT influences link budgets and achievable capacity.

6.1.1.2 Target Service Requirements

A service provider may have specific QoS targets for different services. Examples include a cell-edge reliability target for the link budget for the coverage aspect of the system and a cell throughput target for capacity planning.

6.1.1.3 NTN Configurations

NTN platforms such as GEO and LEO satellites and HAPSs have significantly different altitudes, resulting in widely different device-NTN platform distances and hence overall path losses. Hence,

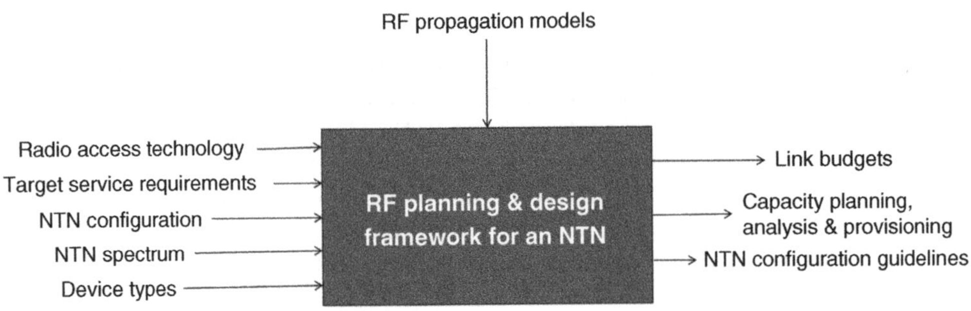

Figure 6.1 RF Planning and Design in an NTN.

for a given target signal-to-interference plus noise ratio (SINR) (or equivalently, a target spectral efficiency in bits per second per Hertz), a GEO satellite-based NTN would require larger power levels and antenna gains compared to an LEO satellite-based NTN. The altitude of the NTN platform also dictates whether atmospheric effects would be significant or not. For example, HAPSs are not affected by ionospheric effects, but satellites are affected by ionospheric effects. Different types of beams, such as Earth-moving beams and quasi-Earth-fixed beams, have different device-NTN platform distances. System capabilities such as the range, the capacity, and the throughput are affected by NTN configurations.

6.1.1.4 NTN Spectrum

An NTN may utilize low-frequency spectrum, such as around 2 GHz, or high-frequency spectrum greater than 10 GHz. At frequencies below 10 GHz, rain effects can often be ignored. However, rain effects start becoming important for frequencies above 10 GHz. Low-frequency spectrum experiences smaller propagation path losses but has limited bandwidth available. In contrast, high-frequency spectrum experiences larger propagation path losses but has more bandwidth available.

6.1.1.5 Device Types

Wireless devices have different RF power and processing power capabilities. A typical 5G smartphone utilizes Power Class 3, which corresponds to the maximum transmit power of 200 mW or 23 dBm. An LTE-M/NB-IoT device may use the maximum transmit power of 23 dBm, 20 dBm, or even 14 dBm. Many smartphones and IoT devices use omnidirectional antennas (and some smartphones, especially the ones operating at mmW spectrum, may use transmit beamforming). Very small aperture terminals (VSATs) use high-gain dish antennas. VSATs can thus have more favorable link budgets and achieve higher throughput than smartphones and IoT devices that are using omnidirectional antennas.

6.1.1.6 RF Propagation Models

RF propagation in an NTN could be quite different from that in a TN. For example, there is typically non-line-of-sight (NLOS) propagation in a TN between the gNB/eNB and the UE in a TN, and propagation path loss exponent n could be quite high, such as $n = 3.5$, leading to significant attenuation as the distance between the transmitter and the receiver increases. In contrast, when line-of-sight (LOS) propagation is assumed between the NTN platform and the NTN UE, propagation path loss exponent n is quite small, such as $n = 2.0$ for free-space propagation path loss. However, transmitter-receiver distances tend to be much larger in an NTN than in a TN, usually leading to weak NTN signals due to significant distance attenuation. Blockages, target deployment scenarios, and service goals in suburban and urban environments need to be considered in RPD. As mentioned in Section 4.10, atmospheric effects such as scintillations could exist. Faraday effects should be considered in the RPD.

6.1.2 RPD Outputs

6.1.2.1 Link Budgets

In a TN, the link budget yields the maximum allowable path loss (MAPL) that can be tolerated between the transmitter and the receiver while meeting a target QoS at the cell edge. For example, a certain cell edge data rate requirement (e.g., X Mbps) and a certain spectral efficiency target (e.g., X bits per modulation symbol) would lead to the determination of the target SINR and the required amount of time-frequency resources to be used in the TN link budget. The MAPL can be translated into cell radius using a suitable propagation path loss model for a TN. However, in an NTN, the size (e.g., diameter) of an NTN cell on the ground is a design parameter that is influenced by the altitude of the NTN platform and the planned beam coverage of an NTN platform or payload, with a GEO satellite typically illuminating relatively large NTN cells and LEO cells illuminating relatively smaller NTN cells (still much larger than TN cells!). NTN cells tend to be hundreds of kilometers or tens of kilometers in diameter. In contrast, TN cells in urban and suburban environments are less than a few kilometers in diameter, and TN cells in a rural environment are tens of kilometers in diameter. Diameters of TN cells are usually much smaller than 100 km (about 62.14 mi). The link budget should be validated based on the NTN platform's maximum distance from the Earth's surface area (computed from the NTN platform's altitude/orbit and the minimum elevation angle) and the NTN cell's diameter. While the link budget directly influences the cell radius in a TN, the link budget directly influences the distance between the device and the NTN platform (or equivalently, the NTN payload, because the NTN payload resides on the NTN platform and performs communications-related functions) instead of the radius of the NTN cell illuminated by the NTN payload in an NTN. Link budgets for both the device-to-the NTN platform link (which is like the traditional 5G NR uplink) and the NTN platform-to-the-device link (which is like the traditional 5G NR downlink) need to be determined in an NTN. Section 6.4 discusses link budgets for the access link for various NTN platforms.

6.1.2.2 Capacity and Throughput Analysis and Planning

Cell capacity/throughput for the uplink and the downlink on the access or service link can be estimated based on the factors including (but not limited to!) expected amount of resources (e.g., available channel bandwidth), supportable modulation and coding scheme (MCS), transmit and receive antenna gains, and noise figures. Using the cell capacity/throughput and target service area capacity/throughput, a required number of NTN cells (with appropriate diameters) and the amount of spectrum per cell can be provisioned.

6.1.2.3 NTN Configuration Guidelines

Parameters such as transmit power levels, transmit/receive antenna gains, supportable types of beams, supportable elevation angle, cell selection/reselection parameters, and random-access parameters can be selected. Guidelines on orbits and/or altitudes, along with the NTN platform's antenna and power amplifier configurations, can also be developed to reflect the outcomes of the RPD of the NTN. Section 6.6.1 provides an overview of NTN configuration guidelines as part of capacity planning.

6.2 NTN Spectrum

An NR-NTN and an IoT-NTN use a variety of spectra. Both spectra below 6 GHz, such as the S-band, and spectra above 6 GHz, such as the Ka-band, may be used. In one configuration, handheld terminals with omnidirectional or semi-directional antennas use the S band, and VSATs with high-gain directional antennas use the Ka band (3GPP 2020).

An NTN may use any suitable spectrum defined per ITU-R or national spectrum allocation policies. The 3GPP has considered the following frequency ranges shown in Table 6.1 for the analysis of the NTN in pre-Release 17 studies and post-Release 17 performance evaluations. The L band ranges

Table 6.1 Example Frequency Ranges for the 3GPP NTN Analysis.

Frequency Band	Downlink on the Access/Service Link (gNB/eNB-to-UE)	Uplink on the Access/Service Link (UE-to-gNB/eNB)	Typical Device Type
S band	2,170–2,200 MHz	1,980–2,010 MHz	Smartphone or IoT Device
K and Ka bands	19.7–21.2 GHz	29.5–30.0 GHz	VSAT

from 1 to 2 GHz, the S band ranges from 2 to 4 GHz, the K band ranges from 18 to 26 GHz, and the Ka band ranges from 26 to 40 GHz (The European Space Agency (ESA) 2024).

Recall that the 3GPP introduced the NTN as a 5G feature in Release 17. In the 3GPP Release 15, 5G was introduced. Two NR frequency bands included in Release 15 are Frequency Range 1 (FR1) and Frequency Range 2 (FR2) for a TN (Tripathi and Reed 2019). While initially FR1 covered from about 450 MHz to about 6 GHz (and hence was often called sub-6 GHz), FR1 was expanded immediately to include up to 7 GHz at the higher end of the spectrum to accommodate a new spectrum above 6 GHz. FR2 was defined to be from about 24 GHz to about 53 GHz.

At the time of this writing, NTN-specific frequency ranges are often referred to as FR1-NTN and FR2-NTN in Release 18, where FR1-NTN corresponds to the frequency range 410–7125 MHz, while FR2-NTN corresponds to the frequency range 17 300–30 000 MHz (3GPP 2024d). Table 6.2 lists 3GPP-specified frequency bands for the NR-NTN use (3GPP 2022a, 2022b, 2024c, 2024d). The initial 3GPP focus is on Frequency Range 1 (FR1), which ranges from 410 to 7125 MHz. The operating bands for the uplink (i.e., the UE to the satellite access node [SAN]) and the downlink (i.e., the SAN to the UE) are defined in Table 6.2. *The SAN is a node that provides NR user plane and control plane protocol terminations toward an NTN satellite-capable UE, connects to the 5GC via the NG interface, and encompasses a transparent NTN payload residing on an NTN platform, an NTN GW (also called satellite-gateway), and gNB functions* (3GPP 2024d). Note that all the NTN frequency bands considered by the 3GPP are FDD bands. FDD is more suitable than TDD in an NTN due to the long propagation delays in an NTN.

Table 6.2 3GPP Operating Bands for the NTN.

NR Operating Band	Frequency Range	Duplex Mode	Uplink Operating Band Frequencies (UE transmit, gNB receive)	Downlink Operating Band Frequencies (gNB transmit, UE receive)
n1 (for HAPS)	FR1	FDD	1,920–1,980 MHz	2,110–2,170 MHz
n254	FR1	FDD	1,610–1,626.5 MHz	2,483.5–2,500 MHz
n255	FR1	FDD	1,626.5–1,660.5 MHz	1,525–1,559 MHz
n256	FR1	FDD	1,980–2,010 MHz	2,170–2,200 MHz
n510	FR2	FDD	27.500–28.350 GHz	17.300–20.200 GHz
n511	FR2	FDD	28.350–30.000 GHz	17.300–20.200 GHz
n512	FR2	FDD	27.500–30.000 GHz	17.300–20.200 GHz

Consider FR1-NTN in Table 6.2. The NR frequency band n1 is available for HAPSs. The NR operating bands n255 (associated with L-band and often called NTN 1.6 GHz) and n256 (associated with S-band and often called NTN 2 GHz) in Table 6.2 have the lowest carrier frequencies and hence the least propagation path losses. These bands are quite helpful to devices with low-gain antennas and/or limited transmit power, such as smartphones and IoT devices. In a typical FDD communication system, a lower frequency is used for the device-to-the-radio network uplink, and a higher frequency is used for the device-to-the-radio network-to-the-device downlink so that there is smaller propagation path loss for the weaker uplink. However, due to the need for coordination with adjacent frequency bands, the band n255 uses a higher frequency for the device to the NTN platform uplink. In contrast, the bands n254 and n256 use a higher frequency for the NTN platform to the device downlink, like a typical FDD communication system.

For FR1, supported Subcarrier Spacing (SCS) values are 15, 30, and 60 kHz. Furthermore, the supported channel bandwidths are 5, 10, 15, 20, and 30 MHz (3GPP 2024d).

Higher-frequency bands in Table 6.2, as discussed by the 3GPP, utilize K band and Ka band frequencies. The K band ranges from 18 to 27 GHz, and the Ka band ranges from 26 to 40 GHz. All the NR operating bands n510, n511, and n512 utilize the same 17.300–20.200 GHz frequency range for the downlink but different frequency ranges for the uplink. Higher frequencies experience larger propagation path losses. Hence, such frequencies are generally usable by devices that can create high gains, such as VSATs with dish antennas and devices with high-gain beamforming.

Table 6.3 lists 3GPP-specified frequency bands for the IoT-NTN use (3GPP 2024b). The Evolved-Universal Terrestrial Radio Access (E-UTRA) operating bands for the uplink (i.e., the UE to the SAN) and the downlink (i.e., the SAN to the UE) are defined in Table 6.3. The SAN for IoT-NTN is a node that provides E-UTRA user plane and control plane protocol terminations toward an NTN satellite-capable UE, connects to the EPC via the S1 interface, and encompasses a transparent NTN payload residing on an NTN platform, an NTN Gateway, and eNB functions (3GPP 2024a, 2024b).

In the case of LTE-M-based IoT-NTN, the UE category M1 operates in the E-UTRA satellite access operating bands defined in Table 6.3 in both half-duplex FDD mode and full-duplex FDD mode. In the case of NB-IoT-based IoT-NTN, Category NB1 and NB2 UEs operate in the E-UTRA satellite access operating bands defined in Table 6.3. Category NB1 and NB2 UEs operate in HD-FDD duplex mode.

LTE-M-based IoT-NTN supports a channel bandwidth of 1.4 MHz. NB-IoT-based IoT-NTN supports the channel bandwidth of 200 kHz corresponding to one PRB with (i) 12 subcarriers with an SCS of 15 kHz and (ii) 48 subcarriers with an SCS of 3.75 kHz (3GPP 2024b).

Table 6.3 3GPP E-UTRA Operating Bands for the IoT-NTN (Satellite Access).

NR Operating Band	Duplex Mode	Uplink Operating Band Frequencies (UE-to-SAN)	Downlink Operating Band Frequencies (SAN-to-UE)
253	FDD	1,668–1,675 MHz	1,518–1,525 MHz
254	FDD	1,610–1,626.5 MHz	2,483.5–2,500 MHz
255	FDD	1,626.5–1,660.5 MHz	1,525–1,559 MHz
256	FDD	1,980–2,010 MHz	2,170–2,200 MHz

6.3 NTN Devices

In traditional satellite communications terminology, an Earth Station refers to a satellite terminal with a high-gain antenna, such as gateways and VSATs (3GPP 2022a). Furthermore, Earth Stations in Motions (ESIMs) are Earth Stations that can move, typically mounted on a moving platform such as a vehicle, a ship, or an aircraft. An ESIM is also referred to as a mobile VSAT.

As mentioned in Section 2.6 in Chapter 2, the NTN supports a variety of devices, such as smartphones, VSATs, and IoT devices. Table 6.4 lists the salient characteristics of these types of NTN devices (3GPP 2020). Different types of NTN devices may emerge in the future, just as new 5G devices have emerged for TN.

High-performance and medium- to high-cost smartphones and VSATs are expected to exploit the high-performance and flexible 5G NR radio interface. Simple and low-cost IoT devices are expected to use IoT technologies such as LTE-M and NB-IoT. A typical 5G NR smartphone utilizes 3GPP-defined Power Class 3, which corresponds to the maximum transmit power of 23 dBm. A VSAT can provide larger transmit power, such as 33 dBm. An IoT device using LTE-M or NB-IoT may operate with 23 dBm transmit power, like a smartphone, or a smaller transmit power level, such as 20 dBm and even 14 dBm.

An omnidirectional antenna is common for smartphones and IoT devices. A high-gain dish antenna is quite common for VSATs. For example, a VSAT may have about 43.2 dBi transmit antenna gain and 39.7 dBi receive antenna gain (3GPP 2020). A VSAT has a low noise figure of about 1.2 dB, while smartphones and IoT devices have a larger noise figure of about 9.0 dB. Furthermore, a VSAT typically uses circular polarization, while linear polarization is common in smartphones and IoT devices. A smartphone may support implementation-dependent beamforming, especially at higher frequencies such as mmW frequencies. An NTN platform typically uses circular polarization. Hence, there is about a 3-dB loss in signal strength for smartphones and IoT devices due to polarization mismatch.

Due to a limited transmit power capability, a lower frequency is more suitable for a low data rate IoT device due to smaller propagation path losses at lower frequencies. As a VSAT utilizes a high-gain antenna, it can overcome large propagation path losses associated with higher frequencies and, hence, can work well even at high frequencies.

Table 6.4 Salient Characteristics of NTN Devices.

Characteristic	Smartphone	VSAT	IoT Device
Technology	5G NR	5G NR	LTE-M and NB-IoT
Transmit power (dBm)	23 dBm	33 dBm	14 dBm, 20 dBm, 23 dBm
Antenna	Omni, beamforming	High-gain dish	Omni
Noise figure	9.0 dB	1.2 dB	9.0 dB
Polarization	Linear	Circular	Linear
Frequency bands	Various (lower more suitable)	Higher workable	Low frequencies more suitable

6.4 RF Propagation Models

The received signal strength experienced by a receiver in a TN can be viewed as the combination of three effects (Tripathi and Reed 2014): (i) propagation path loss, (ii) large-scale fading, and (iii) small-scale fading. The propagation path loss reflects the effects of variables, such as the distance between the transmitter and the receiver and the carrier frequency. Large-scale fading reflects long-term variations in the received signal strength and is often called *shadow fading*. For example, there are variations in the average observed signal strengths for the same distance between the transmitter and the receiver. Large-scale fading reflects long-term signal strength variations (e.g., averaged over a long period such as seconds). Shadow fading is typically modeled by a Gaussian distribution with 0 mean and the environment-based standard deviation (e.g., a rural or suburban environment). Small-scale fading reflects sudden variations in the signal strength as a function of short distances (e.g., even a fraction of a wavelength) or short time intervals (e.g., on the order of milliseconds). Small-scale fading is caused by factors such as relative movement and multipath signals between the transmitter and the receiver. Removing the propagation path loss and large-scale fading effects from the instantaneous signal strength measurements can identify small-scale fading effects.

The received signal strength experienced by a receiver in an NTN can also be viewed as the combination of the same three effects, but NTN-specific radio propagation characteristics need to be considered. The organizations such as the 3GPP and the ITU have specified various radio channel models (3GPP 2020; ITU 2017a, 2017b).

6.4.1 RF Propagation in an NTN: An Overview

Figure 6.2 identifies key RF propagation considerations for an NTN. The propagation path between the device and the NTN platform may be LOS or NLOS. Hence, path loss models are needed for both cases, along with a model to determine the probability of LOS propagation. The NLOS propagation path loss is determined by considering the LOS propagation path loss as the baseline and adding a

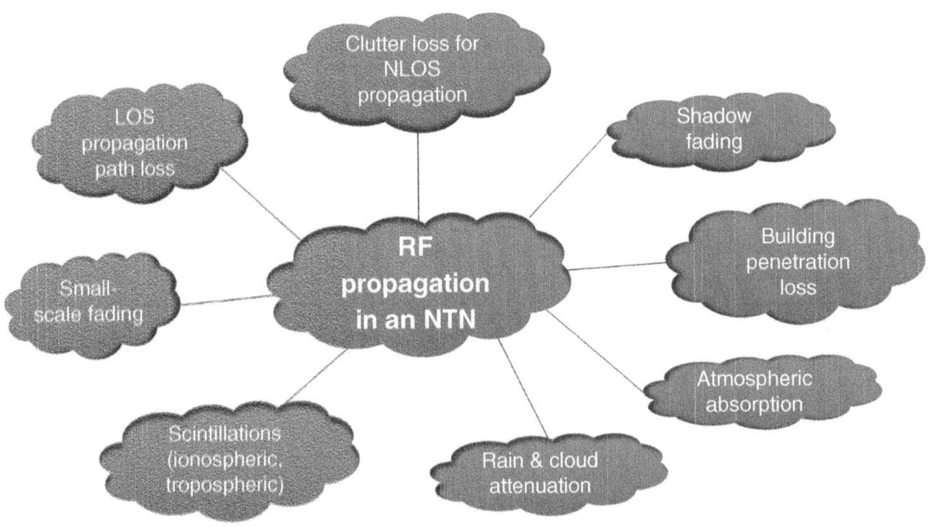

Figure 6.2 RF Propagation Modeling in an NTN.

clutter loss (CL). Shadow fading captures long-term variations in the signal strength. If the device is indoor, building penetration needs to be considered. Atmospheric absorption and rain and cloud attenuation must be considered for higher frequencies. Ionospheric and tropospheric scintillation can affect the signals that pass through the ionosphere and the troposphere. Small-scale fading effects that reflect the influence of Doppler shifts and multipath signals are important for physical layer analysis. These RF propagation factors for the NTN are discussed in Section 6.4.2.

As indoor coverage is challenging for a satellite-based NTN due to a limited link budget and large propagation losses, the 3GPP has focused on outdoor use cases for a satellite-based NTN. However, indoor coverage is considered by the 3GPP for a HAPs-based NTN due to much smaller path losses for the HAPS compared to a satellite. Various propagation environments, including open, rural, suburban, urban, and dense urban, are generally considered based on the frequency band. Furthermore, an Additive White Gaussian Noise (AWGN) channel is assumed for the analysis in open environments, because these environments tend to be relatively static. See 3GPP (2023), 3GPP (2020), and 3GPP (2024d) for examples of channel models used by the 3GPP in performance evaluation and testing of an NTN.

6.4.2 RF Propagation Models for an NTN

6.4.2.1 Free-Space Path Loss

The LOS free-space path loss can be calculated as (Tripathi and Reed 2014)

$$\mathrm{PL(LOS)} = 32.44 + 20 \times \log 10(f_c) + 20^* \log 10(d). \tag{6.1}$$

In Eq. (6.1), f_c is the carrier frequency in MHz, and d is the distance between the transmitter and the receiver in km. The same Eq. (6.1) can also be used if f_c is expressed in GHz and d is expressed in meters. For example, for f_c of 2 GHz and d of 1,000 km, the free-space path loss is 158.5 dB. If the Effective Isotropic Radiated Power (EIRP) of the NTN platform's transmitter is 60 dBm and if the device receiver experiences free-space path loss due to LOS RF propagation toward the NTN platform, the device receiver experiences the received signal strength of (60 dBm − 158.5 dB = −98.5 dBm).

The NLOS path loss can be calculated as

$$\mathrm{PL(NLOS)} = P(\mathrm{LOS}) + \mathrm{CL}. \tag{6.2}$$

In Eq. (6.1), PL(LOS) is the LOS free-space path loss calculated in Eq. (6.1), and CL is the clutter loss that reflects the attenuation associated with the environment experienced by the RF propagation between the transmitter and the receiver. CL is a function of the carrier frequency and the elevation angle. In the case of LOS RF propagation in a typical NTN use case, CL is 0 dB.

Table 6.5 summarizes CL for the S-band and Ka-band in dense urban, urban, suburban, and rural environments (3GPP 2020).

As shown in Table 6.5, CL decreases as the elevation angle increases (i.e., the NTN platform moves away from the horizon and toward the zenith, directly overhead) because there are fewer clutter obstacles in the propagation path between the device and the NTN platform. Furthermore, CL is larger in urban and dense urban environments than in rural and suburban environments because there are more clutter objects in urban and dense urban environments. Finally, CL is larger in the higher frequency Ka band than the lower frequency S band.

The device may experience LOS propagation or NLOS propagation. Due to the significant difference in the path loss for these two types of propagation conditions, it is important to quantify the probability of LOS propagation to be able to estimate the average expected NTN performance.

Table 6.5 Clutter Loss for Different Environments.

	Dense Urban and Urban Environments		Rural and Suburban Environments	
Elevation Angle (Degree)	**S Band (dB)**	**Ka Band (dB)**	**S Band (dB)**	**Ka Band (dB)**
10°	34.3	44.3	19.52	29.5
20°	30.9	39.9	18.17	24.6
30°	29.0	37.5	18.42	21.9
40°	27.7	35.8	18.28	20.0
50°	26.8	34.6	18.63	18.7
60°	26.2	33.8	17.68	17.8
70°	25.8	33.3	16.50	17.2
80°	25.5	33.0	16.30	16.9
90°	25.5	32.9	16.30	16.8

Table 6.6 Probability of LOS RF Propagation.

	Propagation Environment		
Elevation Angle (Degree)	**Rural and Suburban Environments (%)**	**Urban Environment (%)**	**Dense Urban Environment (%)**
10	78.2	24.6	28.2
20	86.9	38.6	33.1
30	91.9	49.3	39.8
40	92.9	61.3	46.8
50	93.5	72.6	53.7
60	94.0	80.5	61.2
70	94.9	91.9	73.8
80	95.2	96.8	82.0
90	99.8	99.2	98.1

As shown in Table 6.6, as the elevation angle increases (i.e., the NTN platform moves away from the horizon and toward the zenith, directly overhead), the LOS probability increases because there is a smaller likelihood of obstacles in the propagation path between the device and the NTN platform (3GPP 2020).

6.4.2.2 Shadow Fading

Shadow fading in both a TN and an NTN can also be modeled as a normal or Gaussian distribution with a mean of 0 dB and a standard deviation of σ dB (or variance of σ^2). Table 6.7 specifies the shadow fading standard deviation for different environments in both LOS and NLOS scenarios for both S and Ka bands (3GPP 2020).

Table 6.7 Shadow Fading Standard Deviation for Different Environments.

	Dense Urban Environment				Urban Environment		Rural and Suburban Environments			
	S Band	S Band	Ka Band	Ka Band	S & Ka Bands	S & Ka Bands	S Band	S Band	Ka Band	Ka Band
Elevation Angle (Degree)	σ (dB) (LOS)	σ (dB) (NLOS)	σ (dB) (LOS)	σ (dB) (NLOS)	σ (dB) (LOS)	σ (dB) (NLOS)	σ (dB) (LOS)	σ (dB) (NLOS)	σ (dB) (LOS)	σ (dB) (NLOS)
10°	3.5	15.5	2.9	17.1	4	6	1.79	8.93	1.9	10.7
20°	3.4	13.9	2.4	17.1	4	6	1.14	9.08	1.6	10.0
30°	2.9	12.4	2.7	15.6	4	6	1.14	8.78	1.9	11.2
40°	3.0	11.7	2.4	14.6	4	6	0.92	10.25	2.3	11.6
50°	3.1	10.6	2.4	14.2	4	6	1.42	10.56	2.7	11.8
60°	2.7	10.5	2.7	12.6	4	6	1.56	10.74	3.1	10.8
70°	2.5	10.1	2.6	12.1	4	6	0.85	10.17	3.0	10.8
80°	2.3	9.2	2.8	12.3	4	6	0.72	11.52	3.6	10.8
90°	1.2	9.2	0.6	12.3	4	6	0.72	11.52	0.4	10.8

More signal strength variations are expected in NLOS scenarios than in LOS scenarios. Hence, the standard deviation is smaller for LOS scenarios and larger for NLOS scenarios in Table 6.7. The dense urban environment exhibits a larger standard deviation compared to all other environments.

6.4.2.3 Building Penetration Loss

As mentioned earlier, outdoor devices are typically assumed by the 3GPP for the NTN. However, a HAPS can support indoor devices. A suitable building penetration loss needs to be used for indoor NTN devices. Building penetration loss varies greatly as a function of the building materials, floor number, and the device location within the building (Tripathi and Reed 2014; ITU 2023b). See ITU (2023a) for details on the building entry/exit-loss model. Two main types of buildings can be defined from the perspective of building penetration loss: thermally efficient modern buildings and traditional buildings. Thermally efficient buildings use materials such as metalized glass and foil-backed panels, causing significantly higher losses than traditional buildings.

Let L be the estimated building penetration or entry loss and P be the probability that the actual building penetration loss does not exceed L. L, expressed in dB, is a function of P and is given by (3GPP 2020)

$$L(P) = 10 \times \log_{10}\left(10^{0.1A(P)} + 10^{0.1B(P)} + 10^{0.1C}\right), \tag{6.3}$$

where

$$\begin{aligned} A(P) &= F^{-1}(P)\sigma_1 + \mu_1 \\ B(P) &= F^{-1}(P)\sigma_2 + \mu_2 \\ C &= -3.0 \end{aligned} \tag{6.4}$$

and

$$\begin{aligned}\mu_1 &= L_h + L_e\\ \mu_2 &= w + x\log(f)\\ \sigma_1 &= u + v\log(f)\\ \sigma_2 &= y + z\log(f)\end{aligned} \tag{6.5}$$

In Eq. (6.5), L_h is the median loss for horizontal paths, and L_e is the correction for the path's elevation angle at the building façade.

L_h is given by

$$L_h = r + s\log(f) + t(\log(f))^2. \tag{6.6}$$

L_e is the correction for the elevation angle of the path at the building façade:

L_e is given by

$$L_e = 0.212|\theta|. \tag{6.7}$$

Furthermore, in Eqs. (6.3)–(6.5), f is the carrier frequency in GHz, θ is the elevation angle of the path at the building façade (degrees), and $F^{-1}(P)$ is the inverse cumulative normal distribution (as a function of probability P).

The coefficients used in Eqs. (6.5) and (6.6) are summarized in Table 6.8 (3GPP 2020).

6.4.2.4 Atmospheric Absorption

Attenuation due to absorption by atmospheric gases depends on the carrier frequency, the elevation angle, the altitude above sea level, and the water vapor density (i.e., absolute humidity). Atmospheric absorption can usually be ignored at frequencies below 10 GHz, especially for the elevation angle above 10°. However, if the elevation angle is below 10°, the 3GPP recommends that the attenuation be determined for a carrier frequency above 1 GHz. The 3GPP considers the approximate method outlined in Annex 2 of ITU (2022) to calculate the atmospheric attenuation, which is given by

$$P_{\text{atm}} = \frac{A_{\text{zenith}}(f)}{\sin(\alpha)}, \tag{6.8}$$

where $A_{\text{zenith}}(f)$ is the frequency-dependent zenith attenuation specified in Figure 6 of ITU (2022), and α is the elevation angle in degrees (3GPP 2020).

6.4.2.5 Rain and Cloud Attenuation

Rain and cloud attenuation can be ignored for frequencies below 6 GHz. The method specified in Section 2.2 of ITU (2017a) can be used to estimate the long-term statistics of location-specific rain

Table 6.8 Propagation Model Coefficients for the Building Penetration Loss.

Building Type	R	s	t	u	v	w	x	y	z
Traditional	12.64	3.72	0.96	9.6	2.0	9.1	−3.0	4.5	−2.0
Thermally Efficient	28.19	−3.00	8.48	13.5	3.8	27.8	−2.9	9.4	−2.1

attenuation. Clear sky conditions can be used as a baseline. For a more accurate analysis, the CDF of the rain attenuation can be used to determine location-specific rain attenuation (ITU 2017a).

6.4.2.6 Scintillation

As discussed in Chapter 4, scintillations are rapid fluctuations of the received signal amplitude and phase due to ionospheric or tropospheric propagation.

6.4.2.7 Ionospheric Scintillation Attenuation

Ionospheric propagation attenuation must only be considered for frequencies below 6 GHz (3GPP 2020). Additionally, these scintillations are relatively more pronounced below 3 GHz. Strong scintillations are generally not routinely observed in mid-latitudes (except during strong geomagnetic storms), but they may occur even daily during post-sunset hours in low-latitude regions. Furthermore, moderate to strong levels of scintillations have been observed at high latitudes such as auroral and polar latitudes.

The 3GPP proposes the use of the Gigahertz scintillation model to estimate the ionospheric scintillation loss. This model is valid for regions approximately 20° north and south of the magnetic equator. However, it is not applicable at high latitudes, such as above 60°. At other latitudes, the ionospheric scintillation can be ignored.

A simplified method to estimate the ionospheric scintillation-related attenuation is summarized below. See (3GPP 2020) for more details.

The ionospheric attenuation, P_{ion}, can be approximated as

$$P_{\text{ion}} = \frac{P_{\text{fluc-target}}}{\sqrt{2}}, \tag{6.9}$$

where $P_{\text{fluc-target}}$ is the attenuation at the target carrier frequency.

$P_{\text{fluc-target}}$ is given by

$$P_{\text{fluc-target}} = \text{FTF} \times P_{\text{fluc-base}}, \tag{6.10}$$

where FTF is the frequency translation factor, and $P_{\text{fluc-base}}$ is the attenuation at the base frequency of 4 GHz.

FTF is given by

$$\text{FTF} = \left(\frac{f_{\text{target}}}{f_{\text{base}}}\right)^{-1.5}, \tag{6.11}$$

where f_{target} is the target frequency at which the attenuation needs to be estimated, and f_{base} is the base frequency at which $P_{\text{fluc-base}}$ graphs are available.

In the P3 curve (which is one of the curves in Figure 6.6.6.1.4-1 in 3GPP (2020) and shows the percentage of time the abscissa is exceeded as a function of the amount of peak-to-peak fluctuation), $P_{\text{fluc-base}}$ at the 99% probability for $f_{\text{base}} = 4$ GHz is 1.1 (3GPP 2020).

As an example, Eq. (6.9) yields P_{ion} of 2.2 dB at 2 GHz when f_{target} is 2 GHz, f_{base} is 4 GHz, and $P_{\text{fluc-base}}$ is 1.1.

6.4.2.8 Tropospheric Scintillation Attenuation

Tropospheric propagation attenuation must be considered only for frequencies above 6 GHz (3GPP 2020). Tropospheric scintillations increase as the carrier frequency increases, with more pronounced effects above 10 GHz. They are caused by sudden changes in the refractive index resulting from variations in temperature, water vapor content, and barometric pressure.

Tropospheric scintillations are more significant at low elevation angles due to longer propagation paths and wider beams of receive antennas of NTN platforms. An algorithm to predict location-based tropospheric scintillation loss is specified in ITU (2017a). For example, the tropospheric scintillation attenuation corresponding to the 99% probability (i.e., there is a 99% probability that the actual attenuation will be less than the estimated value) at 20 GHz ranges from 0.12 dB at 90° elevation angle to about 1 dB at 10° elevation angle at the frequency of 20 GHz in Toulouse, France (3GPP 2020).

6.4.2.9 Small-Scale Fading Models

To model small-scale variations in the signal strength, the ITU two-state model is widely used with the typical assumption of a flat fading radio channel for an NTN (3GPP 2020). The 3GPP describes both the flat fading model and the frequency-selective fading model. The reader is referred to (3GPP 2020) for extensive tables containing parameters of the small-scale models and modeling of time-varying Doppler shifts and Faraday rotation. As RF planning and design consider long-term radio channel effects, small-scaling models are not directly utilized in typical RPD.[1]

6.5 Framework for NTN Link Budgets

Link budget is an RF accounting mechanism that considers (i) relevant characteristics of the transmitter and the receiver, (ii) the RPD performance goals such as target service QoS (e.g., error rate), cell-edge data rate, and cell-edge reliability, and (iii) the RF propagation environment and determines the Maximum Allowable Path Loss (MAPL) (in dB) between the transmitter and the receiver. The MAPL helps meet the target RPD performance goals. In a TN, the MAPL can be translated into the cell radius using a suitable propagation model. This radius can be used to select the gNB/eNB locations to provide RF coverage in the target service area. In an NTN, the MAPL does not help determine the radius of an NTN cell. However, the MAPL can be used to determine certain NTN design parameters (e.g., maximum power amplifier transmit power and antenna gains) to maintain the service link between the device and the NTN platform in its orbit.

NTN link budgets are provided below in multiple tables for different cases. These link budgets are based on typical link budget components of a terrestrial network (Tripathi and Reed 2014) and unique characteristics of an NTN. Example link budgets determined by the 3GPP are discussed in Section 6.7.

Table 6.9 provides an example template for the link budget in an NTN. Each variable is numbered A to Z. Note that the link budget in an NTN is concerned with the farthest distance between the device and the NTN platform rather than the distance between the cell center and the cell edge (which is relevant in a TN).

The focus of Table 6.9 is on the service or access link. The feeder link is implementation specific and beyond the scope of 3GPP specifications. Hence, the link budget for the feeder link is not considered here. Suitable antenna gains on the NTN platform and the NTN-gateway can be used to

1 The impact of small-scale fading resulting from effects such as mobility are indirectly reflected in the Signal-to-Noise Ratio (SNR) value used in the link budget calculation. Small-scale fading effects are typically averaged out in the propagation models used in an RFD tool.

Table 6.9 Key Components of the NTN Link Budget.

ID	Description	Calculation	Example Values/ Explanation
A	Maximum transmit power (dBm)	A	
B	Transmit antenna gain (dB)	B	
C	Transmit EIRP (dBm)	$C = A + B$	
D	Thermal noise density (dBm/Hz)	D	(−174 dBm/MHz)
E	Receiver noise figure (dB)	E	9 dB for the device
F	Subcarrier spacing (SCS) (kHz)	F	SCS = 15 kHz
G	Number of PRBs required for the longest device-NTN platform distance	G	The number of resources needed to meet the data rate requirement using a suitable MCS associated with the target SNR.
H	Noise bandwidth (Hz)	$H = G \times 12 \times F \times 1{,}000$	1 PRB consists of 12 subcarriers.
I	Noise bandwidth (dB-Hz)	$I = 10 \times \log10(H)$	
J	Effective receive noise power (dBm)	$J = D + E + I$	
K	Receive antenna gain (dB)	K	
L	Receiver cable Loss (dB)	L	
M	Required SNR (dB)	M	
N	Receiver sensitivity (dBm)	$N = J - K + L + M$	The receiver needs to detect this signal level.
O	Interference margin (dB)	O	1–3 dB
P	Receiver implementation margin (dB)	P	
Q	Shadow fading margin (dB)	Q	
R	Building penetration loss (dB)	R	

(Continued)

Table 6.9 (Continued)

ID	Description	Calculation	Example Values/ Explanation
S	Clutter loss (dB)	S	
T	Atmospheric absorption loss (dB)	T	
U	Rain and cloud attenuation (dB)	U	
V	Ionospheric scintillation loss (dB)	V	
W	Tropospheric scintillation loss (dB)	W	
X	Head/body loss (dB)	X	
Y	Polarization mismatch loss (dB)	Y	0 dB or 3 dB for circular or noncircular polarization at the device
Z	Total margins (dB)	$Z = O + P + Q + R + S + T + U + V + W + X + Y$	
	MAPL (dB)	$\text{MAPL} = C - N - Z$	

close the feeder link from the NTN platform to the NTN-GW and from the NTN-GW to the NTN platform.

A: Maximum transmit power (dBm). The transmitter's maximum transmit power is expressed in dBm for link budget calculations. For the uplink (i.e., UE-to-NTN platform) link budget, this is the maximum transmit power of the UE. For the downlink (i.e., UE-to-NTN platform) link budget, this is the maximum transmit power of the NTN platform.

B: Transmit antenna gain. For the UE-to-NTN platform uplink, this is the UE's antenna gain. For the UE-to-NTN platform downlink, this is the NTN platform's antenna gain.

C: Transmit EIRP (dBm). Effective isotropic radiated power (EIPRP), C, is calculated as $C = A + B$.

D: Thermal noise density (dBm/Hz). Its value is (−174 dBm/MHz).

E: Receiver noise figure (dB). This is the amount of noise added by the receiver.

F: Subcarrier spacing (SCS). The SCS depends on the configuration determined by the gNB/eNB. For NR-NTN, one candidate SCS value for FR1-NTN is 15 kHz. For NR-NTN, other SCS values for FR1-NTN include 30 kHz and 60 kHz. For IoT-NTN, SCS can be 15 kHz or a smaller value such as 3.75 kHz for the uplink.

G: Number of PRBs required for the longest device-NTN platform distance. A service provider aims to provide a certain minimum data rate at the cell edge. The users at locations other than the cell edge typically experience a higher data rate than the minimum data rate. In a TN, the cell edge distance is equivalent to the cell radius and is the farthest distance from the center of the cell (where the gNB/eNB antenna is located), at which coverage needs to be provided. However, in an NTN, this distance is the distance between the device and the NTN platform when and where the device and the NTN platform are at the farthest distance from each other during the period the NTN platform is designed to serve the device. The number of resources needed to meet the data rate requirement for the longest device-NTN platform distance should be estimated based on a suitable (usually conservative) MCS. Fewer PRBs imply increased power spectral density, which is especially helpful for the uplink. Each MCS has an associated SNR requirement. Generally, a target cell-edge data rate is higher in the downlink and lower in the uplink.

H: Noise bandwidth. The noise bandwidth in Hz is calculated as $H = G \times 12 \times F \times 1{,}000$, where G is the number of PRBs, F is the SCS in kHz, and "12" refers to the number of subcarriers in the PRB.

I: Noise bandwidth (dB-Hz). To facilitate the link budget calculations, the noise bandwidth in dB-Hz is calculated as $I = 10 \times \log10(H)$.

J: Effective receive noise power (dBm). The total received noise power in the receive bandwidth (in dBm) is given by $J = D + E + I$, where D is thermal noise density in dBm/Hz, E is the receiver noise figure in dB, and I is noise bandwidth in dB-Hz.

K: Receive antenna gain. For the uplink, K is the receive antenna gain of the eNB/gNB/NTN platform in dB. For the downlink, K is the receive-antenna gain of the NTN UE in dB.

L: Receiver cable loss (dB). When the receiver and the receiver-antenna connectors are connected via an RF cable and are apart by a few feet or meters, a receiver cable loss of 3 dB can be assumed for the uplink. The NTN UE does not have a long cable between the receiver and the receiver antenna connector. Hence, cable loss at the NTN UE can be ignored for the downlink.

M: Required SNR. This is the required SNR associated with the Modulation and Coding Scheme (MCS) assumed for the parameter G above. The 3GPP has extensively analyzed supportable SNRs, associated MCSs, and achievable throughput for various channel conditions (3GPP 2024c, 2024d).

N: Receiver sensitivity. The receiver sensitivity is the minimum signal level the receiver needs to detect for the target QoS. The receiver sensitivity in dBm can be calculated as $N = J - K + L + M$, where J is the total receive noise power in the receive bandwidth in dBm, K is the receive-antenna gain in dB, L is the receiver cable loss in dB, and M is the required SNR.

O: Interference margin. The receiver must tolerate a certain amount of interference above the total noise power J. Depending on whether the system is a noise-limited system or an interference-limited system, a suitable interference margin needs to be added. In general, field measurements can help determine a good value of interference margin. In an Orthogonal Frequency Division Multiplexing (OFDM) system, a 1 dB margin may be used for low interference scenarios. If high interference is expected, a 3-dB interference margin can be used, especially for the cell edge scenarios. The interference margin should reflect contributions from all sources of interference, including co-channel interference (CCI), adjacent channel interference (ACI), and intermodulation interference. The interference margin reflects how much (power) rise over thermal (RoT) is expected for a heavily loaded system (e.g., during a busy hour).

P: Receiver implementation margin. Different receivers may have different performance capabilities. Minimum performance requirements specified by standards bodies are usually conservative, and commercial receivers often exceed such performance targets by a good margin. A small margin, such as 1 dB, may sometimes be added to account for variables in different receiver implementations.

Q: Shadow fading margin (dB). To account for long-term variations in the signal strength, a shadow fading margin is added to the link budget. Shadow fading is usually modeled with a mean of 0 and the standard deviation of σ dB. Inverse log-normal distribution and the standard deviation of σ dB are used in conjunction with the target cell-edge reliability (e.g., 75% cell-edge reliability or

90% cell-edge reliability) to determine the shadow fading margin. For example, using the Microsoft Excel function norminv (probability = 0.9, mean = 0, standard deviation = 8) yields the shadow fading margin of 10.3 dB.

R: Building penetration loss (dB). The building penetration loss depends on various factors, such as the type of building materials and the floor number. As indoor NTN coverage is quite challenging, an NTN may aim to provide outdoor coverage only. For outdoor coverage, building penetration can be ignored. However, a suitable building penetration margin needs to be added if indoor coverage is targeted for an NTN. In the absence of any specific data, 20 dB can be used as a baseline loss.

S: Clutter loss (dB). Different propagation environments are characterized by clutter, such as trees, open space, and water. Different types of clutter have different propagation path losses. As an example, the clutter loss of 34.3 dB at the S band and 44.3 dB at the Ka band in dense urban and urban environments can be used.

T: Atmospheric absorption loss (dB). TNs typically do not need to consider atmospheric absorption loss. However, NTN signals pass through several layers of the atmosphere, leading to atmospheric absorption loss. The atmospheric absorption loss depends on the frequency and the elevation angle. The atmospheric absorption loss can be assumed to be 0 dB for the elevation angle greater than 10° and the frequency less than 10 GHz. Equation (6.8) can be used to calculate the atmospheric absorption loss for a frequency greater than 10 GHz.

U: Rain and cloud attenuation (dB). *U* can be assumed to be 0 dB for a frequency below 6 GHz. *U* is location dependent for a frequency above 6 GHz. $U = 0$ dB can be used as a baseline for clear skies. The rain margin depends on the rain rate and the total propagation path that experiences rain. There could be 0.02 dB/km attenuation due to rain (ViaSatellite 2023), potentially requiring a large amount of rain margin when the distance between the transmitter and the receiver is long. Clear skies (and hence 0 dB rain attenuation) can be used as a baseline with the understanding that the link performance would degrade under rainy conditions. Alternatively, a moderate value such as 7 dB may be used if rain effects are explicitly considered in the link budget calculations (ViaSatellite 2023).

V: Ionospheric scintillation loss (dB). *V* depends on frequency and latitude. *V* should be considered for frequencies below 6 GHz. For example, $V = 2.2$ dB can be used at 2 GHz as specified in Table 6.10.

W: Tropospheric scintillation loss (dB). *W* depends on the location, the elevation angle, and the frequency. *W* should be considered for the frequency above 6 GHz. For example, $W = 1.0$ dB can be used for the frequency of 20 GHz at the 10° elevation angle for a location similar to Toulouse, France.

X: Head/body loss (dB). Based on how the NTN device is used, a suitable head/body loss should be included. For example, *X* can be about 3 dB for a smartphone and 0 dB for a VSAT.

Y: Polarization mismatch loss (dB). A typical NTN platform uses circular polarization. Hence, when the NTN device uses a noncircular polarization, such as linear polarization, there would be a polarization mismatch. *Y* of 3 dB can be used for a smartphone and IoT device to account for this polarization mismatch.

Z: Total margins (dB). The total margins *Z* is calculated as $Z = O + P + Q + R + S + T + U + V + W + X + Y$.

Finally, the MAPL in dB can be calculated as $\text{MAPL} = C - N - Z$. The MAPL is the maximum path loss that can be tolerated between the transmitter and the receiver while meeting target performance requirements related to the cell-edge data rate and cell-edge reliability.

Table 6.10 summarizes example parameter settings for the uplink link budget for an NTN.

A: Maximum transmit power of the device (dBm). For the uplink, the device or the UE is the transmitter. For a typical smartphone UE using Power Class 3, the maximum transmit power is

Table 6.10 Example Link Budget Parameter Settings for the Uplink.

ID	Description	Example Values/Explanation
A	Maximum transmit power of the device (dBm)	23 dBm for a smartphone. 23, 20, or 14 dBm for an IoT device. 33 dBm for a VSAT.
B	Transmit antenna gain (dB)	0 dBi for a smartphone and an IoT device. 43.2 dBi for a VSAT.
E	Receiver noise figure (dB)	8 dB for the gNB/eNB receiver
F	Subcarrier spacing (SCS) (kHz)	SCS = 15 kHz for a smartphone and an LTE-M device. SCS = 15 kHz or 3.75 kHz for an NB-IoT device.
G	Required number of PRBs	Variable—a function of the target minimum data rate and the target MCS.
K	Receive antenna gain (dB)	The gain of the receive antenna at the gNB/eNB/NTN platform. Design parameter.
L	Receiver cable loss (dB)	The loss due to an RF cable between the antenna and the receive processor (3 dB).
M	Required SNR (dB)	PUSCH SNR for a robust target MCS.
O	Interference margin (dB)	0 dB for a noise-limited system and 3 dB for an interference-limited system
P	Receiver implementation margin (dB)	1 dB (to account for different receivers)
Q	Shadow fading margin (dB)	Function of the environment (e.g., rural vs. urban) and the edge-of-coverage reliability.
R	Building penetration loss (dB)	Function of the building material. Ex: 20 dB. A more accurate attenuation uses the procedure summarized in Section 6.4.2.
S	Clutter loss (dB)	Function of the environment, frequency, and elevation angle. For example: 34.3 dB at the S band and 44.3 dB at the Ka band in dense urban and urban environments.
T	Atmospheric absorption loss (dB)	Function of the frequency and the elevation angle. 0 dB for the elevation angle >10 and frequency <10 GHz. Equation (6.8) for frequency >10 GHz.
U	Rain and cloud attenuation (dB)	0 dB for frequency below 6 GHz. Location dependent for frequency above 6 GHz. 0 dB as a baseline for clear skies.
V	Ionospheric scintillation loss (dB)	Function of frequency and latitude. Consider for frequency <6 GHz. Ex: 2.2 dB at 2 GHz.
W	Tropospheric scintillation loss (dB)	Function of location, elevation angle, and frequency. Consider for frequency >6 GHz. Ex: 1.0 dB at 20 GHz for the 10° elevation angle in Toulouse, France.
X	Head/body loss (dB)	3 dB for a smartphone, 0 dB for a VSAT, application dependent for an IoT device
Y	Polarization mismatch loss (dB)	3 dB for a smartphone and IoT device due to the use of linear polarization when an NTN platform (e.g., a satellite) uses circular polarization, 0 dB for a VSAT due to the use of circular polarization when an NTN platform (e.g., a satellite) uses circular polarization.

23 dBm. An LTE-M/NB-IoT IoT device may use the maximum transmit power of 23, 20, or 14 dBm for an IoT device. A VSAT may use the maximum transmit power of 33 dBm (3GPP 2020).

B: Transmit antenna gain (dB). An omnidirectional antenna can be assumed for a smartphone and an IoT device, leading to the ideal transmit antenna gain of 0 dBi. As a high-gain dish antenna is common for a VSAT, the transmit antenna gain of 43.2 dBi can be assumed for a VSAT (3GPP 2020).

E: Receiver noise figure (dB). A gNB/eNB receiver usually has better noise performance than a UE receiver. Hence, the noise figure of 8 dB can be assumed for the gNB/eNB receiver (3GPP 2020).

F: Subcarrier spacing (SCS) (kHz). SCS = 15 kHz can be assumed for a smartphone and an LTE-M IoT device. SCS = 15 kHz or 3.75 kHz can be assumed for an NB-IoT device.

G: Required number of PRBs. This is a design variable. It is a function of the target minimum uplink data rate at the cell edge for the farthest distance between the device and the NTN platform and the MCS.

K: Receive antenna gain (dB). This is a design variable. It is the gain of the receive antenna at the gNB/eNB/NTN platform.

L: Receiver cable loss (dB). This loss is due to an RF cable between the antenna and the receive-processor. A typical value is 3 dB.

M: Required SNR (dB). The SNR for the Physical Uplink Shared Channel (PUSCH) for a robust target MCS and a suitable Doppler frequency should be used. For geostationary NTN platforms such as a GEO satellite and a HAPS, a low Doppler shift can be used to reflect the expected NTN UE mobility. A large Doppler shift needs to be used for LEO satellites to reflect the NTN platform mobility.

O: Interference margin (dB). The interference margin can be set to 0 dB for a noise-limited system and 3 dB for an interference-limited system.

P: Receiver implementation margin (dB). An optional implementation margin of 1.0 dB can be used to account for different receivers.

Q: Shadow fading margin (dB). As discussed earlier in conjunction with Table 6.9, the shadow fading margin is based on the expected service environment (e.g., rural vs. urban) and the target edge-of-coverage reliability. An NTN usually serves different types of environments, and hence, the worst-case environment needs to be used to determine the applicable shadow fading margin.

R: Building penetration loss (dB). If the NTN is designed to provide outdoor coverage only, the link budget can ignore the building penetration loss. The building penetration loss can be determined for indoor coverage using the procedure summarized in Section 6.4.2. A baseline building penetration loss of 20 dB can be assumed.

S: Clutter loss (dB). Clutter loss is a function of the environment, the carrier frequency, and the elevation angle. For example, the CL can be assumed to be 34.3 dB at the S band and 44.3 dB at the Ka band in dense urban environments (3GPP 2020).

T: Atmospheric absorption loss (dB). The atmospheric absorption loss is a function of the frequency and the elevation angle. It can be assumed to be 0 dB for the elevation angle greater than 10° and frequency less than 10 GHz. Equation (6.8) can be used to estimate the atmospheric absorption loss for frequency >10 GHz.

U: Rain and cloud attenuation (dB). Rain and cloud attenuation can be assumed to be 0 dB for frequencies below 6 GHz. Rain and cloud attenuation of 0 dB can be used as a baseline for clear skies. Rain and cloud attenuation is location dependent for frequencies above 6 GHz.

V: Ionospheric scintillation loss (dB). *V* depends on frequency and latitude. Specifically, *V* should be considered for frequencies below 6 GHz. For example, $V = 2.2$ dB can be used at 2 GHz.

W: Tropospheric scintillation loss (dB). As mentioned earlier in Table 6.9, *W* depends on the location, the elevation angle, and the frequency. In particular, *W* should be considered for the frequency above 6 GHz. For example, $W = 1.0$ dB can be used as tropospheric scintillation loss for the frequency of 20 GHz at the 10° elevation angle for a location similar to Toulouse, France.

X: Head/body loss (dB). As mentioned earlier in Table 6.9, a suitable head/body loss needs to be included based on how the NTN device is used. For example, X can be assumed to be 3 dB for a smartphone and 0 dB for a VSAT.

Y: Polarization mismatch loss (dB). As mentioned earlier in Table 6.9, when an NTN platform (e.g., a satellite) uses circular polarization, but the NTN UE uses a noncircular polarization, there is a polarization mismatch. Hence, $Y = 3$ dB can be used for smartphones and IoT devices.

Table 6.11 provides example parameter settings for the downlink link budget in an NTN.

Table 6.11 Example Link Budget Parameter Settings for the Downlink.

ID	Description	Example Values/Explanation
A	Maximum transmit power of the gNB/eNB/NTN platform (dBm)	Design parameter.
B	Transmit antenna gain (dB)	Design parameter.
E	Receiver noise figure (dB)	9 dB for a smartphone and an IoT device, 1.2 dB for a VSAT
F	Subcarrier spacing (SCS) (kHz)	SCS = 15 kHz for a smartphone, an LTE-M device, and an NB-IoT device.
G	Required number of PRBs	Variable—a function of the target minimum data rate, worst-case SNR, and associated MCS.
K	Receive antenna gain (dB)	0 dBi for a smartphone, an LTE-M device, and an NB-IoT device, 39.7 dBi for a VSAT.
L	Receiver cable loss (dB)	0 dB
M	Required SNR (dB)	PDSCH SNR for a robust target MCS.
O	Interference margin (dB)	0 dB for a noise-limited system and 3 dB for an interference-limited system
P	Receiver implementation margin (dB)	1 dB (to account for different receivers)
Q	Shadow fading margin (dB)	Function of the environment (e.g., rural vs. urban) and the edge-of-coverage reliability.
R	Building penetration loss (dB)	Function of the building material. For example: 20 dB. A more accurate attenuation, use the procedure summarized in Section 6.4.2.
S	Clutter loss (dB)	Function of the environment, frequency, and elevation angle. Ex: 34.3 dB at the S band and 44.3 dB at the Ka band in dense urban and urban environments.
T	Atmospheric absorption loss (dB)	Function of the frequency and the elevation angle. 0 dB for the elevation angle >10 and frequency <10 GHz. Equation (6.8) for frequency >10 GHz.
U	Rain and cloud attenuation (dB)	0 dB for frequency below 6 GHz. Location dependent for frequency above 6 GHz. 0 dB as baseline for clear skies.
V	Ionospheric scintillation loss (dB)	Function of frequency and latitude. Consider for frequency <6 GHz. Ex: 2.2 dB at 2 GHz.
W	Tropospheric scintillation loss (dB)	Function of location, elevation angle, and frequency. Consider for frequency >6 GHz. Ex: 1.0 dB at 20 GHz for the 10° elevation angle in Toulouse, France.
X	Head/body loss (dB)	3 dB for a smartphone, 0 dB for a VSAT, application dependent for an IoT device
Y	Polarization mismatch loss (dB)	3 dB for a smartphone and IoT device due to the use of linear polarization when an NTN platform (e.g., a satellite) uses circular polarization, 0 dB for a VSAT due to the use of circular polarization when an NTN platform (e.g., a satellite) uses circular polarization.

A: Maximum transmit power of the gNB/eNB/ NTN payload (dBm). In the case of a regenerative payload, this is the maximum transmit power of the gNB, the eNB, or the gNB-DU. For a transparent payload, this is the maximum transmit power of the NTN payload that is serving as a repeater. This is a design parameter to help close the downlink for communications.

B: Transmit antenna gain (dB). This is the transmit antenna gain of the gNB, the eNB, or the gNB-DU depending on the type of regenerative payload. For a transparent payload, this is the transmit antenna gain of the NTN payload that is serving as a repeater. This is a design parameter to help close the downlink for communications.

E: Receiver noise figure (dB). For the downlink, the receiver is at the NTN UE. *E* can be assumed to be 9 dB for a smartphone and an IoT device and 1.2 dB for a VSAT.

F: Subcarrier spacing (SCS) (kHz). For the downlink, F can be assumed to be 15 kHz for an NR smartphone, an LTE-M device, and an NB-IoT device. 5G NR also supports additional SCSs such as 30 and 60 kHz. These SCSs can also be used per NTN design configuration.

G: Required number of PRBs. This is a design variable and is a function of the target minimum data rate in the downlink and the target supportable MCS. This MCS is associated with the target supportable SNR in the downlink.

K: Receive antenna gain (dB). For the downlink, *K* is the receive antenna gain at the UE. *K* can be assumed to be 0 dBi for a smartphone, an LTE-M IoT device, and an NB-IoT device. *K* can be set to 39.7 dBi for a VSAT (3GPP 2020).

L: Receiver cable loss (dB). As the receive antenna and the receiver processor are quite close in the UE, the receiver cable loss can be ignored and set to 0 dB.

M: Required SNR (dB). The SNR for the Physical Downlink Shared Channel (PDSCH) for a robust target MCS and a suitable Doppler frequency should be used. For geostationary NTN platforms such as a GEO satellite and a HAPS, a low Doppler shift can be used to reflect the expected NTN UE mobility. For LEO satellites, a large Doppler shift needs to be used to reflect the NTN platform mobility.

O: Interference margin (dB). The interference margin can be set to 0 dB for a noise-limited system and 3 dB for an interference-limited system.

P: Receiver implementation margin (dB). An optional implementation margin of 1.0 dB can be used to account for different receivers.

Q: Shadow fading margin (dB). As discussed earlier in conjunction with Table 6.9, the shadow fading margin based on the expected service environment (e.g., rural vs. urban) and the target edge-of-coverage reliability. An NTN usually serves different types of environments and hence the worst-case environment needs to be used to determine the applicable shadow fading margin.

R: Building penetration loss (dB). If the NTN is designed to provide outdoor coverage only, the link budget can ignore the building penetration loss. For indoor coverage, the building penetration loss can be determined using the procedure summarized in Section 6.4.2. A baseline building penetration loss of 20 dB can be assumed.

S: Clutter loss (dB). Clutter loss is a function of the environment, the carrier frequency, and the elevation angle. As an example, the CL can be assumed to be 34.3 dB at the S band and 44.3 dB at the Ka band in dense urban and urban environments (3GPP 2020).

T: Atmospheric absorption loss (dB). The atmospheric absorption loss is a function of the frequency and the elevation angle. It can be assumed to be 0 dB for the elevation angle greater than 10° and frequency less than 10 GHz. Equation (6.8) can be used to estimate the atmospheric absorption loss for frequency >10 GHz.

U: Rain and cloud attenuation (dB). Rain and cloud attenuation can be assumed to be 0 dB for frequencies below 6 GHz. Rain and cloud attenuation of 0 dB can be used as a baseline for clear skies. Rain and cloud attenuation is location dependent at frequencies above 6 GHz.

V: Ionospheric scintillation loss (dB). *V* depends on frequency and latitude. Specifically, *V* should be considered for frequencies below 6 GHz. For example, $V = 2.2$ dB can be used at 2 GHz.

W: Tropospheric scintillation loss (dB). As mentioned earlier for Table 6.9, *W* depends on the location, the elevation angle, and the frequency. In particular, W should be considered for the frequency above 6 GHz. For example, $W = 1.0$ dB can be used as tropospheric scintillation loss for the frequency of 20 GHz at the 10° elevation angle for a location similar to Toulouse, France.

X: Head/body loss (dB). As mentioned earlier in Table 6.9, a suitable head/body loss needs to be included based on how the NTN device is used. For example, *X* can be assumed to be 3 dB for a smartphone and 0 dB for a VSAT.

Y: Polarization mismatch loss (dB). As mentioned earlier in Table 6.9, when an NTN platform (e.g., a satellite) uses circular polarization, but the NTN UE uses a noncircular polarization, there is a polarization mismatch. Hence, $Y = 3$ dB can be used for a smartphone and IoT devices.

6.5.1 A Note on NTN Link Budgets

A typical link budget provides the MAPL between the transmitter and the receiver, which can be used with a suitable propagation model to estimate the supportable longest distance between the NTN UE and the NTN platform or payload. If the target worst-case (i.e., the longest) distance is greater than the supportable longest distance estimated by the link budget, it means that the design configuration (e.g., transmit power, antenna gains, and target data rate) cannot support the target worst-case (i.e., the longest) distance. In such case, one or more design configuration parameters would need to be adjusted. If the target worst-case (i.e., the longest) distance is much shorter than the supportable longest distance estimated by the link budget, it means that the design configuration (e.g., transmit power, antenna gains, and target data rate) can support a longer distance than the target worst-case (i.e., the longest) distance; equivalently, the design configuration can achieve better performance than the performance targeted by the link budget. The performance target can then be adjusted to reflect the enhances performance that the link budget can accommodate.

An iterative link budget calculation process can be used to finalize the link budget and design configuration parameters. *A word of caution...the link budget is concerned with a single UE; it is unrelated to the average user throughput in the downlink or the uplink and the average cell throughput in the downlink and the uplink.* Prior to the deployment, average user throughput and cell throughput can be estimated analytically and/or via simulations using the design configurations finalized in the link budget and capacity planning processes.

In another approach to the link budget calculations, the worst-case or farthest distance is used to calculate the propagation path loss. Such distance corresponds to the minimum elevation angle between the NTN UE and the NTN platform. EIRP, estimated propagation path loss, and various losses and margins specified in Table 6.9 are used to estimate carrier-to-interference-plus-noise ratio (CINR). If CINR is inadequate to meet target throughput, the design configuration needs to be modified.

Note that the key components specified in Table 6.9 are still applicable to this CINR approach to the link budget. See Section 6.7 for a simplified example that determines CNR[2] based on 3GPP-estimated link budgets.

2 In a Code Division Multiple Access (CDMA) system, SNR is (Processing Gain ∗ CNR). In a typical non-CDMA system including an OFDM system, SNR = CNR.

6.6 NTN Capacity Planning

6.6.1 Capacity Planning in a Nutshell

Capacity planning aims to determine the number, locations, and configurations of gNBs/eNBs to provide adequate capacity and throughput in the target service area. Capacity planning in a cellular network involves various aspects such as estimation of cell capacity and throughput, user traffic analysis, and suitable provisioning of network equipment. Figure 6.3 illustrates the key stages of the capacity planning process in a cellular system. These stages are based on typical capacity planning considerations for a terrestrial network (Tripathi and Reed 2014) and unique characteristics of an NTN.

6.6.1.1 Step 1: Capacity/Throughput Estimation

Capacity or throughput must be estimated before deploying a new radio access technology. *Due to the different characteristics of the downlink and the uplink, throughput needs to be estimated separately of the downlink and the uplink.* For example, higher transmit power, a more aggressive MCS, and more spatial multiplexing layers are available in the downlink compared to the uplink. If Time Division Duplex (TDD) is used, the distribution of the time between the downlink and the uplink must be considered when estimating throughput. As a starting point in the planning process, the target service area is divided into cells. The cell size is typically dictated by the link budget.

Example approaches to estimate capacity/throughput include an *analytical approach*, a *simulation approach*, and an *RPD tool approach*. In general, the analytical approach can provide a rough estimate of throughput, the simulation approach provides a more accurate estimate of throughput than the analytical approach, and the RPD tool approach can provide the most accurate estimate of throughput. In one extreme, the analytical approach is the simplest and requires the least amount of time and effort. On the other hand, the RPD tool approach requires the most time and effort and is widely used in commercial cellular networks. Note that the RPD tool utilizes more accurate propagation models (typically tuned based on real-world measurements) and includes simulations that model different types of traffic and user mobility.

In Figure 6.4, an example analytical approach to estimating cell capacity or throughput shows the cell divided into multiple radial segments.

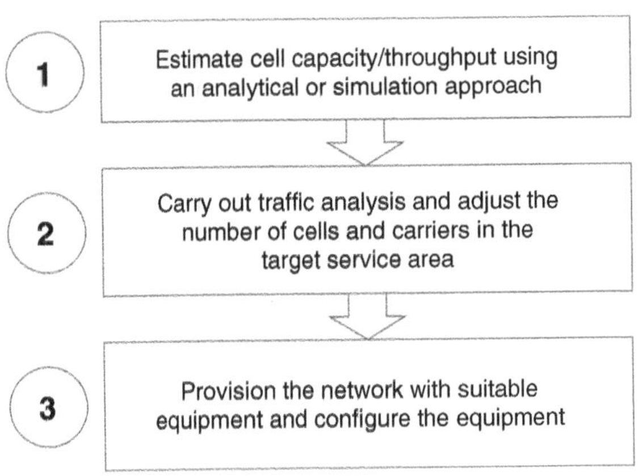

Figure 6.3 Capacity Planning in a Cellular Network.

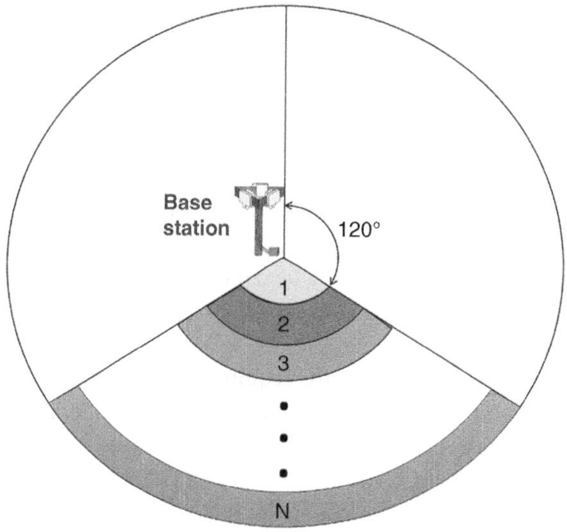

Figure 6.4 Capacity Planning in a Cellular Network.

In Figure 6.4, a hexagonal sector or cell is approximated as a circle to simplify calculations. A 120° sector or cell is divided into N circular (or partial doughnut/donut-shaped) segments located at different distances from the center of the sector. In each of the N segments, a suitable MCS (based on the expected average SINR) and a suitable number of spatial multiplexing layers are assumed, and segment throughput is calculated using the planned amount of radio resources. The amount of radio resources should reflect the number of PRBs usable for user traffic on planned carrier frequencies and any overhead, such as control channels and synchronization signals and the Physical Broadcast Channel. The estimated segment throughput and the segment area for all segments are considered to estimate the overall cell throughput.[3]

In an enhancement to the approach described above, an elliptical area instead of a circular area can be used to model elliptical beams of an NTN.

In an example *simulation approach*, a layout of multiple cells with at least two tiers of interference around the center cell is used. Users are uniformly distributed in the center cell. Suitable propagation models relevant to deployment environments, such as urban, suburban, and rural environments, are used to estimate user location-specific SINRs and subsequently supportable MCSs and spatial multiplexing layers. These location-specific MCSs, spatial multiplexing layers, and the number of radio resources based on the PRBs available for user traffic are translated into an overall cell throughput estimate.

In an example *RPD tool approach*, RF propagation analysis is carried out to predict SINRs at different locations or bins in the target service area. The SINRs in beans are mapped to supportable throughput based on the mapping with MCS and assumptions on the number of spatial multiplexing layers. In addition, depending on the RPD toll capability, user traffic is modeled based on selected types of services. Various performance metrics, including overall throughput, are quantified.

3 After a new technology is initially launched in limited areas, field testing or drive testing can be carried out to refine such estimates for RPD in other geographic areas.

6.6.1.2 Step 2: Traffic Analysis and Adjustment of Cells and Carriers

The initial network design is typically coverage driven, aiming to ensure contiguous coverage that meets minimum cell-edge performance (e.g., the data rate of X Mbps and certain cell-edge reliability, such as 75% or 90%). The link budget and a suitable propagation model are used to determine the cell radius. In a greenfield deployment scenario where the operator does not have any deployed radio network infrastructure (e.g., cell sites), cell site locations are selected to meet the cell radius constraint. In a nongreenfield scenario, the cell radius can be used to select cell site locations for the new technology deployment from the available cell sites. In some cases, all coverage-driven cell sites are selected as a starting point, and more cell sites (which may be completely new cell sites or the existing cell sites that have been deployed to provide increased capacity instead of basic coverage) are added to address any coverage holes.

The coverage-driven deployment is modified to meet the user traffic requirements. For example, traffic engineering is carried out to estimate the traffic in the next few years. Traffic simulations are carried out to check if the selected cell sites can meet the expected traffic demand. If the selected cell sites are inadequate, new cell sites are added, or additional carrier frequencies are added based on the business objectives and plans. After the traffic demands are met, *the selected cell sites satisfy both coverage requirements and capacity/throughput requirements.*

6.6.1.3 Step 3: Provisioning and Configuration of Network Equipment

The gNBs/eNBs at the selected cell sites must be configured with suitable parameter settings. For example, cell selection and neighbor cell parameters must be determined to broadcast appropriate System Information Blocks (SIBs). The cell edge based on the link budget, or the planned coverage area, can be used to determine an optimal value for the cell selection threshold. Serving cell parameters such as the Physical Cell Identifier (PCI) for 5G NR, LTE-M, and NB-IoT need to be determined. Random-access parameters must also be properly determined to minimize interference on the Physical Random-Access Channel (PRACH) and ensure reasonable random-access delay.

The backhaul bandwidth between the eNBs/gNBs and between the gNBs/eNBs and the core network must be adequately provisioned. Furthermore, when virtualization is used, suitable cloud resources must be provisioned to ensure the expected performance of the radio (and core) network. When disaggregated gNBs are used, the midhaul bandwidth between the gNB-DU and the gNB-CU must also be provisioned properly.

6.6.2 Complexities of Capacity Planning for an NTN

An NTN has distinct features that make capacity planning much more complex than a TN. These unique NTN features include large cells, cell mobility, variable cell size, variable device to NTN payload distance, and unique interfaces.

6.6.2.1 Large Cells

Figure 6.5 shows TN and NTN cells. In a TN, cells are relatively small, and hence, multiple cells have a similar large-area propagation environment, such as an urban environment, and a similar local-area environment, such as a certain type of clutter (e.g., trees vs. buildings). In contrast, an NTN cell is relatively large and could contain multiple types of local-area clutter and large-area propagation environments.

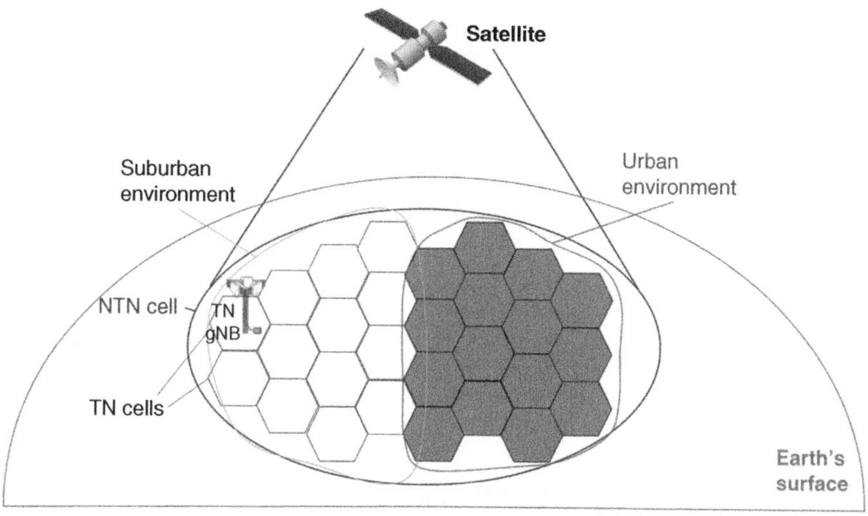

Figure 6.5 Consideration of Large Cells for NTN Capacity Planning.

6.6.2.2 Cell Mobility

The NTN cell of a given GSO satellite covers essentially the same geographic area and the NTN cell of a given HAPS covers essentially the same geographic area. The same NTN cell covers different geographic areas in the case of an NGSO satellite at tine t_1 and t_2 as shown in Figure 6.6. Hence, the RF propagation environment can differ for a given NTN cell at different instants, leading to different achievable capacities during different periods.

6.6.2.3 Variable Cell Size

As illustrated in Figure 6.7, elliptical beam coverage areas may not always be the same size for NGSO satellites. For example, for quasi-Earth-fixed beams, the elliptical beams, and hence NTN

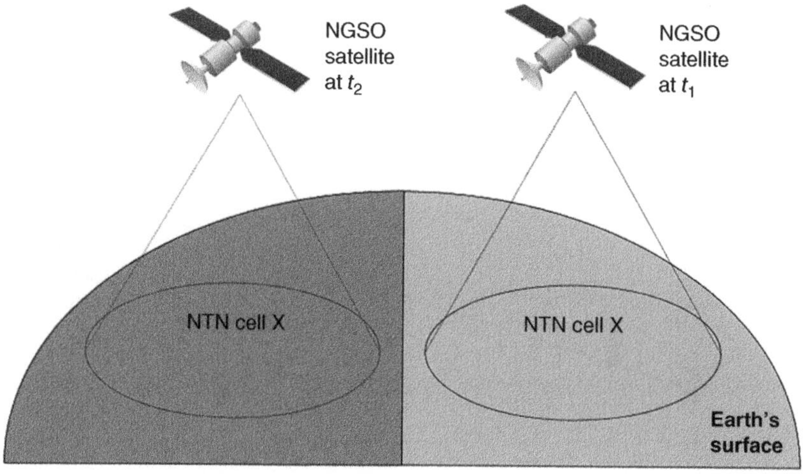

Figure 6.6 Consideration of Cell Mobility for NTN Capacity Planning.

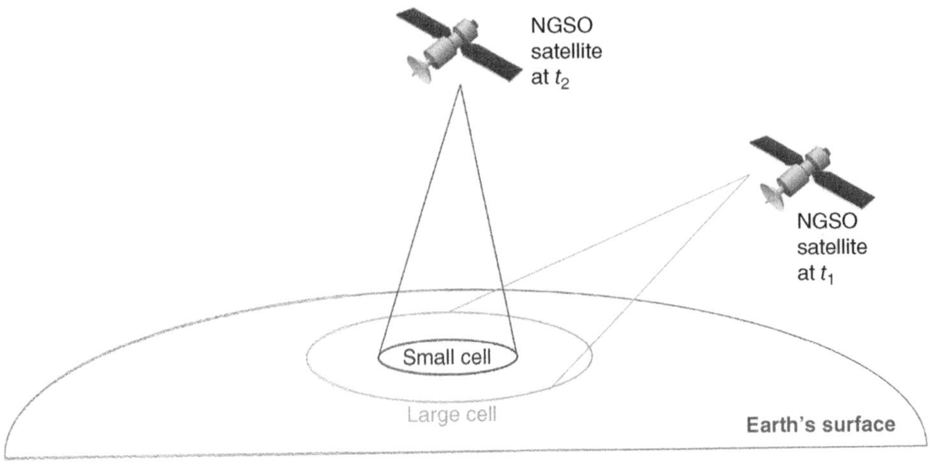

Figure 6.7 Consideration of a Variable Cell Size for NTN Capacity Planning.

cells, are larger when the NGSO satellite is near the horizon, as shown by the scenario at time t_1 in Figure 6.7. Similarly, the elliptical beams, and hence NTN cells, are smaller when the NGSO satellite is just above the device, as shown by the scenario at time t_2 in Figure 6.7. Elliptical beam coverage areas are of similar sizes in the case of Earth-moving beams.

6.6.2.4 Variable Device-NTN Payload Distance

The farthest distance between the NTN payload and the device at the edge of coverage can vary significantly from one period to the next in the case of a quasi-Earth-fixed beam, as shown in Figure 6.8. The farthest distance between the device and the NTN payload would be relatively short when the NTN platform is directly above the center of the NTN cell, as shown by the scenario at time t_2 in Figure 6.8. In contrast, the farthest distance between the device and the NTN payload is

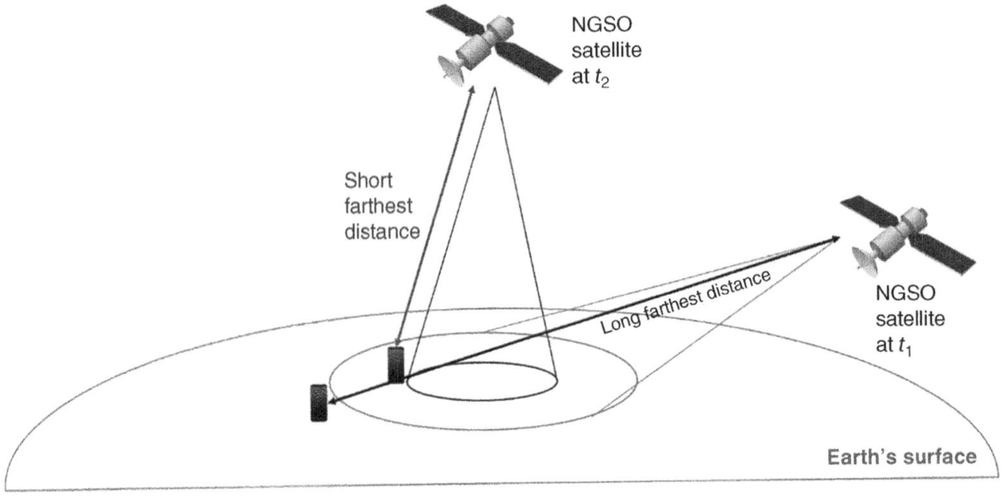

Figure 6.8 Consideration of Variable Device–NTN Payload Distances for NTN Capacity Planning.

relatively long when the NTN platform is at the lowest elevation angle, and the NTN UE is at the far cell edge of the NTN cell, as shown by the scenario at time t_1 in Figure 6.8.

6.6.2.5 Unique Interfaces

Figure 6.9 focuses on unique NTN interfaces. The feeder link between the NTN payload and the NTN-GW is a unique interface in an NTN. While the feeder link specifications are beyond the scope of the 3GPP standardization, adequate bandwidth (along with a suitable link budget) needs to be provisioned on the feeder link so that the feeder link does not become a capacity bottleneck. Furthermore, when the regenerative NTN payload is used, inter-satellite links (ISLs) would need to be provisioned with adequate bandwidth to carry the expected amount of traffic between the satellites.

6.6.3 Addressing the Complexities of NTN Capacity Planning

The NTN capacity planning would need to consider all the radio environments that the moving NTN cell would pass through to determine the supportable capacity/throughput. In one strategy, the environment with the most challenging radio environment and the heaviest traffic demand can be considered for planning. This strategy ensures adequate capacity is available in any area where the NTN cell provides coverage.

When estimating the traffic demand, the largest expected cell size can be considered so that the maximum number of users are considered, and the maximum amount of user traffic is generated per cell for a given user density (e.g., a certain number of users per square kilometer area) and user traffic intensity (e.g., a certain number of bytes per unit time). This strategy ensures adequate capacity or throughput is always available in the NTN cell.

The worst-case farthest distance between the device and the NTN payload should be used in planning to ensure that the most conservative MCS is assumed while estimating the supportable throughput.

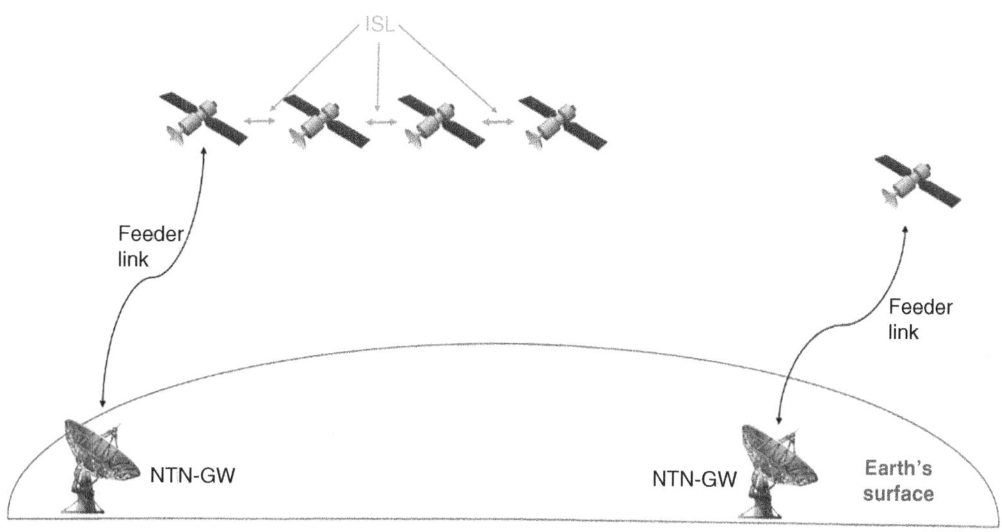

Figure 6.9 Consideration of Unique Interfaces for NTN Capacity Planning.

6.7 3GPP-Estimated Link Budgets and Throughput for an NTN

The 3GPP identified 30 study cases (SCs) to analyze performance of the NTN (3GPP 2023). These study cases correspond to combinations of the following scenarios: (i) GEO satellites, LEO satellites at the altitude of 600 km, and LEO satellites at the altitude of 1200 km; (ii) frequency reuse factor (FRF) of 1 and FRF of 3; and (iii) handheld devices (i.e., smartphones) and VSATs. FRF = 1 means that all the beams use the same carrier frequency or the same set of carrier frequencies. FRF = 3 means that three adjacent beams use three different carrier frequencies or three different sets of carrier frequencies.

The 3GPP made several assumptions for the NTN analysis. See 3GPP (2023) for the details of assumptions; selected assumptions are mentioned in self-explanatory Table 6.12 and Table 6.13 (3GPP 2023). There are two sets of satellite EIRP densities in Table 6.12. Set 1 has higher EIRP density than Set 2. Several parameters depend on the satellite EIRP density as shown in Table 6.12.

Section 6.7.1 summarizes 3GPP-determined link budgets for selected SCs, and Section 6.7.2 summarizes 3GPP-estimated UE throughput for selected SCs.

Table 6.12 Satellite Parameters for the NTN Analysis.

Parameter	Settings
Satellite	GEO (GEO satellites at the altitude of 35,786 km), LEO-1200 (LEO satellites at the altitude of 1200 km), LEO-600 (LEO satellites at the altitude of 600 km)
Frequency band	S-band (2 GHz; channel bandwidth: 30 MHz DL and 30 MHz UL)
	Ka-band (20 GHz DL and 30 GHz UL; channel bandwidth: 400 MHz DL and 400 MHz UL)
Satellite EIRP density	Set 1: (GEO: 59 dBW/MHz for S-band and 40 dBW/MHz for Ka-band), (LEO-1200: 40 dBW/MHz for S-band and 10 dBW/MHz for Ka-band), (LEO-600: 34 dBW/MHz for S-band and 4 dBW/MHz for Ka-band)
	Set 2: (GEO: 53.5 dBW/MHz for S-band and 32 dBW/MHz for Ka-band), (LEO-1200: 34 dBW/MHz for S-band and 2 dBW/MHz for Ka-band), (LEO-600: 28 dBW/MHz for S-band and −4 dBW/MHz for Ka-band)
Satellite beam diameter	Set 1: (GEO: 250 km for S-band and 110 km for Ka-band), (LEO-1200: 90 km for S-band and 40 km for Ka-band), (LEO-600: 50 km for S-band and 20 km for Ka-band)
	Set 2: (GEO: 450 km for S-band and 280 km for Ka-band), (LEO-1200: 190 km for S-band and 90 km for Ka-band), (LEO-600: 90 km for S-band and 50 km for Ka-band)
Maximum transmit and receive antenna gain	Set 1: (GEO: 51 dBi for S-band and 58.5 dBi for Ka-band), (LEO-1200: 30 dBi for S-band and 38.5 dBi for Ka-band), (LEO-600: 30 dBi for S-band and 38.5 dBi for Ka-band)
	Set 2: (GEO: 45.5 dBi for S-band and 50.5 dBi for Ka-band), (LEO-1200 : 24 dBi for S-band and 30.5 dBi for Ka-band), (LEO-600: 24 dBi for S-band and 30.5 dBi for Ka-band)

Table 6.13 NTN UE Parameters for the NTN Analysis.

Parameter	VSAT UE	Smartphone UE
Antenna type	Directive antenna with a 60 cm aperture diameter	Omnidirectional antenna
Transmit power	2 W (33 dBm)	200 mW (23 dBm)
Carrier frequency	Ka band (20 GHz DL and 30 GHz UL)	S band (2 GHz)
Channel bandwidth	400 MHz DL, 400 MHz UL	30 MHz DL, 30 MHz UL
Polarization	Circular	Cross-pol
Antenna gain	Tx antenna gain: 43.2 dBi, Rx antenna gain: 39.7 dBi	0 dBi
Noise figure	1.2 dB	7 dB

6.7.1 3GPP Link Budget Analysis

Table 6.14 summarizes the link budget-dictated CNR for the downlink and the uplink with certain losses per 3GPP analysis (3GPP 2023).

In Table 6.14, SC1 corresponds to a GEO satellite-based NTN, FRF = 1, Ka band, and a VSAT UE. SC4 corresponds to a GEO satellite-based NTN, FRF = 1, S band, and a smartphone UE. SC6 corresponds to an LEO satellite-based NTN (at the altitude of 600 km), FRF = 1, Ka band, and a VSAT UE. SC9 corresponds to an LEO satellite-based NTN (at the altitude of 600 km), FRF = 1, S band, and a smartphone UE.

Furthermore, in Table 6.14, FSPL is free space path loss, AL is atmospheric loss, SFM is shadow fading margin, and SL is scintillation loss.

Additionally, in Table 6.14, (G/T) is the ratio of the antenna power gain-to-system noise temperature (ITU 2000). The (G/T) ratio is usually expressed in dB/K, where K is temperature in Kelvin. The system noise temperature affects the overall noise power in the system. The (G/T) ratio is an important configuration parameter in a satellite system. The (G/T) ratio indicates the ability of the antenna to retrieve a desired signal relative to the noise. A higher G/T value implies better signal reception performance.

Table 6.14 CNR for Selected 3GPP Study Cases for an NTN (3GPP 2023).

Study case	Link	Frequency (GHz)	EIRP (dBm)	(G/T) (dB/K)	Bandwidth (MHz)	FSPL (dB)	AL (dB)	SFM (dB)	SL (dB)	CNR (dB)
SC1	DL	20.0	96.0	15.9	400	210.6	1.2	0.0	1.1	11.6
SC1	UL	30.0	76.2	28.0	400	214.1	1.1	0.0	1.1	0.5
SC4	DL	2.0	103.8	−31.6	30	190.6	0.2	3.0	2.2	0.0
SC4	UL	2.0	23.0	19.0	0.4	190.6	0.2	3.0	2.2	−10.9
SC6	DL	20.0	60.0	15.9	400	179.1	0.5	0.0	0.3	8.5
SC6	UL	30.0	76.2	13.0	400	182.6	0.5	0.0	0.3	18.4
SC9	DL	2.0	76.8	−31.6	30	159.1	0.1	3.0	2.2	6.6
SC9	UL	2.0	23.0	1.1	0.4	159.1	0.1	3.0	2.2	2.8

As seen from Table 6.14, FSPL in the Ka band is about 20 to 24 dB larger than FSPL in the S band due to the Ka band frequency being 10 or 15 times greater than the S band frequency. Larger channel bandwidths are feasible in the Ka band compared to the S band due to the higher carrier frequency of the Ka band. Note that (G/T) ratios are much better (i.e., higher) for the cases of satellite receivers and VSAT UE receivers due to large receive antenna gains. In contrast, the (G/T) ratio is low for the case of the smartphone UE receiver due to low receive antenna gain.

Here is the calculation of downlink CNR for SC1. CNR can be calculated as

$$\text{CNR} = \text{EIRP} - \text{Losses} + (G/T) - k - B, \tag{6.12}$$

where CNR is in dB, EIRP is in dBm, losses are the total losses including propagation path losses and all types of margins including shadow fading margin (in dB), (G/T) is in dB/K, k is Boltzmann's constant in dBm/K/Hz, and B is the noise bandwidth in dB-Hz.

In Eq. (6.12), k is (−198.6 dBm/K/Hz). k represents Boltzmann's constant (=1.38 × 10–23 J/K), which is equivalent to −228.6 dBW/K/Hz or (−198.6 dBm/K/Hz.

In Eq. (6.12), B is 10 × log10(400,000,000) = 86.02 dB-Hz.

Hence, downlink CNR for SC1 is

$$\begin{aligned}\text{CNR(dB)} &= 96\,\text{dBm} - (210.6\,\text{dB} + 1.2\,\text{dB} + 1.1\,\text{dB}) \\ &\quad + (15.9\,\text{dB/K}) + 198.6\,\text{dBm/K/Hz} - 86.02\,\text{dB} - \text{Hz},\end{aligned} \tag{6.13}$$

$$\text{CNR} = 96 - 212.9 + 15.9 + 198.6 - 86.02 = 11.58\,\text{dB}. \tag{6.14}$$

Note that the link budget components specified in Table 6.9 can be used in conjunction with Table 6.14 to determine any additional losses and margins and calculate an updated CNR value and the CINR value. When the interference margin is defined to be the rise-over-thermal noise, CINR can be calculated as CNR minus interference margin.

6.7.2 3GPP UE Throughput Analysis

Table 6.15 summarizes 5 percentile, 50 percentile, and 95 percentile UE throughput (in Mbps) for four study cases (3GPP 2023).

In Table 6.14, SC1 corresponds to a GEO satellite-based NTN, FRF = 1, Ka band, and a VSAT UE. SC2 corresponds to a GEO satellite-based NTN, FRF = 3, Ka band, and a VSAT UE. SC6 corresponds to an LEO satellite-based NTN (at the altitude of 600 km), FRF = 1, Ka band, and a VSAT UE. SC9 corresponds to an LEO satellite-based NTN (at the altitude of 600 km), FRF = 1, S band, and a smartphone UE. Table 6.15 assumes that there are 10 UEs per NTN cell. Table 6.15 further assumes that proportional fair scheduling is used. See (3GPP 2023) for additional details on simulation assumptions.

For example, the median (i.e., 50 percentile) UE throughput is 2.7 Mbps, 2.05 Mbps, 4.08 Mbps, and 0.31 Mbps for SC1, SC2, SC6, and SC9, respectively, when resource utilization (RU) is 20%. RU = 100% means that all physical resource blocks (PRBs) in the channel bandwidth are used.

In Table 6.15, 5 percentile UE throughput can be viewed as the 5% worst throughput, because 5% of the UE throughout values are below the 5% value. Similarly, 95 percentile throughput can be viewed as 5% best throughput, because (100% − 95% = 5%) values of the UE throughput are greater than the 95% value.

Table 6.15 shows that the UE throughput for a VSAT UE is significantly larger than that for a smartphone UE due to much higher antenna gain of a VSAT UE compared to a smartphone UE.

Table 6.15 UE Throughput for Selected 3GPP Study Cases for an NTN (3GPP 2023).

Study case	Percentile	UE Throughput for Four Study Cases (Mbps)			
		SC1	SC2	SC6	SC9
RU = 20%	5%	0.96	0.48	2.04	0.11
RU = 20%	50%	2.7	2.05	4.08	0.31
RU = 20%	95%	4.93	3.9	6.43	0.52
RU = 50%	5%	5.34	4.03	6.91	0.44
RU = 50%	50%	8.53	6.61	10.03	0.74
RU = 50%	95%	11.99	9.99	13.7	1.09

6.8 Major Takeaways

Radio frequency (RF) planning and design (RPD) aims to meet coverage, capacity, throughput, and quality of service (QoS) requirements in a wireless network. As the non-terrestrial network (NTN) possesses quite distinct characteristics compared to the TN, RPD for an NTN differs from RPD for a TN. Here are the major takeaways from this chapter.

- The RPD framework needs to consider various inputs, including radio access technology, target service requirements, available NTN configurations, NTN spectrum, and device types.
- Key outputs of the RPD framework are the link budget, capacity planning, and configurations.
- While spectrum below 6 GHz (e.g., the L band between 1 and 2 GHz and the S band between 2 and 4 GHz) is initially specified by the 3GPP, spectrum above 6 GHz (e.g., the K band from 18 GHz to 26 GHz and the Ka band from 26 GHz to 40 GHz) is expected to be supported as well. Examples of 3GPP-supported NR frequency bands include bands n1, n254, n255, and n256. Examples of 3GPP-supported E-UTRA frequency bands include bands 253, 254, 255, and 256.
- The NTN supports various devices such as smartphones, VSATs, and IoT devices.
- The propagation path between the device and the NTN platform may be LOS or NLOS. Hence, path loss models are needed for both cases, and a model is needed to determine the probability of LOS propagation.
- The NLOS propagation path loss is determined by considering the LOS propagation path loss as the baseline and adding a clutter loss (CL).
- Shadow fading captures long-term variations in the signal strength. If the device is indoor, building penetration needs to be considered.
- For higher frequencies, atmospheric absorption and rain and cloud attenuation must be considered.
- Ionospheric and tropospheric scintillation can affect the signals that pass through the ionosphere and the troposphere.
- Small-scale fading effects that reflect the influence of Doppler shifts and multipath signals are important for physical layer analysis.
- Link budget is an RF accounting mechanism that considers (i) relevant characteristics of the transmitter and the receiver, (ii) the RPD performance goals such as target service QoS (e.g., error rate), cell-edge data rate, and cell-edge reliability, and (iii) the RF propagation environment and

determines the Maximum Allowable Path Loss (MAPL) (in dB) between the transmitter and the receiver.

- The MAPL helps meet the target RPD performance goals. In an NTN, the MAPL does not help determine the radius of an NTN cell. However, the MAPL can be used to determine certain NTN design parameters (e.g., maximum power amplifier transmit power and antenna gains) to maintain the service link between the device and the NTN platform in its orbit.
- Approaches to estimating capacity/throughput include an analytical approach, a simulation approach, and an RPD tool approach.
- An NTN has distinct features that make capacity planning much more complex than a TN. These unique NTN features include large cells, cell mobility, variable cell size, variable device-to-NTN payload distance, and unique interfaces.

References

3GPP (2020). *Study on New Radio (NR) to support non-terrestrial networks*. TR 38.811. https://portal.3gpp.org/desktopmodules/Specifications/SpecificationDetails.aspx?specificationId=3234 (accessed 5 September 2024).

3GPP (2022a). *Ka-band satellite NTN band definition*. R4-2219076. https://www.3gpp.org/ftp/tsg_ran/WG4_Radio/TSGR4_105/Docs (accessed 5 September 2024).

3GPP (2022b). *WF for above 10 GHz band definition and system parameters*. R4-2220239. https://www.3gpp.org/ftp/tsg_ran/WG4_Radio/TSGR4_105/Docs (accessed 5 September 2024).

3GPP (2023). *Solutions for NR to support non-terrestrial networks (NTN)*. TR 38.821. https://portal.3gpp.org/desktopmodules/Specifications/SpecificationDetails.aspx?specificationId=3525 (accessed 5 September 2024).

3GPP (2024a). *User Equipment (UE) radio transmission and reception for satellite access*. TS36.102. https://portal.3gpp.org/desktopmodules/Specifications/SpecificationDetails.aspx?specificationId=4063 (accessed 5 September 2024).

3GPP (2024b). *Satellite Access Node radio transmission and reception*. TS 36.108. https://portal.3gpp.org/desktopmodules/Specifications/SpecificationDetails.aspx?specificationId=4064 (accessed 5 September 2024).

3GPP (2024c). *User Equipment (UE) radio transmission and reception; Part 5: Satellite access Radio Frequency (RF) and performance requirements*. TS 38.101-5. https://portal.3gpp.org/desktopmodules/Specifications/SpecificationDetails.aspx?specificationId=3982 (accessed 5 September 2024).

3GPP (2024d). *Satellite Access Node radio transmission and reception*. TS 38.108. https://portal.3gpp.org/desktopmodules/Specifications/SpecificationDetails.aspx?specificationId=3934 (accessed 5 September 2024).

The European Space Agency (ESA) (2024). *Satellite frequency bands*. https://www.esa.int/Applications/Connectivity_and_Secure_Communications/Satellite_frequency_bands (accessed 5 September 2024).

ITU (2000). *Determination of the G/T Ratio for Earth Stations Operating in the Fixed-Satellite Service*. Recommendation ITU-R S.733-2. https://www.itu.int/dms_pubrec/itu-r/rec/s/R-REC-S.733-2-200001-I!!PDF-E.pdf (accessed 5 September 2024).

ITU (2017a). *Propagation data and prediction methods required for the design of Earth-space telecommunication systems*. Recommendation ITU-R P.618-14. https://www.itu.int/rec/R-REC-P.618-14-202308-I/en (accessed 5 September 2024).

ITU (2017b). *Propagation data required for the design of earth-space land mobile telecommunication systems*. Recommendation ITU-R P.681-10. https://www.itu.int/rec/R-REC-P.681-10-201712-S (accessed 5 September 2024).

ITU (2022). *Attenuation by atmospheric gases and related effects*. Recommendation ITU-R P.676. https://www.itu.int/rec/R-REC-P.676 (accessed 5 September 2024).

ITU (2023a). *Prediction of building entry loss*. Recommendation ITU-R P.2109-2. https://www.itu.int/rec/R-REC-P.2109-2-202308-I/en (accessed 5 September 2024).

ITU (2023b). *Compilation of measurement data related to building entry loss*. Recommendation ITU-R P.2346-5. https://www.itu.int/pub/R-REP-P.2346 (accessed 5 September 2024).

Tripathi, N.D. and Reed, J.H. (2014). *Cellular Communications - A Comprehensive and Practical Guide*. Wiley. https://www.wiley.com/en-kr/Cellular+Communications%3A+A+Comprehensive+and+Practical+Guide-p-9780470472071 (accessed 5 September 2024).

Tripathi, N.D. and Reed, J.H. (2019). *5G Cellular Communications - Journey and Destination*. The Wireless University. https://thewirelessuniversity.com/ (accessed 5 September 2024).

ViaSatellite (2023). *Cover Story: Rain Degradation: Its Implications For Satellite Communication*. https://www.satellitetoday.com/connectivity/2003/02/01/cover-story-rain-degradation-its-implications-for-satellite-communication/ (accessed 5 September 2024).

7

Pre-Data Transfer Operations in an NTN

Before an NTN UE can exchange data with the radio and core network of the NTN, certain pre-data transfer operations occur between the NTN UE and the NTN radio and core networks. These operations include cell search, network acquisition, random access, RRC connection setup, registration or attach, and PDU session or EPS bearer setup. These operations are explained in the context of the NTN, and key enhancements for the NTN are highlighted that make the NTN possible. These operations are discussed for three radio access technologies: NR, LTE-M, and NB-IoT. Once these operations are successfully completed, the NTN UE is ready for the really exciting phase of data transfer!

7.1 Overview of Pre-Data Transfer NTN Operations

Recall from Section 3.1 in Chapter 3 that 5G supports three usage scenarios envisioned by the International Telecommunication Union (ITU): enhanced Mobile Broadband (eMBB), Ultra-Reliable Low-Latency Communications (URLLC), and massive Machine Type Communications (mMTC). The eMBB usage scenario is characterized by high data rates. The URLLC usage scenario focuses on both high reliability and low latency. The mIoT usage scenario is characterized by support for many delay-tolerant IoT devices requiring only low data rates. The NR-NTN is best suited for eMBB, while IoT-NTN is best suited for mMTC. The NR-NTN utilizes the 5G NR air interface between the NTN UE and the gNB, whereas the IoT-NTN utilizes an LTE-M or NB-IoT air interface.

At a high level, certain operations between the UE and the network depend on the radio access technology, while others are generic. The exact process by which UE-network operations occur depends on the radio access technology (e.g., LTE-M, NB-IoT, or NR) and the core network (e.g., EPC or 5GC). Figure 7.1 summarizes key pre-data transfer operations that occur after UE power-up. 3GPP **NTN-specific enhancements supporting these operations are discussed in detail in Sections 7.2 and 7.3 for NR-NTN and IoT-NTN using LTE-M and NB-IoT**.

7.1.1 Step 1: Network and System Information Acquisition

The NTN UE, upon power-up, detects a cell by observing Primary Synchronization Signal (PSS) and Secondary Synchronization Signal (SSS) and determines the Physical Cell ID (PCI) of the cell using both signals (3GPP 2024s). An NR-NTN UE, an IoT-NN UE using LTE-M, or an NB-IoT has different PSS and SSS structures. The NTN UE then processes the Physical Broadcast Channel

3GPP-based Non-Terrestrial Networks in 5G and 6G: Expanding the Frontiers of Wireless Communications, First Edition.
Nishith D. Tripathi and Jeffrey H. Reed.

Companion website: www.wiley.com/go/tripathi5g

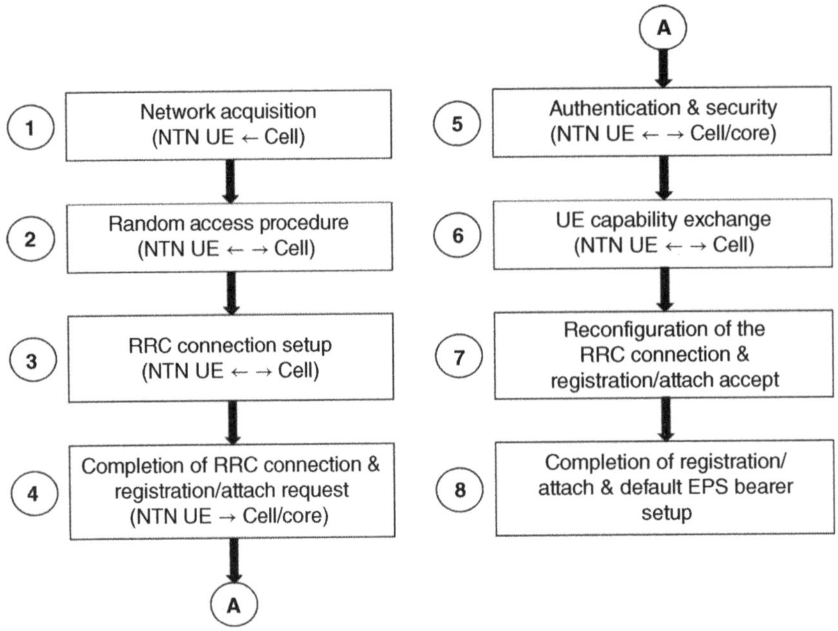

Figure 7.1 Pre-Data Transfer Operations upon NTN UE Power-Up in an NTN.

(PBCH) to obtain the Master Information Block (MIB) containing critical information about the system (3GPP 2024v). The NR-NTN UE observes the spectrum equivalent to twenty Physical Resource Blocks (PRBs) or 240 subcarriers in the channel bandwidth used by the NTN cell to obtain the MIB. The IoT-NTN UE using LTE-M observes the spectrum equivalent to 6 PRBs or 1.08 MHz with the subcarrier spacing of 15 kHz to obtain the MIB. The IoT-NTN UE observes the spectrum equivalent to 1 PRB or 180 kHz with the subcarrier spacing of 15 kHz to obtain the MIB-Narrowband (MIB-NB).

After processing the MIB, the NTN UE acquires information about the system by observing one or more System Information Blocks (SIBs). For example, the IoT-NTN UE using LTE-M obtains SIB1-Bandwidth reduced (SIB1-BR), the IoT-NTN UE using NB-IoT obtains SIB1-Narrowband (SIB1-NB), and the NR-NTN UE obtains SIB1.

SIB1 provides random-access parameters to an NR-NTN UE. However, an IoT-NTN UE using LTE-M must obtain SIB2, and an IoT-NTN UE using NB-IoT must obtain SIB2-Narrowband (SIB2-NB) to learn about the random-access parameters.

The NTN UE acquires new NTN-specific SIBs in addition to typical SIBs (3GPP 2024v). These SIBs are discussed in Sections 7.2.2 and 7.3.2, respectively.

7.1.2 Step 2: Random-Access Procedure

The random-access procedure has been enhanced for the NTN to accommodate much longer propagation delays compared to a TN. The NTN UE sends a Random-Access (RA) preamble to the cell and receives a Random-Access Response (RAR). Using the UL grant specified by the RAR, the NTN UE initiates the establishment of a Radio Resource Control (RRC) signaling connection. The RA procedures for NR-NTN and IoT-NTN are discussed in Sections 7.2.3 and 7.3.3, respectively.

7.1.3 Step 3: RRC Connection Setup

The RRC connection is a dedicated signaling connection on the radio interface between the UE and the serving cell (3GPP 2024v). The NTN UE sends an RRC message to the cell toward the end of the RA procedure to request the setup of the RRC connection. The IoT-NTN UE may also provide some of its IoT-related capabilities to the network. The NTN cell responds with an RRC message to set up the RRC connection and provides configuration of the radio interface from the serving cell. The RRC connection setup procedures for NR-NTN and IoT-NTN are discussed in Sections 7.2.4 and 7.3.4, respectively.

7.1.4 Step 4: Completion of the RRC Connection Setup and Registration/Attach Request

The NTN UE completes the RRC connection setup by sending an RRC message in response to the RRC setup-related message received from the NTN cell in Step 3. In this RRC message, the NTN UE also initiates connectivity with the core network by sending a Non-Access Stratum (NAS) message in the last RRC message of the RRC connection setup procedure. An IoT-NTN UE sends an Attach Request to the EPC via the cell/eNB, and NR-NTN UE sends a Registration Request to the 5GC via the cell/gNB.

7.1.5 Step 5: Mutual Authentication and NAS Security

No NTN-specific security changes have been made in Release 17 and Release 18. The IoT-NTN UE and the EPC/MME exchange authentication-related NAS signaling messages to execute mutual authentication. At the end of successful authentication, NAS security is established between the IoT-NTN UE and the MME. For NR-NTN, the NR-NTN UE and the UASF exchange authentication-related NAS signaling messages via the AMF to execute mutual authentication. At the end of successful authentication, NAS security is established between the NR-NTN UE and the AMF.

7.1.6 Step 6: UE Capability Transfer

The eNB/gNB NR cell requests the NTN UE to provide its radio capabilities by sending an RRC message called UE Capability Enquiry. The NTN UE conveys its radio access capabilities to the NTN cell (managed by the eNB/gNB) by replying with an RRC message called UE Capability Information. The UE capability transfer procedures for NR-NTN and IoT-NTN are discussed in Sections 7.2.6 and 7.3.6, respectively.

7.1.7 Step 7: Reconfiguration of the RRC Connection and Registration/Attach Accept

An IoT-NTN cell sends an RRC message that encapsulates the NAS Attach Accept message received from the MME. Similarly, an NR-NTN cell sends an RRC message that encapsulates the NAS Registration Accept message received from the AMF. The NAS Attach/Registration Accept message specifies the core network assigned UE identity and other core network-related parameters, such as Discontinuous Reception (DRX) cycle parameters and tracking area update/registration area update-related timers. In the case of IoT-NTN, the Attach Accept message contains a NAS message to set up a default EPS bearer with specific QoS parameters for the UE.

7.1.8 Step 8: Completion of Registration/Attach and Default EPS Bearer Setup

The NTN UE replies to the Attach/Registration Accept message with the Attach/Registration Complete message. The IoT-NTN UE also includes a NAS message that conveys acceptance of the EPS bearer setup.

After the successful Attach procedure and the associated default EPS bearer setup, the IoT-NTN UE can exchange traffic with the NTN. Additional EPS bearers can also be set up to facilitate data transfer.

In the case of NR-NTN, one or more PDU sessions are established after successful registration, summarized in Figure 7.1. The PDU session enables the NR-NTN UE to exchange traffic with the NTN.

7.2 Pre-Data Transfer Operations in NR-NTN: A Closer Look

Section 7.2.1 provides an overview of key pre-data transfer operational enhancements in NR-NTN, and Section 7.2.2 provides details of these enhancements.

7.2.1 Operational Enhancements in NR-NTN: An Overview

Figure 7.2 highlights key enhancements in operations made by the 3GPP to support NR-NTN.

NR-NTN reuses the overall frameworks of these operations and makes enhancements related to the NTN.

Cell Selection and SI Acquisition. Several new IEs have been introduced in the SI, such as (i) cellBarredNTN, to convey accessibility to the 5GC via the NTN access and (ii) trackingAreaList to support the broadcast of multiple TAs per NTN cell. Two new NTN-specific SIBs, SIB19 and SIB25, have been introduced in Release 17 and 18 (3GPP 2024v).

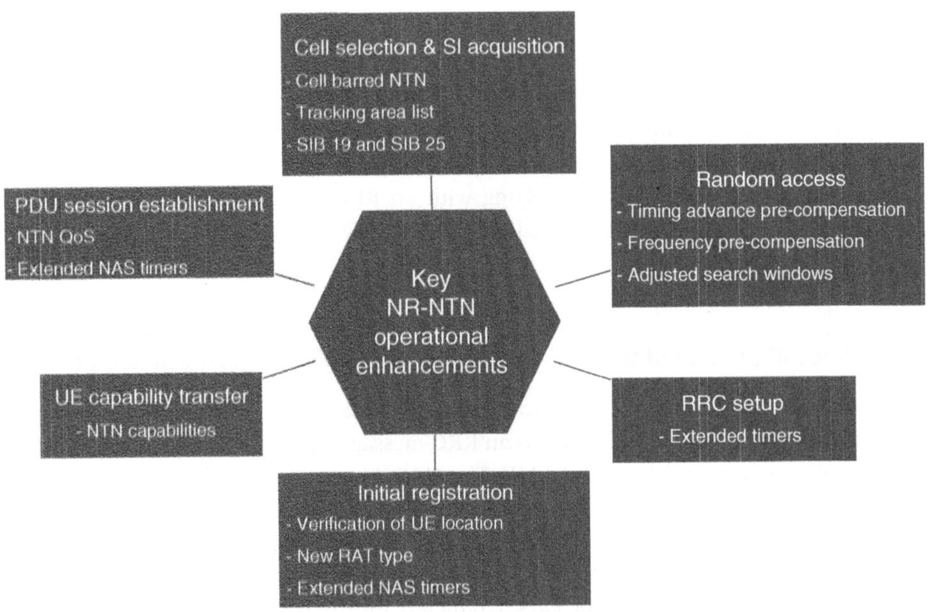

Figure 7.2 Key NR-NTN Enhancements in Pre-Data Transfer Operations.

Random-Access Procedure. To accommodate long delays and high Doppler shifts, the NTN UE carries out Timing Advance and frequency pre-compensation while transmitting during the random-access procedure. The NTN UE may also support TA reporting during the random-access procedure.

RRC Connection Setup. During the establishment of the RRC connection between the NTN UE and the NTN cell, the gNB can configure the NTN UE at suitable extended times to address long propagation delays.

Initial Registration. The AMF may verify the NTN UE's location to ensure that the NTN UE can access the NTN-related PLMN at its current location. Extended NAS timers can be configured. The NTN-specific RAT type is used. Discontinuous coverage-related parameters can be configured for the NTN UE.

UE Capability Transfer. The NTN UE specifies its typical and NTN-specific capabilities to the NTN gNB.

PDU Session Establishment. After registration, the NTN can initiate the establishment of the PDU session. If applicable, the NTN-specific QoS is used for the PDU session. Extended NAS timers, as appropriate, are also utilized.

7.2.2 Cell Selection and SI Acquisition in NR-NTN

A UE may only support an NTN or may support both a TN and an NTN. For example, a certain VSAT (e.g., in a rural area without any TN coverage) may only support connectivity to an NTN. In contrast, a smartphone may support connectivity to both.[1] When an NTN UE powers up, it searches for an NTN cell, selects a cell, and acquires System Information (SI) as shown in Figure 7.3.

7.2.2.1 NTN Cell Search and Cell Selection

During NR-NTN cell search, the NR-NTN UE utilizes Primary Synchronization Signal (PSS), the Secondary Synchronization Signal (SSS), and the Physical Broadcast Channel (PBCH) Demodulation Reference Signal (DMRS) to acquire time and frequency synchronization with the NR-NTN cell and detects the Physical layer Cell ID (PCI) of the NR-NTN cell (3GPP 2024s).

Figure 7.3 Cell Selection and SI Acquisition in an NTN.

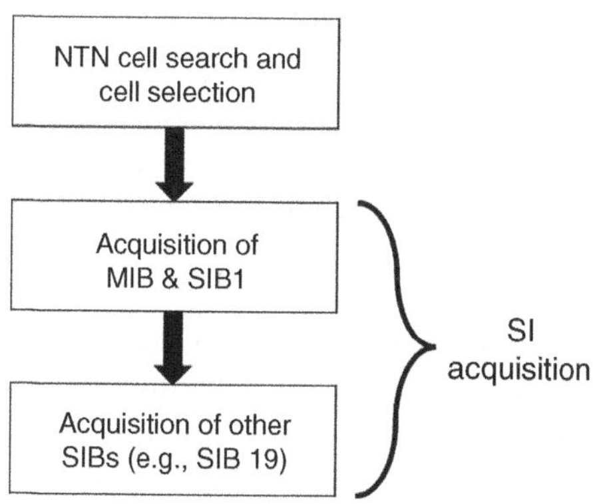

1 In Release 17 and Release 18, a UE connects to a TN or an NTN for communications. Simultaneous connectivity of a UE with both a TN and an NTN may be supported in future 3GPP releases.

The same cell selection process is used for a TN and an NTN. When an NTN UE is switched on, a Public Land Mobile Network (PLMN)[2] is selected by the Non-Access Stratum (NAS) (3GPP 2024t). Suitable UE policies may define the association between the selected PLMN and associated RAT(s) (3GPP 2024a). The Access Stratum (AS) in the UE uses the list of equivalent PLMNs for cell selection and reselection.

Cell selection may be *initial cell selection* or *cell selection by leveraging stored information*. In the case of initial selection, the UE does not know which Radio Frequency (RF) channels are NR frequencies. In such a case, the NTN UE scans all RF channels in the NR-NTN bands. On a given RF channel or frequency, the NTN UE searches for the strongest NTN cell and selects a suitable cell. A suitable cell is the cell that meets the cell selection criterion "S" (3GPP 2024t). In the case of cell selection by leveraging stored information, the UE utilizes stored information of frequencies and detected cells (and potentially cell parameters).

To evaluate whether the currently monitored NTN cell meets the cell selection criterion, the NTN UE needs to make measurements such as Reference Signal Received Power (RSRP) and Reference Signal Received Quality (RSRQ) of a cell. Hence, the NTN UE searches for the PSS and the SSS. The NTN UE determines the Physical Cell Identifier (PCI) of the NTN cell based on the combination of the PSS and the SSS. The PSS and the SSS are part of the time-frequency structure called the Synchronization Signal/Physical Broadcast Channel (SS/PBCH) Block. The NTN UE uses the SSB resources to estimate RSRP and RSRQ and compares measured RSRP and RSRQ to evaluate the cell selection criterion S.

The cell selection criterion S is said to be fulfilled when ((Srxlev >0) AND (Squal >0)), where Srxlev is the difference between the measured RSRP and a signal threshold (which is Q_{rxlevmin} potentially adjusted by additional parameters), and Squal is the difference between the measured RSRQ and a signal quality threshold (which is Q_{qualmin} potentially adjusted by additional parameters). In essence, if the measured RSRP and RSRQ exceed relevant thresholds, the cell is considered suitable. See 3GPP (2024t) for details of the formula and associated parameters.

In general, an NR-NTN UE at a given location can observe the SS/PBCH Block (SSB) at least once in a 20 ms period. The gNB transmits a set of provisioned SSB beams within a 5 ms half-frame in a given NTN cell identified by the PCI. One or more half-frames (e.g., four half-frames) may be used in a 20 ms period by the gNB to transmit all the planned beams in the 5G NR cell. In one example scenario, one beam created by the NTN payload may correspond to one SSB and one PCI. In another deployment configuration, multiple beams may correspond to different SSB beams, and a set of these SSB beams would belong to one NTN cell associated with one PCI.

After finding a suitable NTN cell that meets the cell selection criterion S, the NTN UE carries out the *SI acquisition*. The SI consists of the Master Information Block (MIB), System Information Blocks (SIBs), and positioning SIBs (posSIBs). Depending on the NTN UE capabilities and the network design, only certain SIBs and posSIBs are transmitted by the NTN gNB and processed by the NTN UE.

7.2.2.2 Acquisition of MIB and SIB1

After performing cell selection, the NR-NTN UE processes the PBCH to obtain the MIB and additional PBCH payload (which is in addition to the MIB) containing critical information about the system such as the System Frame Number (SFN) and cell barring information (3GPP 2024v). The MIB also specifies the information, such as the SCS and the PDCCH configuration, to enable the UE to retrieve SIB1. The MIB is sent with a periodicity of 80 ms, and repetitions are made within this

2 A private network may also be selected. Such a network is called a Standalone Non-Public Network (SNPN).

80 ms period. The NTN UE ignores the cell barring information of the MIB; there is a separate NTN cell barring information in SIB1.

An NR-NTN UE acquires SIB1 to determine whether the cell it has been observing is an NTN cell (3GPP 2024v). Note that the NR-NTN UE knows where to look for SIB1 after processing the MIB. Specifically, the NR-NTN UE processes PDCCHs and obtains SIB1 from the PDSCH that utilizes the radio resources specified by the associated PDCCH; such PDCCH has its CRC[3] masked with System Information-Radio Network Temporary Identifier (SI-RNTI). SIB1 contains information used by the UE to determine if it can access this cell based on the list of prohibited PLMNs, NTN cell availability, and signal strength and quality-based cell selection criterion S mentioned above.

The NR-NTN utilizes its location capability to select a PLMN that can operate in the current NR-NTN UE location (3GPP 2024a, 2024d). Specifically, the NR-NTN UE checks if the PLMN in "satellite NG-RAN access technology" is part of the list of "PLMNs not allowed to operate at the present UE location." The NR-NTN UE can develop a list of prohibited PLMNs if it has received such an indication from the 5GC in past NAS signaling exchange(s). This is, for instance, used to make sure that reception in an authorized country is allowed.

SIB1 also includes radio resource configuration information applicable to all UEs. Furthermore, it contains numerous parameters, including scheduling information about other SIBs and random-access parameters. SIB1 is transmitted with a periodicity of 160 ms, and repetitions are made within such a period.

SIB1 contains an NTN-specific parameter called *cellBarredNTN*. If *cellBarredNTN* is set to "barred," the NR-NTN UE considers that the cell is not allowed to have connectivity to NTN (i.e., barred). Furthermore, if the *cellBarredNTN* is absent in SIB1, the NTN UE also considers the cell to be barred. When the cell currently being monitored is barred, the NR-NTN UE performs cell reselection to another cell at the same frequency as the barred cell (3GPP 2024t). If *cellBarredNTN* is set to "notBarred," the cell being monitored can connect to NTN.

Another SIB1 enhancement for the NTN is that the field "trackingAreaCode" is absent for an NTN cell. Furthermore, the field "trackingAreaList" in SIB1 plays an important role; it contains a list of Tracking Areas (TAs) associated with the cell indicated by cell Identity. The total number of different TACs across different PLMN-IdentityInfo fields cannot exceed maxTAC.

Note that the network may broadcast multiple Tracking Area Codes (TACs) per PLMN in an NR NTN cell in an NTN. The network changes the TACs in the SI in an implementation-specific manner; the TACs being broadcast by an NR-NTN cell may not exactly reflect the ground area illuminated by the NR-NTN cell's beam(s) at a given instant, because the beams illuminated by the NTN payload have a reasonably regular geometric shape (e.g., elliptical), but the geographic area that is designated as the TAC during the network planning process may have an irregular geometric shape or the beam may be sweeping a geographic area continuously.

In support of enhanced cell reselection for LEO-based NTN, the Information Element (IE)[4] *ServingCellConfigCommonSIB* IE in SIB1 may contain the IE enhanced Measurement LEO. If enhanced Measurement LEO is present and if the NR-NTN UE supports the enhanced cell reselection requirements for NTN LEO in RRC_IDLE and RRC_INACTIVE states, the NR-NTN UE applies the enhanced cell reselection requirements for the NTN LEO-based NTN specified in (3GPP 2024p).

3 CRC stands for Cyclic Redundancy Check and is used by the receiver to evaluate if the received information has any error or not.

4 The terms "field" and "Information Element (IE)" are used interchangeably in this book and 3GPP specifications. Both refer to an element or part of a signaling message.

Table 7.1 System Information Enhancements for the NR-NTN.

System Information (SI)	NR-NTN-Related Changes
MIB	Cell barring information in the MIB is ignored by the NR-NTN UE.
SIB1	Contains cell Barred NTN to enable the NTN UE to determine if this is an NTN cell or not. Also, includes the field "trackingAreaList" containing a list of TAs to support multiple TAs per cell. Also, indicates support for enhanced cell reselection for LEO-based NTN SIB1 in the form of enhanced Measurement LEO.
SIB2	The absence of p-Max indicates that the NR-NTN UE can use the maximum transmit power per UE class.
SIB4	The absence of p-Max indicates that the NR-NTN UE can use the maximum transmit power per UE class in intra-frequency neighboring NR cells. The IE *smtc4list* is facilitates measurements of NTN neighbor cells by providing an SSB Measurement Timing Configuration list.
SIB9	Contains time information to enable the NR-NTN UE to determine the UTC, the GPS and the local time.
SIB19	Contains numerous parameters providing satellite assistance information for NTN access.
SIB25	Contains TN coverage information to assist neighbor cell measurements for the UEs in an NTN cell.

SIB 19 is especially important for the NTN. Indeed, SIB19 has been introduced by the 3GPP to support the NTN. Hence, in SIB1, si-BroadcastStatus[5] for the SIB19 is set to "broadcasting" for an NTN cell. SIB25 is also introduced by the 3GPP for the NR-NTN to influence monitoring of the TN by an NR-NTN UE.

7.2.2.3 Acquisition of Additional SIBs

The NR-NTN gNB and the NR-NTN UE can reuse several SIBs previously used for a TN (e.g., at least SIB2 to SIB4) and SIB1. Furthermore, NTN-specific SIB19 and SIB25 contain important NTN parameters. Table 7.1 summarizes SI enhancements for the NTN (3GPP 2024v). The MIB and SIB1 enhancements have already been discussed earlier. The remaining enhancements are briefly outlined in Table 7.1.

7.2.2.4 SIB2

SIB2 contains cell reselection information (i) common to intra-frequency, inter-frequency, and/or inter-RAT cell reselection and (ii) non-neighbor cell intra-frequency cell reselection information (3GPP 2024v). The IE p-Max in SIB2 indicates the maximum power that a UE can use in intra-frequency neighboring NR cells. If this IE is absent, the NR-NTN UE considers the maximum power as defined in (3GPP 2024o).

7.2.2.5 SIB4

SIB4 contains information relevant for inter-frequency cell reselection for the current cell inter-frequency neighbor cells (3GPP 2024v). Cell reselection parameters that are common for multiple cells associated with a given carrier frequency, as well as cell-specific reselection parameters, are

5 The IE si-BroadcastStatus indicates if the associated SI message is being broadcasted or not.

included in SIB4. As in the case of SIB2, if the IE p-Max is absent, the NR-NTN UE considers the maximum power to be as defined in 3GPP (2024o).

An NR-NTN UE may be configured to make neighbor cell measurements during specific periods through measurement gaps if the NR-NTN UE needs such gaps. Data transfer between the serving cell and the UE does not occur during such gaps. If these gaps are not chosen properly, the NR-NTN UE may be unable to detect the neighbor cells' target reference signals. Hence, to facilitate measurements of neighbor cells that may have widely different propagation delays with the NR-NTN UE, the IE *smtc4list* is introduced for the NTN, which provides an SSB Measurement Timing Configuration (SMTC) list for NTN deployments. The offset of each SSB-MTC4 in smtc4list assumes that the NTN gNB- NTN UE propagation delay difference between the serving cell and neighbor cells is 0 ms. The NR-NTN UE can adjust the offset based on the estimated propagation delay difference between the serving and neighbor cells. If an NR-NTN UE supports fewer SMTCs than the number of SMTCs provided by the IE *smtc4list* in this list, the NR-NTN UE selects a subset of SMTCs.

7.2.2.6 SIB9

SIB9 contains GPS and Coordinated Universal Time (UTC) information (3GPP 2024v). The NR-NTN UE may utilize SIB9 to determine the UTC, the GPS, and the local time. The UE can use such time information to serve different purposes, such as assisting with GPS initialization and synchronizing the NR-NTN UE clock. Specifically, the IE timeInfoUTC represents UTC corresponding to the SFN boundary at or immediately after the ending boundary of the SI-window in which SIB9 is transmitted. In an NTN cell, this indicated time is referenced at the uplink time synchronization Reference Point (RP). The NR-NTN considers the propagation delay between the NR-NTN UE and the RP to determine the UTC at the NR-NTN UE.

7.2.2.7 SIB19

As mentioned earlier, the 3GPP has introduced SIB19 specifically for the NR-NTN. SIB19 provides satellite assistance information for NTN access to NR-NTN UEs. Table 7.2 lists key IEs of SIB19 (3GPP 2024v).

NTN-Config-r17. This IE is equivalent to ntn-Config. This IE provides information such as ephemeris data to the NR-NTN UE so the NR-NTN UE can communicate with the gNB via NR-NTN access. See Table 7.3 for details of this IE.

t-Service-r17. This IE indicates the time information on when an NR-NTN will stop serving the current geographic area. This IE applies to both (i) the service link switch for NR-NTN using quasi-Earth-fixed beams and (ii) the feeder link switches for both NTN quasi-Earth fixed and Earth-moving systems. The field indicates a time in multiples of 10 ms after 00:00:00 on the Gregorian calendar date January 1, 1900 (i.e., midnight between Sunday, December 31, 1899, and Monday, January 1, 1900). The exact stop time instant is between (i) the time indicated by the value of this field minus 1 and (ii) the time indicated by the value of this field. Note that the reference point for t-Service is the uplink time synchronization reference point of the NTN cell.

referenceLocation-r17. This IE specifies the reference location of the serving NTN cell using quasi-Earth-fixed beams. It is used in location-based measurement initiation in the NR-NTN UE's RRC_IDLE and RRC_INACTIVE states (3GPP 2024t).

Table 7.2 SIB 19 in NR-NTN: A Closer Look.

Information Element (IE)	Brief Description
NTN-Config-r17	Provides information such as ephemeris data to the NR-NTN UE to facilitate access to NR-NTN (see Table 7.3 for details).
t-Service-r17	Indicates the time information on when an NR-NTN will stop serving the current geographic area.
referenceLocation-r17	Reference location of the serving NTN cell using quasi-Earth-fixed beams.
distanceThresh-r17	A threshold specifying the distance of the NTN-UE from the serving cell reference location.
ntn-NeighCellConfigList-r17	A list of up to "maxCellNTN-r17" instances of the "NTN-NeighCellConfig-r17" IE. NTN-NeighCellConfig-r17 contains ntn-Config-r17, carrierFreq-r17, and physCellId-r17. ntn-Config-r17: Provides information such as ephemeris data to the NR-NTN UE to facilitate access to NR-NTN. carrierFreq-r17: Carrier frequency of the neighbor cell. physCellId-r17: PCI of the neighbor cell.
movingReferenceLocation-r18	Reference location of the serving NR-NTN cell using Earth-moving beams at a time reference.
satSwitchWithReSync-r18	Provides parameters for the target satellite associated with satellite switch with re-synchronization. Contains ntn-Config-r18, t-ServiceStart-r18, and ssb-TimeOffset-r18. ssb-TimeOffset: Indicates the time offset between (i) the SSB from the source satellite and (ii) the SSB from the target satellite at the uplink time synchronization reference point. t-ServiceStart: Indicates the time information on when the target satellite will start serving the area currently served by the serving satellite.

distanceThresh-r17. This IE represents a threshold specifying the distance of the NTN-UE from the serving cell reference location. It is used in location-based measurement initiation in the NR-NTN UE's RRC_IDLE and RRC_INACTIVE states (3GPP 2024t). Each step represents 50 m.

ntn-NeighCellConfigList-r17. This is a list of up to maxCellNTN-r17 instances of the IE "NTN-NeighCellConfig-r17." NTN-NeighCellConfig-r17 contains ntn-Config-r17, carrierFreq-r17, and physCellId-r17. The IE ntn-Config-r17 provides information such as ephemeris data to the NR-NTN UE to facilitate access to NR-NTN. The IE carrierFreq-r17 is the carrier frequency of the neighbor cell. The IE physCellId-r17 is the Physical Cell Identifier (PCI) of the neighbor cell.

movingReferenceLocation-r18. This is the reference location of the serving NR-NTN cell using Earth-moving beams at a time reference. It is used in location-based measurement initiation in the NR-NTN UE's RRC_IDLE and RRC_INACTIVE states (3GPP 2024t). The time reference of this IE is indicated by the epochTime in ntn-Config of the serving cell.

satSwitchWithReSync-r18. This IE provides parameters for the target satellite associated with the satellite switch with re-synchronization. This IE indicates that a satellite switch without PCI change is supported in the cell. This IE contains ntn-Config-r18, t-ServiceStart-r18, and ssb-TimeOffset-r18. The IE ssb-TimeOffset indicates the time offset (expressed in subframes) between (i) the SSB from the source satellite and (ii) the SSB from the target satellite at the uplink time synchronization Reference Point. The IE t-ServiceStart indicates the time information on when the target satellite will start serving the area currently served by the serving satellite. The field indicates a time in multiples of 10 ms after 00:00:00 on the Gregorian calendar date

Table 7.3 NTN-Config in SIB 19 in NR-NTN: A Closer Look.

Information Element (IE)	Brief Description
epochTime-r17	Indicates the epoch time for the NTN assistance information such as serving and/or neighbor satellite ephemeris and Common Timing Advance parameters. Contains sfn-r17 and subFrameNR-r17. sfn-r17: Specifies the SFN. subFrameNR-r17: Specifies the subframe number.
ntn-UlSyncValidityDuration-r17	The duration for which the previously acquired network assistance information is valid.
cellSpecificKoffset-r17	This is the scheduling offset used for the timing relationships in the NR-NTN.
kmac-r17	This is the scheduling offset if downlink and uplink frame timing are not aligned at the NR-NTN gNB.
ta-Info-r17	Provides parameters related to Timing Advance (TA). Contains ta-Common-r17, ta-CommonDrift-r17, and ta-CommonDriftVariant-r17. ta-Common: Indicates network-controlled common TA value. ta-CommonDrift: Indicates the drift rate of the common TA. ta-CommonDriftVariant: Indicates the drift rate variation of the common TA.
ntn-PolarizationDL-r17	Indicates polarization of the downlink transmission on the service link, such as Right Hand Circular Polarization (RHCP), Left Hand Circular Polarization (LHCP), and linear polarization.
ntn-PolarizationUL-r17	Indicates polarization of the uplink transmission on the service link.
ephemerisInfo-r17	Specifies satellite ephemeris in the format of (i) position and velocity state vector or (ii) orbital parameters.
ta-Report-r17	This field indicates that TA reporting is enabled during the random-access procedure, RRC connection establishment or resume, and RRC connection reestablishment. When this field is included in ServingCellConfigCommon within dedicated signaling, it indicates that TA reporting is enabled during Random Access due to reconfiguration with sync.

1st January 1900 (i.e., midnight between Sunday, December 31, 1899, and Monday, January 1, 1900). The exact start time instant is between (i) the time indicated by the value of this field minus 1 and (ii) the time indicated by the value of this field.

Table 7.3 lists key IEs of the IE NTN-Config (3GPP 2024v). As mentioned earlier, NTN-Config provides the parameters needed for the UE to access NR-NTN.

epochTime-r17. This IE indicates the epoch time for the NTN assistance information, such as serving and/or neighbor satellite ephemeris and Common Timing Advance parameters, and contains sfn-r17 and subFrameNR-r17. The IE epochTime-r17 is the starting time of a DL subframe, indicated by the SFN (i.e., sfn-r17) and the subframe number (i.e., subFrameNR-r17) signaled together with the assistance information. For the serving NR-NTN cell, the IE sfn-r17 indicates the current SFN or the next upcoming SFN after the frame where the message indicating the IE epochTime-r17 is received. For a neighbor NR-NTN cell, sfn-r17 indicates the SFN nearest the frame where the message indicating the IE epochTime-r17 is received. If this IE is absent for the serving NR-NTN cell, the epoch time is the end of the SI window where this SIB19 is scheduled.

ntn-UlSyncValidityDuration-r17. This IE specifies the validity duration (in seconds) for assistance information. It indicates the maximum time duration from epochTime-r17, during which the NR-NTN UE can utilize the assistance information without acquiring new assistance information.

cellSpecificKoffset-r17. This is the scheduling offset used for the timing relationships in the NR-NTN (3GPP 2024q). It is expressed in the units of the number of slots for the subcarrier spacing of 15 kHz. If this IE is 0, the NR-NTN UE assumes its value to be 0.

kmac-r17. This is the scheduling offset when downlink and uplink frame timing are not aligned at the NR-NTN gNB. If this IE is absent, the NR-NTN UE assumes the IE value to be 0. In FR1, this IE is expressed in the unit of the number of slots for the subcarrier spacing of 15 kHz.

ta-Info-r17. This IE provides Timing Advance (TA) parameters and contains ta-Common-r17, ta-CommonDrift-r17, and ta-CommonDriftVariant-r17. The IE ta-Common indicates network-controlled common TA value and has the granularity of 4.072×10^{-3} µs. The IE ta-CommonDrift indicates the drift rate of the common TA and has a granularity of 0.2×10^{-3} µs/s. The IE ta-CommonDriftVariant indicates the drift rate variation of the common TA and has a granularity of $0.2 \times 10\text{^}^{-4}$ µs/s^2.

ntn-PolarizationDL-r17. This IE indicates the polarization of the DL transmission on the service link, such as Right-Hand Circular Polarization (RHCP), Left-Hand Circular Polarization (LHCP), and linear polarization.

ntn-PolarizationUL-r17. This IE indicates polarization of the UL transmission on the service link. If this IE is absent but the IE ntn-PolarizationDL is present, the NR-NTN UE assumes the same polarization for the UL and the DL.

ephemerisInfo-r17. This IE specifies satellite ephemeris in the format of (i) position and velocity state vector or (ii) orbital parameters.

ta-Report-r17. This IE indicates that the reporting of TA is enabled during (i) the random-access procedure executed due to RRC connection establishment or RRC connection resume and (ii) RRC connection re-establishment.

7.2.2.8 SIB25

The 3GPP introduced SIB 25 in Release 18 to support NTN-TN mobility. The network broadcasts suitable information in NR TN and EUTRA TN coverage areas in SIB25 for Earth-Fixed, Quasi-Earth-fixed, and Earth-Moving cells. SIB25 contains information related to the TN coverage to assist the NR-NTN UEs with TN neighbor cell measurements. The coverage information essentially consists of a list of geographical TN areas. The NR-NTN UE can skip TN measurements based on the broadcast TN coverage information. Table 7.4 lists the key IEs of SIB25 (3GPP 2024v).

Table 7.4 SIB 25 in NR-NTN: A Closer Look.

Information Element (IE)	Brief Description
coverageAreaInfoList-r18	This IE lists up to maxTN-AreaInfo-r18 instances of the IE "CoverageAreaInfo-r18." The NR-NTN UE can use this IE in the RRC_IDLE or RRC_INACTIVE state to skip TN measurements. Contains tn-AreaId-r18, tn-ReferenceLocation-r18, and tn-DistanceRadius-r18. tn-AreaId-r18: Area identity in the TN coverage area. tn-ReferenceLocation-r18: TN coverage area reference location. tn-DistanceRadius: Distance from the TN coverage area reference location.

coverageAreaInfoList-r18. This IE lists up to maxTN-AreaInfo-r18 instances of the IE "CoverageAreaInfo-r18." The NR-NTN UE can use this IE in the RRC_IDLE or RRC_INACTIVE state to skip TN measurements (3GPP 2024t), saving battery power and reducing processing requirements. This IE contains tn-AreaId-r18, tn-ReferenceLocation-r18, and tn-DistanceRadius-r18. The IE tn-AreaId-r18 is the area identity in the TN coverage area. The IE tn-ReferenceLocation-r18 is the TN coverage area reference location. Finally, the IE tn-DistanceRadius is the distance of the NR-NTN UE from the TN coverage area reference location.

7.2.2.9 Beyond the SI Acquisition

After acquiring the MIB, SIB1, and other SIBs, such as SIB19, the NTN UE tries to register with the network. The NTN UE first contacts the radio network by executing the random access and RRC connection setup procedures. Once the RRC connection is set up, it exchanges registration-related signaling with the 5GC.

The SI acquisition procedure explained above is executed in various situations: cell selection (e.g., after power on), cell reselection, return from out of coverage, completion of reconfiguration with sync, entering the network from another RAT, receiving an indication that the system information has changed, receiving a Public Warning System (PWS) notification, receiving the request (e.g., a positioning request) from upper layers, and whenever the UE does not have a valid version of a stored SIB or posSIB (3GPP 2024v).

7.2.3 Random-Access Enhancements in NR-NTN

As mentioned in Section 7.2.1, before initiating the RA procedure, the NR-NTN UE completes network acquisition by synchronizing with the 5G NR cell using the PSS and the SSS and acquiring the relevant SI relevant to the NR-NTN. The overall RA procedure must be enhanced in NR-NTN due to NTN-specific challenges such as long and variable propagation delays and high NTN cell mobility.

Figure 7.4 illustrates timing relationship parameters such as service link Round Trip Time (RTT), Feeder link RTT, Common TA, k_{mac}, and T_{TA} for an NR-NTN using a transparent NTN payload (3GPP 2024s).

The uplink time synchronization Reference Point (RP) is the entity where the downlink frame and the uplink frame are aligned through the offset $N_{TA,offset}$ (3GPP 2024q, 2024s). Common TA is a configured timing offset that specifies the RTT between the RP and the NTN payload. K_{offset} is a configured scheduling offset that needs to be larger or equal to the sum of the service link RTT and the Common TA. k_{mac} is a configured timing offset that specifies the RTT between the RP and the NTN gNB. k_{mac} is used to delay the application of a downlink configuration indicated by a MAC CE command on PDSCH (3GPP 2024q) and in estimating the UE-gNB RTT (3GPP 2024u). The NTN gNB may provide k_{mac} when the downlink and the uplink frame timing are not aligned at the NTN gNB. During the RA procedure in the NR-NTN, k_{mac} is used to determine the start time of the RAR window/MsgB window after a Msg1/MsgA transmission (3GPP 2024q). The scheduling offset K_{offset} provides the UE sufficient processing time between a downlink reception and an uplink transmission (3GPP 2024q).

Figure 7.5 illustrates a simplified view of the 4-step RA procedure in an NR-NTN. See 3GPP (2024u) for a comprehensive view of the 4-step RA procedure.

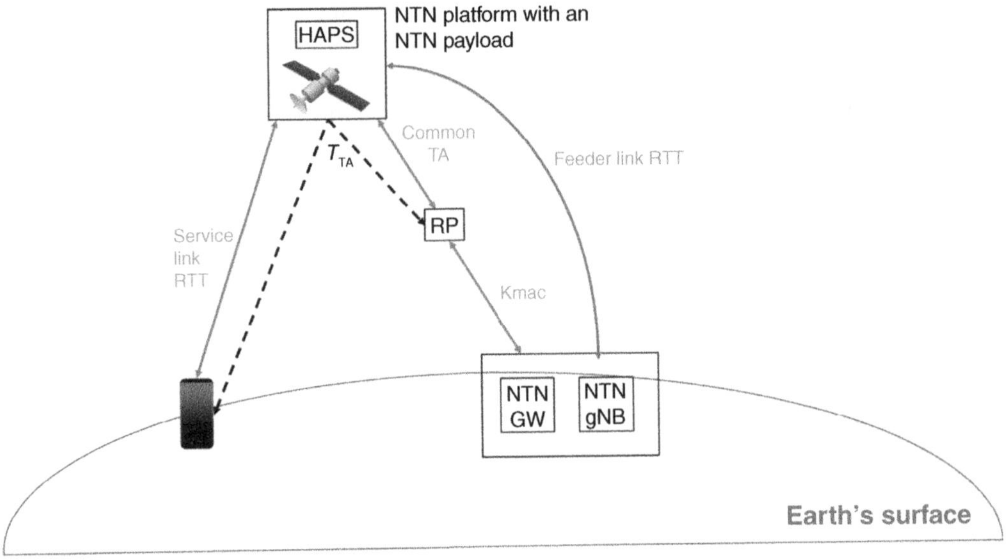

Figure 7.4 Timing Relationship in an NR-NTN. *Source:* hobrath / Adobe Stock Photos.

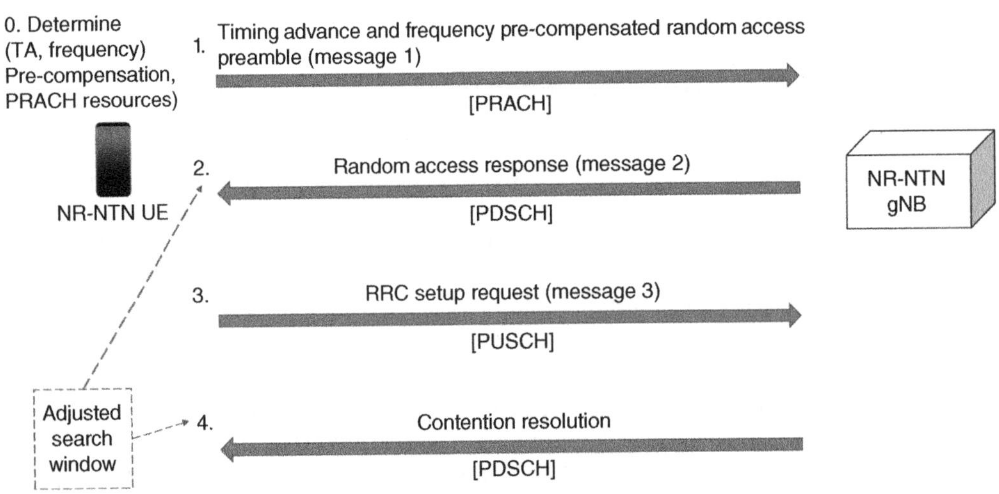

Figure 7.5 Simplified 4-Step Random-Access Procedure in NR-NTN. *Source:* hobrath / Adobe Stock Photos.

7.2.3.1 Step 0: Prior to RA Preamble Transmission

In *Step 0*, before the transmission of an RA preamble on the PRACH, the NR-NTN UE determines a suitable Timing Advance to pre-compensate the propagation delay so that its preamble arrives at the RP (approximately) at the target time and the NR-NTN UE's signal does not cause timing misalignment related interference in the uplink. To determine timing pre-compensation, the NR-NTN UE utilizes its location (determined using the GNSS) and the NTN platform's ephemeris from SIB1/NTN-Config-r17/ephemerisInfo-r17[6] to estimate the service link RTT. The UE also obtains the

6 SIB1 contains the IE NTN-Config-r17. The IE NTN-Config-r17 contains ephemerisInfo-r17.

Common TA from SIB1/NTN-Config-r17/ta-Info-r17. The NR-NTN UE computes the RTT between itself and the RP using the service link RTT and Common TA to pre-compensate for T_{TA}, as shown in Figure 7.4.

Furthermore, the NR-NTN UE needs to pre-compensate for the Doppler shift so that the NR-NTN UE's signal does not cause frequency misalignment-related interference in the uplink. Hence, the NR-NTN UE computes the Doppler shift of the service link based on its position and the ephemeris data from SIB1/NTN-Config-r17/ephemerisInfo-r17 and determines frequency pre-compensations for its RA preamble transmission so that the NR-NTN UE's signal does not cause frequency misalignment-related interference in the uplink.

Finally, as in a 5G NR TN, since the 5G NR is a beamformed radio interface, the NR-NTN UE utilizes the best SSB beam, and determines the PRACH configuration associated with this SSB beam and randomly selects an RA preamble.

7.2.3.2 Step 1: RA Preamble Transmission

In Step 1, the NR-NTN UE sends a TA and Doppler shift pre-compensated RA preamble (also known as message 1 or Msg1) on the PRACH using the beam-specific PRACH configuration specified in SIB1.

7.2.3.3 Step 2: Random-Access Response

In *Step 2*, the NR-NTN gNB looks for the RA preambles from NR-NTN UEs during certain periods and in specific SSB beams based on the mapping between the SSB beam and the PRACH configuration. The NR-NTN gNB responds with the RAR, considered to be message 2, that contains typical configuration parameters for the UE, such as the Timing Advance Command to facilitate UL synchronization for the NR-NTN UE, UL grant for message 3 transmission, and (Temporary) C-RNTI that identifies the NR-NTN UE on the radio interface in the current NR-NTN cell.

Note that the NR-NTN UE looks for the Random-Access Response in the RA Response window, denoted by ra-ResponseWindow configured in the IE RACH-ConfigCommon in SIB1, after sending the RA preamble. To accommodate long propagation delays in NR-NTN, the NR-NTN UE delays the start of the ra-ResponseWindow by ($T_{\text{TA}} + k_{\text{mac}}$) ms (3GPP 2024q).

7.2.3.4 Step 3: Msg3 Transmission

In *Step 3*, using the allocated UL grant, the NR-NTN UE sends Message 3/Msg 3 such as the RRC Setup Request message. This message contains the NR-NTN UE's identity, such as 5G-S-TMSI, if an AMF has previously allocated it a 5G-GUTI. 5G-S-TMSI is a short form of 5G-GUTI. If the NR-NTN UE does not have 5G-S-TMSI, it generates a random number and includes it in the RRC Setup Request message. Since more than one NR-NTN UE might have used the same RA preamble ID, more than one 5G NR UE may send the RRC Setup Request message using the UL grant specified by the NR-NTN gNB in the RAR. The UE ID, such as 5G-S-TMSI or a random number, enables the NR-NTN gNB to resolve the conflict among the NR-NTN UEs that might have used the same RA preamble simultaneously.

Note that the NR-NTN UE looks for the Contention Resolution message after sending Msg3 within a certain period. This period corresponds to the timer ra-ContentionResolutionTimer. To accommodate long propagation delays in NR-NTN, the NR-NTN UE starts the ra-ContentionResolutionTimer in the first symbol after the end of all repetitions (if any per configuration) of the Msg3 transmission plus the RTT between the NR-NTN UE and the NR-NTN gNB (3GPP 2024u).

7.2.3.5 Step 4: Contention Resolution

The NR-NTN gNB sends the Contention Resolution message containing the UE identity to the NR-NTN UE that the NR-NTN gNB has successfully detected on the PDSCH. The location of this PDSCH is identified by the PDCCH associated with the C-RNTI allocated to the NR-NTN UE in Step 2.

A successful RA procedure enables the RRC connection setup procedure to continue. In case the RA procedure fails, the NR-NTN UE retries the RA procedure by sending another RA preamble after a random wait time.

7.2.3.6 Random Access in NR-NTN: Additional Considerations

While a 4-step RA procedure is shown in Figure 7.5, a 2-step RA procedure may also be used in an NR-NTN. The 2-stepRA procedure involves message A from the NR-NTN UE and message B from the NR-NTN gNB. The NR-NTN gNB specifies resources for the 2-step RA if it wants NR-NTN UEs to use the 2-step RA. See 3GPP (2024u) for details of the 2-step RA procedure.

The NR-NTN UE reports Timing Advance during the RA procedure using Timing Advance Report MAC CE (3GPP 2024u) on the PUSCH if SIB1/ta-Report-r17 indicates that the TA report is enabled (3GPP 2024v). The Timing Advance reporting procedure is used in the NTN so that the NR-NTN UE can provide the gNB with an estimate of the UE's Timing Advance value (i.e., T_{TA}).

The TA reporting procedure is also used in the NTN after the RA procedure. For example, the TA reporting procedure is triggered if the NR-NTN gNB configures the NR-NTN UE with offsetThresholdTA. This procedure is also triggered if the UE has not previously reported the Timing Advance value to the current Serving Cell. Furthermore, if the variation between the current estimate of the Timing Advance value and the last reported Timing Advance value is equal to or larger than offsetThresholdTA, the NR-NTN UE sends a TA report to the NR-NTN gNB.

Note that the NR-NTN UE performs the pre-compensation of the Doppler shift on the service link; the management of (i) the Doppler shift on the feeder link, and (ii) transponder frequency error is network implementation-specific.

7.2.4 RRC Connection Setup in NR-NTN

The RRC connection setup process is similar for a TN and an NTN. Figure 7.6 summarizes key steps of a 3-way handshake between the NR-NTN UE and the NR-NTN gNB to set up the RRC connection in an NR-NTN (3GPP 2024v).

In *Step 1*, like a TN, the RRC Setup Request message includes the UE ID, such as 39 rightmost bits of 5G-S-TMSI or a 39-bit random number. The NR-NTN UE indicates the purpose behind the message by specifying the establishment cause (e.g., mobile-originated signaling). While not inside the RRC Setup Request message, a Timing Advance Report MAC CE accompanies the RRC Setup Request message if the NR-NTN gNB enables such reporting.

In *Step 2*, the NR-NTN gNB responds with the RRC Setup message to set up NR-NTN UE-specific SRB1. The RRC Setup message includes IEs such as for SRB1 (i) masterCellGroup containing RLC-BearerConfig and CellGroupConfig (with mac-CellGroupConfig, physicalCellGroupConfig, and spCellConfig) and (ii) radioBearerConfig.

The IE RLC-BearerConfig sets up RLC parameters for SRB1 as an RLC bearer. The reassembly timer has been extended to accommodate longer propagation delays. For example, the IE T-ReassemblyExt-r17 supports the following values: 210, 220, 340, 350, 550, 1,100, 1,650, and 2,200 ms.

The IE mac-CellGroupConfig contains MAC parameters such as Discontinuous Reception (DRX) parameters, Scheduling Request (SR) parameters, and Buffer Status Report (BSR) parameters.

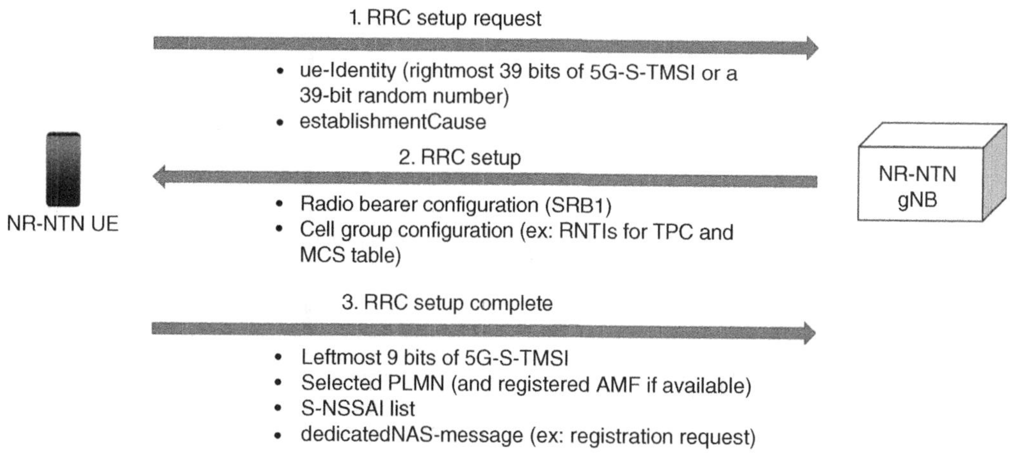

Figure 7.6 RRC Connection Setup in an NR-NTN. *Source:* hobrath / Adobe Stock Photos.

The IE physicalCellGroupConfig configures cell-group-specific physical layer parameters such as maximum transmit power, HARQ feedback configurations, and mcs-C-RNTI to indicate the use of qam64LowSE for grant-based transmissions.

spCellConfig contains an index of the serving cell (e.g., 0 for the PCell) and RLF timers.

SRB1 is specific to the NR-NTN UE. This message also specifies 5G NR-specific cell group configuration such as C-RNTIs for Transmit Power Control (TPC) and Modulation and Coding Scheme (MCS) table. In general, due to the large path losses in an NTN and limited power at the NR-NTN UE and the NTN payload, SINRs are not expected to be high in an NTN, reducing the likelihood of the use of aggressive coding and modulation schemes defined for a CQI table. Furthermore, due to long propagation delays, CQIs reported by an NR-NTN UE may not reflect the prevailing radio channel conditions. Conservative coding and modulation would be more appropriate for an NR-NTN.

In *Step 3*, the NR-NTN UE uses SRB1 to send the RRC Setup Complete to the NR-NTN gNB, which is the response to the RRC Setup message. The RRC Connection Setup Complete message conveys the leftmost 9 bits of 5G-S-TMSI, selected PLMN (and registered AMF if available), S-NSSAI list, and dedicated NAS message (e.g., NAS Registration Request).

7.2.5 Initial Registration via NR-NTN

The overall steps executed for initial registration via an NR-NTN are the same as those for an NR-based TN. Recall that pre-requisites such as registration and PDU session setup need to be met before an NR-NTN UE can exchange traffic. Figure 7.7 shows major steps of initial registration for a 5GS using NR-NTN (3GPP 2024e). PDU session setup is summarized in Section 7.2.7.

7.2.5.1 Step 1: Network Acquisition

As discussed in Section 7.2.2, the NR-NTN UE, upon power-up, detects an NR-NTN cell by observing the PSS and the SSS and determines the PCI of the NR-NTN cell. The NR-NTN UE then acquires MIB, SIB1, and other SIBs, including NTN-specific SIBs SIB19 and SIB25.

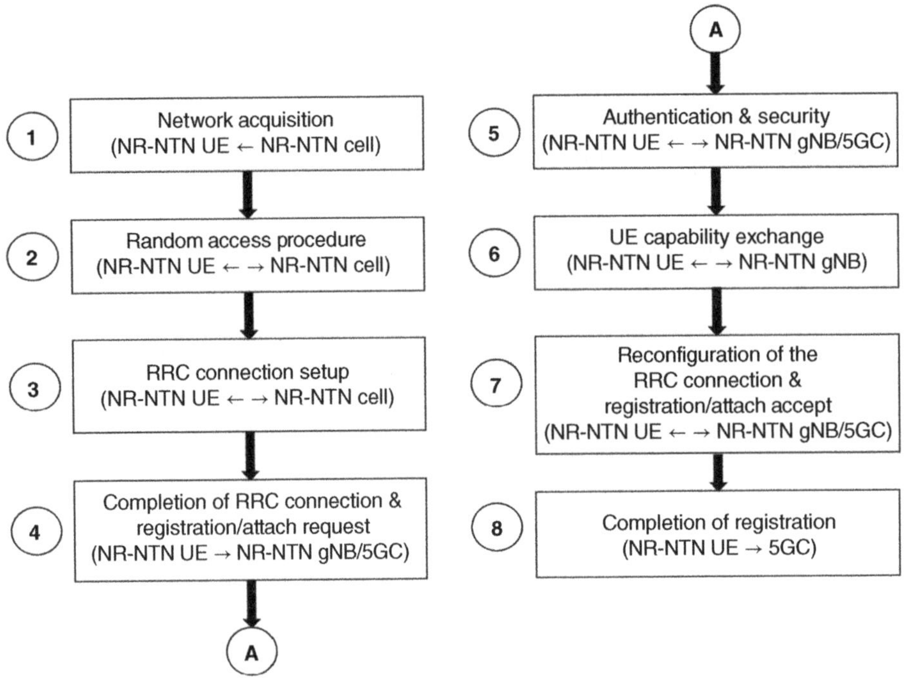

Figure 7.7 Initial Registration in the 5GS via NR-NTN.

7.2.5.2 Step 2: Random-Access Procedure

After acquiring SIBs (especially SIB1), the NR-NTN UE executes the RA procedure in the selected NR-NTN cell, as discussed in Section 7.2.3.

7.2.5.3 Step 3: RRC Connection Setup

The NR-NTN UE sends an RRC Setup Request message toward the end of the RA procedure and obtains the configuration of SRB1 from the serving NR-NTN cell via the RRC Setup message. Furthermore, the NR-NTN UE initiates connectivity with the 5GC by sending a NAS message in the RRC Setup Complete message. Section 7.2.4 provides details of the RRC connection setup procedure.

7.2.5.4 Step 4: Registration Request

The NR-NTN UE sends the (NAS) Registration Request message to the AMF via the NR-NTN gNB (3GPP 2024e). The NR-NTN UE encapsulates this NAS message in the RRC Setup Complete message. The NR-NTN gNB selects a suitable AMF for the NR-NTN UE and sends the (NGAP) INITIAL UE MESSAGE to the selected AMF. This message contains the NAS message sent by the NR-NTN UE and IEs, such as 5G-STMSI, RRC Establishment Cause (i.e., the reason such as mobile-originated signaling), and User Location Information (3GPP 2024w). In support of the NR-NTN access and any associated restriction for the NR-NTN UE, the User Location Information (ULI) includes (i) TN-like parameters such as NR Cell Global Identity (CGI) and Age of Location (i.e., the UTC time when the location information was generated) and (ii) an NTN-specific IE "NR NTN TAI Information" that includes (a) the identity of the serving PLMN, (b) all the Tracing Area

Codes (TACs)[7] broadcast in the NR-NTN cell where the NR-NTN UE is currently located, and (c) (optionally) UE Location Derived TAC in NR NTN (i.e., the specific TAC derived by the NR-NTN gNB based on the actual location of the NR-NTN UE, if available).

The Registration Request message from an NR-NTN UE contains parameters that are also present for the 5GS access via a TN. These parameters include Registration Type (e.g., initial registration or mobility/periodic registration update), UE ID (e.g., SUCI[8]), security parameters, requested Network Slice Selection Assistance Information (NSSAI),[9] requested DRX parameters, and selected UE capabilities. Like a TN UE, the NR-NTN UE also indicates a follow-on request if it intends to exchange additional signaling after registration (e.g., PDU session setup signaling) (3GPP 2024e).

In support of discontinuous NR-NTN coverage, in the Registration Request message, the NR-NTN UE may include the IE unavailability information that consists of (i) the unavailability type (e.g., due to UE reasons or discontinuous coverage), (ii) unavailability period duration presence indication (indicating whether the unavailability period duration field is present or absent, (iii) unavailability period duration (which is the unavailability period duration in seconds), and (iv) the start of unavailability period (which is the time until the start of the next unavailability period in seconds) (3GPP 2024f, 2024g).

The AMF determines the RAT Type for NR satellite access; supported RAT Types include NR(LEO), NR(MEO), NR(GEO), and NR(OTHERSAT). The NR-NTN cells of each NR satellite RAT Type need to be deployed in Tracking Areas different from (i) the Tracking Areas for other NR satellite RAT Types and (ii) the Tracking Areas supporting terrestrial access RAT Types.

To meet the regulatory requirements, the service provider may configure the network to enforce regulatory requirements that prohibit a UE at certain locations from selecting a PLMN. Hence, upon receiving an NGAP message containing User Location Information for an NR-NTN UE using NR satellite access, the AMF may decide to verify the UE location. If the AMF evaluates if the NR-NTN UE can access the selected PLMN via the NR-NTN based on the Selected PLMN ID and ULI (which includes CGI and NR NTN TAI Information). If the selected PLMN is prohibited from operating at the current UE location, the AMF rejects a NAS request with a suitable cause value. If the AMF is unable to determine the NR-NTN UE's location with sufficient accuracy to determine whether to allow the access or not, the AMF proceeds with the Mobility Management or Session Management procedure and may initiate UE location procedure to determine the NR-NTN UE's location after the Mobility Management or Session Management procedure is complete (3GPP 2024b, 2024d). The AMF deregisters the NR-NTN UE if the information obtained from the Location Management Function (LMF) indicates that the NR-NTN UE is registered to a PLMN that is not allowed to operate in the NR-NTN UE's current location. In the case of a NAS procedure, the AMF (i) rejects a NAS request targeted toward a disallowed PLMN and indicates a suitable cause value or (ii) accepts the NAS procedure and initiates the deregistration procedure (3GPP 2024a, 2024d, 2024g).

7.2.5.5 Step 5: Authentication and Security Activation

Mutual authentication between the NR-NTN UE and the 5GC occurs just like between a TN UE and the 5GC. Specifically, the AUSF obtains authentication information from the UDM and

7 As of Release 18, up to 12 (specified by "maxnoofTACsinNTN") TACs can be included in one NR-NTN cell (3GPP 2024w).

8 SUCI is Subscription Concealed Identifier and can be used to hide the Subscription Permanent Identifier.

9 NSSAI contains the network slices requested by the UE. Examples of these network slices in Release 15 include eMBB, URLLC, and mIoT.

exchanges authentication messages with the NR-NTN UE using the NAS signaling connection between the UE and the AMF. Upon successful mutual authentication, the NAS security between the NR-NTN UE and the AMF and the AS security between the NR-NTN UE and the NR-NTN gNB are set up. Such NAS and AS security setup in an NTN is similar to that in a TN. The RRC and NAS messages can be ciphered and integrity protected. The AMF obtains relevant policy information (e.g., allowed NSSAI) from the PCF. The AMF provides the (NAS) Registration Accept message to the NR-NTN gNB.

7.2.5.6 Step 6: UE Capability Exchange

The NR-NTN UE and the NR-NTN gNB exchange messages so that the NR-NTN UE can convey its 5G NR and NR-NTN capabilities to the NR-NTN gNB. Section 7.2.6 for details of the UE capability exchange.

7.2.5.7 Step 7: Registration Accept

The NR-NTN gNB sends an RRC message, such as RRC Reconfiguration or DL Information Transfer, that encapsulates the (NAS) Registration Accept message received from the AMF. The (NAS) Registration Accept message specifies parameters such as 5G-GUTI, Registration Area (equivalent to the TAI List containing one or more Tracking Areas), allowed NSSAI, periodic registration update timer, and accepted DRX parameters (3GPP 2024e).

In the Registration Accept message, the AMF may indicate the TAIs that are forbidden for the NR-NTN UE for NR-NTN access in one or both IEs: (i) Forbidden TAI(s) for the list of "5GS forbidden tracking areas for roaming" and (ii) Forbidden TAI(s) for the list of "5GS forbidden tracking areas for regional provision of service."

Furthermore, in support of discontinuous NR-NTN coverage, in the Registration Accept message, the AMF may include the IE Unavailability configuration that consists of (i) End of unavailability report (which indicates whether the UE is required to perform the registration procedure when the unavailability period has ended) and (ii) Unavailability period duration (which specifies the availability period duration in seconds).

7.2.5.8 Step 8: Registration Complete

The NR-NTN UE replies to the Registration Accept message with the Registration Complete message by encapsulating such NAS message in the RRC UL Information Transfer message (3GPP 2024e).

Note that no PDU session is set up during registration for a TN and an NTN. If the NR-NTN UE wishes to set up a PDU session, it can initiate such a procedure at any time after successful registration.

7.2.5.9 Coarse UE Location Reporting

Upon request by the network, an NR-NTN UE reports its coarse UE location information (which is specified as the most significant bits of the GNSS coordinates to achieve location accuracy on the order of 2 km) to the NG-RAN if such location information is available at the UE. Such reporting occurs after AS security is established when the NR-NTN UE is in RRC_CONNECTED state (3GPP 2024s).

7.2.5.10 Verification of the UE Location

The network must verify the UE-reported location to meet the regulatory requirements. Hence, for the UE location verification based on multi-RTT with a single satellite in the NTN, the following UE and gNB measurements are reported (3GPP 2024r): (i) gNB receive-transmit time difference at the uplink time synchronization reference point, (ii) UE receive-transmit time difference, (iii) UE receive-transmit time difference subframe offset, and (iv) DL timing drift (3GPP 2024s).

The assistance information that may be provided to the 5GC may include (i) ephemeris information, including accurate satellite position and velocity at the time of multi-RTT measurement, (ii) common TA parameters (e.g., ta-Common, ta-CommonDrift, and ta-CommonDriftVariant), and (iii) Epoch time (3GPP 2024s).

7.2.6 UE Capability Exchange in NR-NTN

The NR-NTN gNB learns about the capabilities of the NR-NTN UE through the UE capability transfer procedure (3GPP 2024v). Figure 7.8 summarizes the UE capability transfer-related RRC signaling message exchange between the NR-NTN UE and the NR-NTN gNB.

In *Step 1*, the NR-NTN gNB sends a UE Capability Enquiry message to the NR-NTN UE. For example, the NR-NTN gNB may ask the UE to provide capabilities related to one or more RATs (e.g., NR or E-UTRA-NR) in its UE Capability Enquiry message to the UE by including the IEs such as UE-CapabilityRAT-RequestList and UE-CapabilityEnquiryExt containing UE-CapabilityRequestFilterCommon (contains IEs such as mrdc-Request to obtain the UE's multi-RAT capabilities and codebookTypeRequest to obtain the UE's MIMO codebook capabilities) (3GPP 2024v). The IE UE-CapabilityRAT-RequestList includes RAT-Type (e.g., nr for 5G NR, e-utra-nr for LTE and NR, eutra for LTE, and utra-fdd-v1610 for UMTS FDD) and capabilityRequestFilter (by which the gNB requests the UE to filter the UE capabilities; defined by UE-CapabilityRequestFilterNR when rat-Type is NR). The IE UE-CapabilityRequestFilterNR contains parameters such as FreqBandList. *Since NR-NTN is a special case of NR, the NR-NTN gNB can set RAT-Type = nr to obtain the UE's NR-NTN capabilities.*

In *Step 2*, the NR-NTN UE responds with a UE Capability Information message conveying its capabilities related to the RAT type specified by the NR-NTN gNB and includes the IE UE-CapabilityRAT-ContainerList. The IE UE-CapabilityRAT-ContainerList, as the name suggests, a list of "UE-CapabilityRAT-Container" IEs, with each UE-CapabilityRAT-Container specifying the RAT-Type and associated capabilities such as UE-NR-Capability for RAT-Type=nr and UE-MRDC-Capability for RAT-Type=eutra-nr. *Since NR-NTN is a special case of NR, the NR-NTN UE provides its typical NR capabilities and NR-NTN-specific capabilities in the IE UE-NR-Capability.*

In the UE Capability Information message, the NR-NTN UE includes the IE UE-NR-Capability/UE-NR-Capability-v1700 to convey the NR UE Radio Access Capability Parameters that include

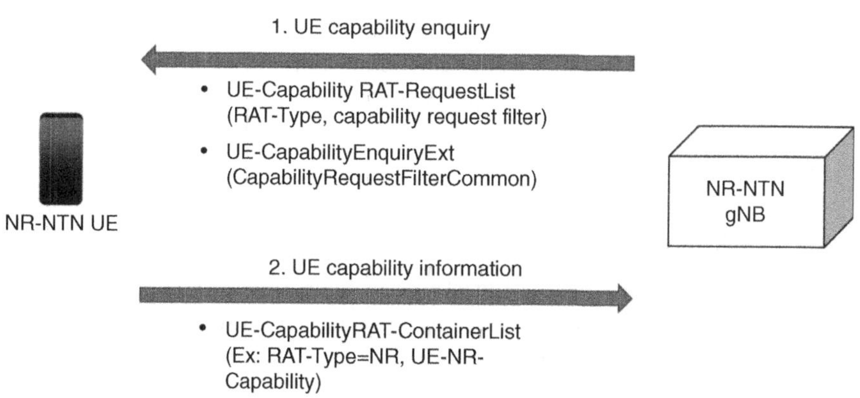

Figure 7.8 UE Capability Transfer in NR-NTN. *Source:* hobrath / Adobe Stock Photos.

(i) various capabilities that are common to both a TN and an NTN, (ii) nonTerrestrialNetwork-r17 (whether the UE supports the NTN or not), (iii) ntn-ScenarioSupport-r17 (indicating support for GSO satellites or NGSO satellites), and (iv) NTN-specific capabilities in the form of the IE NTN-Parameters-r17. Furthermore, the IE UE-NR-Capability-v1800 defined in Release 18 includes NTN-specific capabilities such as support for soft satellite switch with resync (via the IE softSatelliteSwitchResyncNTN-r18) and support for hard satellite switch with resync (via the IE hardSatelliteSwitchResyncNTN-r18) (3GPP 2024v).

Examples of UE capabilities common to both a TN and an NTN include numerous capabilities of protocols such as PHY, MAC, RLC, and PDCP. The NR UE also provides RF parameters, measurement parameters, and mobility parameters. For example, the NR UE specifies supported SCS values, modulation schemes (e.g., QPSK, 16-QAM, and 64-QAM), bandwidth, and power class (e.g., power class 3, which corresponds to 23 dBm EIRP).

The IE NTN-Parameters have been introduced to convey a set of UE Radio Access Capability Parameters that apply to the NR-NTN access when the NR-NTB UE's TN and NTN capabilities differ. The IE NTN-Parameters-r17 includes the following parameters (3GPP 2024v): (i) inactiveStateNTN-r17 (to indicate if the RRC_INACTIVE state is supported by the NR-NTN UE on the NR-NTN access), (ii) ra-SDT-NTN-r17 (to indicate if the RRC_INACTIVE state is supported by the NR-NTN UE on the NR-NTN access), (iii) srb-SDT-NTN-r17 (to indicate if SRB for SDT is supported by the NR-NTN UE on the NR-NTN access), (iv) measAndMobParametersNTN-r17 (to specify measurement and mobility parameters supported by the NR-NTN UE on the NR-NTN access but different from TN access), (v) mac-ParametersNTN-r17 (to specify MAC layer capabilities supported by the NR-NTN UE on the NR-NTN access but different from TN access), (vi) phy-ParametersNTN-r17 (to specify physical layer capabilities supported by the NR-NTN UE on the NR-NTN access but different from TN access), (vii) fdd-Add-UE-NR-CapabilitiesNTN-r17 (contains UE-NR-CapabilityAddXDD-Mode supported by the NR-NTN UE on the NR-NTN access but different from TN access), (viii) fr1-Add-UE-NR-CapabilitiesNTN-r17 (contains UE-NR-CapabilityAddFRX-Mode supported by the NR-NTN UE on the NR-NTN access but different from TN access), (ix) ue-BasedPerfMeas-ParametersNTN-r17 (contains UE-BasedPerfMeas-Parameters-r16; specifies UE-based performance measurements and parameters supported by the NR-NTN UE on the NR-NTN access but different from TN access), and (x) son-ParametersNTN-r17 (contains SON-Parameters-r16; specifies SON parameters supported by the NR-NTN UE on the NR-NTN access but different from TN access).

If any of fdd-Add-UE-NR-CapabilitiesNTN, fr1-Add-UE-NR-CapabilitiesNTN, mac-ParametersNTN, measAndMobParametersNTN, phy-ParametersNTN, son-ParametersNTN, and ue-BasedPerfMeas-ParametersNTN are absent, their NR counterpart (which is specified for the NR-based TN) applies to the NR-NTN.

Note that the procedure shown in Figure 7.8 will be carried out after AS security has been activated. The NR-NTN gNB can forward the UE capabilities to the 5GC via suitable NGAP signaling toward the AMF.

7.2.7 PDU Session Establishment in NR-NTN

Recall that prerequisites such as registration and PDU session setup must be met before an NR-NTN UE can exchange traffic. The overall steps executed for PDU session setup via an NR-NTN are the same as those for an NR-based TN (Tripathi and Reed 2019; 3GPP 2024e). The main differences between the TN and the NTN are (i) the availability and use of new NTN QoS for a GSO-based

NTN, (ii) verification of the NR-NTN UE location by the AMF, and (iii) the use of extended session management timers.

Section 5.7 in Chapter 5 mentions that the 3GPP has introduced a new 5G QoS Identifier (5QI) for the NR-NTN- 5QI=10. 5QI=10 has a Packet Delay Budget (PDB) of 1,100 ms that can accommodate the worst-case propagation delay for a GEO satellite-based NTN. Additionally, 5QI=10 has the resource type = Non-Guaranteed Bit Rate (Non-GBR), the default priority of 90, and the packet error rate of 10^{-6} (3GPP 2024d).

When the AMF receives ULI from the NR-NTN gNB as part of a NAS procedure, it may verify that the NR-NTN UE can still use the selected PLMN in its current location. Recall that the service provider may configure the network to enforce regulatory requirements that prohibit a UE at certain locations from selecting a PLMN. Hence, upon receiving an NGAP message containing ULI for an NR-NTN UE using NR satellite access, the AMF may decide to verify the UE location. The AMF evaluates if the NR-NTN UE can access the selected PLMN via the NR-NTN based on the Selected PLMN ID and ULI (which includes CGI and NR NTN TAI Information). If the selected PLMN is not allowed to operate at the current UE location, the AMF rejects a NAS request with a suitable cause value.

NAS signaling messages use a set of NAS timers that govern the NAS behavior at the UE and the 5GC. Since propagation delays can be much longer in an NTN compared to a TN, these NAS timers have been extended (3GPP 2024g). The NR-NTN UE and the SMF can utilize extended NAS timers for 5GS session management in the NR-NTN when the NTN RAT Type is NR(MEO) and NR(GEO). For example, if the SMF wants to execute PDU session authentication, a typical value for the NAS timer T3590 for a TN is 15 seconds. This timer is started after transmission of the PDU SESSION AUTHENTICATION COMMAND message. However, if this timer expires, the PDU SESSION AUTHENTICATION COMMAND message is retransmitted. For the NR(MEO) and NR(GEO), T3590 is extended to 21 seconds to accommodate a longer propagation delay.

Figure 7.9 shows the major steps of PDU session setup for a 5GS using NR-NTN (Tripathi and Reed 2019).

Recall that a PDU session is the association between the UE and a Data Network (DN) for a given network slice. A PDU session enables the exchange of PDUs containing user traffic. The User Plane connection for a PDU session consists of (i) the Data Radio Bearer (DRB) on the NR air interface between the UE and the gNB in a given cell and (ii) the N3 connection (which is a GTP-U tunnel) between the gNB and the UPF.

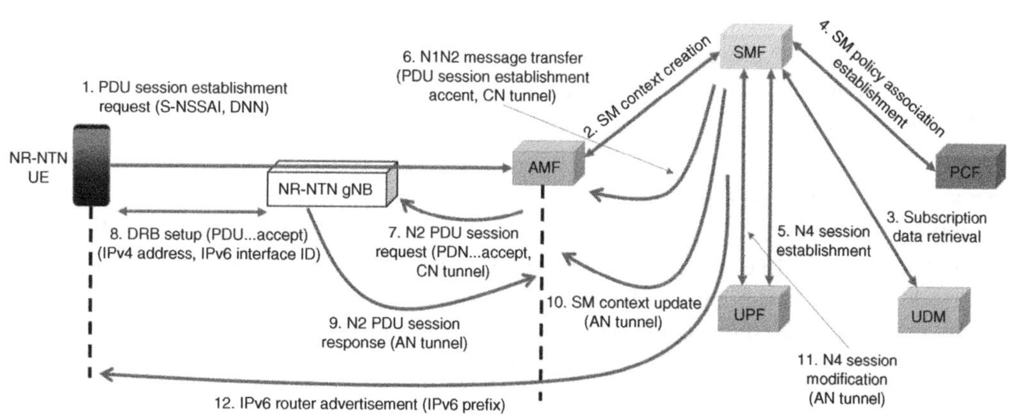

Figure 7.9 PDU Session Setup in 5GS via NR-NTN. *Source:* hobrath / Adobe Stock Photos.

7.2.7.1 Step 1: PDU Session Establishment Request

The NR-NTN UE, any time after successful registration, sends a PDU Session Establishment Request to the 5GC to establish connectivity with a DN (3GPP 2024e). Specifically, the NR-NTN UE sends the NAS UL NAS Transport message encapsulated in the RRC UL Information Transfer message. The NAS UL NAS Transport message has the NAS payload container type of "N1 SM information" with the N1 SM Payload being the PDU Session Establishment Request message (3GPP 2024g). The NAS UL NAS Transport includes S-NSSAI to identify the Network Slice and the Data Network Name (DNN) to identify the DN toward which the NR-NTN UE requests the PDU session. The PDU Session Establishment Request message contains the PDU Session ID determined by the UE, requested PDU Session Type (e.g., IPv6 or IPv4), requested SSC mode, and 5GSM Capability (e.g., support for reflective QoS). The AMF selects a suitable SMF based on the SMF capabilities, S-NSSAI, and the DNN (3GPP 2024g). The AMF may decide to verify the NR-NTN UE's location to meet regulatory requirements for the NR-NTN.

7.2.7.2 Step 2: SM Context Creation

The AMF sends Nsmf_PDUSession_CreateSMContext Request to the SMF to create the NR-NTN UE-specific Session Management (SM) context (3GPP 2024e). Along with the N1 SM payload, this message includes parameters such as SUPI, S-NSSAI, DNN, PDU Session ID, AMF ID (i.e., GUAMI), and RAT Type.[10] For the 3GPP access, the AMF determines the RAT type (e.g., NR and LTE-M) based on the Global RAN Node ID and the Tracking Area indicated by NG-RAN. Furthermore, the AMF determines the specific satellite RAT type, such as NR(LEO), NR(MEO), NR(GEO), and NR(OTHERSAT).

7.2.7.3 Step 3: Subscription Data Retrieval

The SMF obtains subscriber data from the UDM. Examples of subscription data retrieved by the SMF from the UDM include allowed PDU session types, allowed SSC modes, default 5G QoS Identifier (5QI) and Allocation and Retention Priority (ARP), and Session-AMBR. The SMF can subscribe to the UDM for notifications. The SMF then replies to the AMF with Nsmf_PDUSession_CreateSMContext Response containing the SM Context ID and cause. The new NTN-specific 5QI=10 is helpful in the case of GSO satellite-based NTN (3GPP 2024e).

7.2.7.4 Step 4: SM Policy Association Establishment

The SMF exchanges signaling messages with the PCF to establish the association of the SM policy (3GPP 2024e). The PCF provides the SMF with policy information such as authorized Session-AMBR and authorized default 5QI and AP. The new NTN-specific 5QI=10 is helpful in the case of GSO satellite-based NTN.

7.2.7.5 Step 5: N4 Session Establishment

The SMF selects a suitable UPF as a PDU Session Anchor (PSA) for the NR-NTN UE's PDU session toward the DN. The SMF establishes the NR-NTN UE-specific N4 session with the UPF. The SMF specifies the packet detection rule to the UPF and the Core Network (CN) Tunnel Info (e.g., Tunnel Endpoint ID [TEID]) for the NR-NTN UE-specific N3 Tunnel between the NR-NTN gNB and the UPF (3GPP 2024e). More specifically, the CN Tunnel Info specifies the UPF side of the GTP-U

10 The AMF, per configuration, knows if the 5G NR UE is accessing the TN gNB that is using satellite backhaul. In such a case, the AMF provides satellite backhaul related information to the SMF.

tunnel between the NR-NTN gNB and the UPF on the N3 interface. The UPF or the SMF may determine the CN Tunnel Info.

7.2.7.6 Step 6: N1N2 Message Transfer

The SMF sends the Namf_Communication_N1N2MessageTransfer message to the AMF to (i) transport the N1 SM container with the PDU Session Establishment Accept message and (ii) provide the PDU session related information such as CN Tunnel information, PDU Session ID, QoS Flow Indicator (QFI), QoS Profile, S-NSSAI, Session-AMBR, and PDU Session Type (3GPP 2024e). The PDU Session Establishment Accept message itself, to be sent to the UE by the AMF, contains QoS Rules and QoS Flow level parameters, selected SSC Mode, S-NSSAI, DNN, IPv4 address and/or IPv6 interface ID, Session-AMBR, Selected PDU Session Type, and P-CSCF address.

7.2.7.7 Step 7: N2 PDU Session Request

The AMF sends an NGAP N2 PDU Session Resource Setup Request message to the NR-NTN gNB. This message contains (i) PDU Session Resource Setup Request Transfer with details of QoS parameters to enable the NR-NTN gNB to perform admission control and (ii) the NAS PDU (3GPP 2024e). The NAS PDU carries the DL NAS Transport message, including the PDU Session ID and the N1 SM information containing the PDU Session Establishment Accept message.

7.2.7.8 Step 8: DRB Setup and PDU Session Establishment Accept

The NR-NTN gNB considers the QoS requirements for the DRB associated with the PDU session and sets up a DRB for the UE. The NR-NTN gNB also transports the NAS PDU Session Establishment Accept message in the RRC message. The NR-NTN gNB can utilize the RRC Reconfiguration message to set up the DRB and encapsulate the DL NAS Transport message containing the PDU Session Establishment Accept message.

7.2.7.9 Step 9: N2 PDU Session Response

After setting up the DRB and sending the PDU Session Establishment Accept message, the NR-NTN gNB sends N2 PDU Session Resource Setup Response to the AMF and specifies the NR-NTN gNB side of the GTP-U tunnel for the UE-specific N3 connection between the NR-NTN gNB and the UPF in the form of the Access Network (AN) Tunnel Information (3GPP 2024e).

7.2.7.10 Step 10: SM Context Update

The AMF sends the SM context update to the SMF and provides the AN tunnel information as part of N2 SM information.

7.2.7.11 Step 11: N4 Session Modification

The SMF modifies the N4 session toward the UPF by providing the AN Tunnel Information. The GTP-U Tunnel between the NR-NTN gNB and the UPF is now complete. Since the DRB has already been set up in Step 8, the UP connection for the PDU session between the UE and the UPF is fully established.

7.2.7.12 Step 12: IPv6 Router Advertisement

If the NR-NTN UE is allocated an IPv6 address, the SMF sends an IPv6 Router Advertisement message through the UPF and the NR-NTN gNB to the UE in the form of an IP layer message to specify the network prefix portion of the IPv6 address.

7.3 Selected Operations in IoT-NTN: A Closer Look

Nomenclature in This Book[11]

Certain terms are defined here for enhanced clarity; these terms are not used in 3GPP NTN specifications. In this book, an NTN-capable UE that supports LTE-M or NB-IoT is called an IoT-NTN UE. An NTN-capable eNB that supports LTE-M or NB-IoT is referred to as an IoT-NTN eNB. An NTN cell that supports LTE-M or NB-IoT is called an IoT-NTN cell. The NTN that utilizes LTE-M on the radio interface is called LTE-M NTN, and the NTN that utilizes NB-IoT on the radio interface is called NB-IoT NTN.

An LTE-M NTN UE utilizes (i) an LTE-M radio interface to communicate with an LTE-M NTN cell/eNB and (ii) EPS signaling to communicate with the EPC. Similarly, an NB-IoT NTN UE utilizes (i) NB-IoT radio interface to communicate with an NB-IoT NTN cell/eNB and (ii) EPS signaling to communicate with the EPC.

7.3.1 Operational Enhancements in IoT-NTN: An Overview

Figure 7.10 highlights key operations enhancements made by the 3GPP to support IoT-NTN. These enhancements are briefly mentioned below and discussed in more detail in Sections 7.3.2–7.3.6.

IoT-NTN reuses the overall frameworks of these operations and makes enhancements related to the NTN.

Cell Selection and SI Acquisition. Several new IEs have been introduced in the SI, such as (i) cellBarredNTN to convey accessibility to the EPC via the NTN access and (ii) trackingAreaList to support broadcasting multiple TAs per IoT-NTN cell. Three new NTN-specific SIBs, SIB31, SIB32, and SIB33, have been introduced in Release 17 and 18.

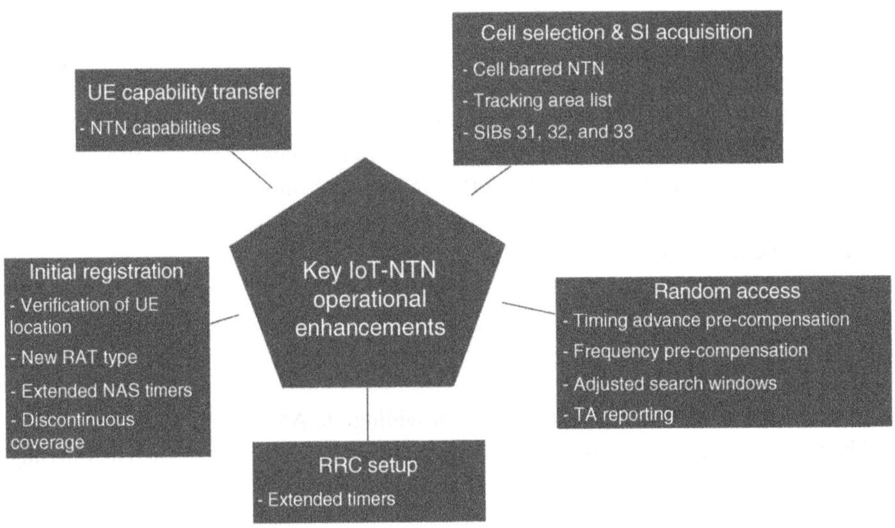

Figure 7.10 Key IoT-NTN Enhancements in Pre-Data Transfer Operations.

11 Certain terms are defined here in this book for enhanced clarity; these terms are not used in 3GPP NTN specifications. These terms include IoT-NTN UE, IoT-NTN eNB, IoT-NTN cell, LTE-M NTN, NB-IoT NTN, LTE-M NTN UE, and NB-IoT NTN UE.

Random-Access Procedure. To accommodate long delays and high Doppler shifts, the IoT-NTN UE carries out Timing Advance and frequency pre-compensation while transmitting during the random-access procedure. Additionally, the IoT-NTN UE may also support TA reporting during the random-access procedure.

RRC Connection Setup. During establishing the RRC connection between the IoT-NTN UE and the IoT-NTN cell, the IoT-NTN gNB can configure the IoT-NTN UE with suitable extended times to address long propagation delays.

Initial Registration. The MME may verify the IoT-NTN UE's location to ensure that it can access the NTN-related PLMN at its current location. Extended NAS timers can be configured. The NTN-specific RAT type is used. Discontinuous coverage-related parameters can be configured for the IoT-NTN UE.

UE Capability Transfer. The IoT-NTN UE specifies its typical and NTN-specific capabilities to the IoT-NTN gNB.

7.3.2 Cell Selection and SI Acquisition in IoT-NTN

The same cell selection and SI acquisition process is used for an IoT-TN and an IoT-NTN. See Chapter 3 for an overview of the overall cell selection process used in a TN in LTE-M and NB-IoT. When an IoT-NTN UE powers up, it searches for an IoT-NTN cell, selects a suitable cell, and acquires the SI as shown in Figure 7.11.

7.3.2.1 Step 1: IoT-NTN Cell Search and Cell Selection

The same cell selection process is used for a TN and an NTN. When an NTN UE is switched on, the NAS layer selects a PLMN (3GPP 2024k). UE policies may define the association between the selected PLMN and associated RAT(s) (3GPP 2024a). The AS in the NTN UE uses the list of equivalent PLMNs for cell selection and reselection.

As summarized in the case of NR-NTN, cell selection in IoT-NT may also be *initial cell selection* or *cell selection by leveraging stored information*. In the case of initial selection, the IoT-NTN UE does not have any prior knowledge of which RF channels are LTE-M/NB-IoT frequencies. In such a case, the IoT-NTN UE scans all RF channels in the IoT-NTN bands. On a given RF channel or frequency, the IoT-NTN UE searches for the strongest IoT-NTN cell and selects a suitable cell. A suitable cell is

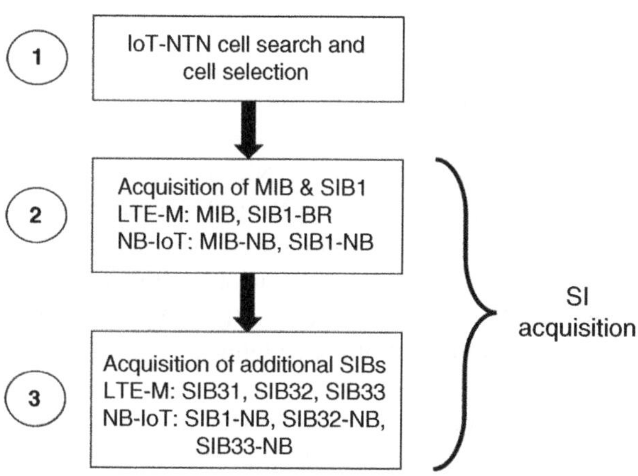

Figure 7.11 Cell Selection and SI Acquisition in IoT-NTN.

a cell that meets the cell selection criterion S (3GPP 2024k); a brief description of the IoT-NTN cell selection criterion S is given below. In general, to evaluate whether the currently monitored IoT-NTN cell meets the cell selection criterion, the IoT-NTN UE must make measurements of a cell, such as its RSRP and RSRQ. In the case of cell selection by leveraging stored information, the IoT-NTN UE utilizes stored information of frequencies and detected cells (and potentially cell parameters).

7.3.2.2 Step 2 in LTE-M NTN: Acquisition of MIB and SIB1-BR

Upon power-up, the LTE-M NTN UE detects an LTE-M NTN cell by observing the PSS and the SSS and determines the PCI of the LTE-M NTN cell just like an LTE-M TN UE. The LTE-M NTN UE then processes the PBCH to obtain (i) the MasterInformationBlock (MIB) containing critical information about the system and (ii) schedulingInfoSIB1-BR indicating whether SystemInformationBlockType1-Bandwidth Reduced (SIB1-BR) is scheduled or not. In support of the NTN, the MIB contains the IE earfcn-LSB that indicates the two least significant bits of the EARFCN[12] for the NTN bands that are using a 100 kHz raster (3GPP 2024h, 2024n). This IE facilitates the detection of IoT-NTN signals.

The structure of the SystemInformationBlockType1-BR is the same as the structure of the SystemInformationBlockType1 (i.e., SystemInformationBlockType1-BR-r13 = SystemInformationBlockType1) (3GPP 2024n). A SIB1-BR enhancement for the NTN is that the field "trackingAreaList-r17" in SIB1-BR plays an important role; it contains a list of Tracking Areas (TAs) associated with the cell indicated by cellIdentity. Note that the network may broadcast multiple Tracking Area Codes (TACs) in an LTE-M NTN cell in an NTN. The network changes the TACs in the SI in an implementation-specific manner; the TACs broadcast by an LTE-M NTN cell may not exactly reflect the ground area illuminated by the LTE-M NTN cell's beam(s) at a given instant. Additionally, a pre-Release 17 IE multiBandInfoList can be used to specify a list of prioritized NTN frequency bands (3GPP 2024h). Furthermore, a pre-Release 17 IE freqBandInfo can be used to define a list of additionalPmax and additionalSpectrumEmission values (3GPP 2024h, 2024j). These values affect the transmit power of the NTN UE.

After acquiring the MIB and schedulingInfoSIB1-BR, the LTE-M NTN UE obtains SystemInformationBlockType1-BR that provides important LTE-M parameters such as the signal threshold for Coverage Enhancement (CE), the starting OFDM symbol number for the MPDCCH, and schedule of other SIBs. The LTE-M IoT UE evaluates the cell selection criterion S and proceeds to acquire additional SIBs if the cell selection criterion S is met.

7.3.2.3 Step 2 in NB-IoT NTN: Acquisition of MIB-NB and SIB1-NB

The NB-IoT NTN UE, upon power-up, detects an NB-IoT NTN cell by observing the Narrowband Primary Synchronization Signal (NPSS) and the Narrowband Secondary Synchronization Signal (NSSS) and determines the Narrowband Physical Cell ID (NPCI) of the NB-IoT NTN cell using the NSSS. The NPSS and the NSSS are transmitted using a single PRB in NB-IoT because the maximum transmission bandwidth on a given NB-IoT carrier is one PRB in NB-IoT. The NB-IoT NTN UE then processes the Narrowband Physical Broadcast Channel (NPBCH) to obtain the MasterInformationBlock-NB containing critical information about the system such as (i) the System Frame Number (SFN) and Hyper SFN, and (ii) SIB1-Narrowband (SIB1-NB) scheduling information such

---

12 EARFCN represents E-UTRA Absolute Radio Frequency Channel Number, where E-UTRA represents LTE radio access and stands for Evolved Universal Terrestrial Radio Access. ERAFCN designates a specific location of the RF channel.

as Transport Block Size (TBS) and number of repetitions, (iii) Operation mode information, and (iv) Access Barring Enabled or Not. Like the MIB in LTE-M NTN, the MIB-NB contains the IE earfcn-LSB that specifies the two least significant bits of the EARFCN for the NTN bands that are using a 100 kHz raster (3GPP 2024h, 2024n).

After processing the MIB-NB, the NB-IoT NTN UE obtains SystemInformationBlockType1-NB (SIB1-NB) that provides NB-IoT cell-related information such as the cell selection threshold, schedule of other SIBs, and Frequency Band Indicator. Like SIB1-BR in LTE-M, SIB1-NB also includes the field "trackingAreaList-r17," which contains a list of TAs. The network may broadcast multiple TACs in an NB-IoT NTN cell. The network changes the TACs in the SI in an implementation-specific manner. The TACs being broadcast by an NB-IoT NTN cell may not exactly reflect the ground area illuminated by the NB-IoT NTN cell's beam(s) at a given instant. A pre-Release 17 IE freqBandInfo can be used to define a list of additionalPmax and additionalSpectrumEmission values (3GPP 2024h, 2024j). These values affect the transmit power of the NB-IoT NTN UE.

After processing the SIB1-NB, the NB-IoT NTN UE evaluates the cell selection criterion S and acquires additional SIBs if it is met.

7.3.2.4 Cell Selection Criterion S in IoT-NTN

The cell selection criterion from the TN is reused in the IoT-NTN. To evaluate whether the currently monitored IoT-NTN cell meets the cell selection criterion S, the IoT-NTN UE must make measurements such as RSRP and RSRQ of an IoT-NTN cell. The IoT-NTN UE utilizes measured RSRP and RSRQ to evaluate the cell selection criterion S. The cell selection criterion S in normal coverage[13] is said to be met when ((Srxlev >0) AND (Squal >0)), where Srxlev is the difference between the measured RSRP and a signal threshold (which is $Q_{rxlevmin}$, potentially adjusted by additional parameters), and Squal is the difference between the measured RSRQ and a signal quality threshold (which is $Q_{qualmin}$, potentially adjusted by additional parameters). See 3GPP (2024k) for details of the formula and associated parameters.

7.3.2.5 NTN Cell Indication and NTN Cell Barring in IoT-NTN

The IoT-NTN UE ignores the cell-barring information of the SystemInformationBlockType1-BR and SystemInformationBlockType1-NB. SystemInformationBlockType1-BR and SystemInformationBlockType1-NB contain an NTN-specific parameter called *cellBarredNTN*. If *cellBarredNTN* is set to "barred," the IoT-NTN UE considers that the cell is not allowed (i.e., barred) to connect to the EPC via the NTN. When the cell currently being monitored is barred, the IoT-NTN UE performs cell reselection to another cell at the same frequency as the barred cell (3GPP 2024k). If *cellBarredNTN* is set to "notBarred," the cell being monitored is allowed to connect to the EPC via the NTN. Hence, the IE *cellBarredNTN* can use the NTN-capable UE to infer that the cell being monitored is indeed an NTN cell.

7.3.2.6 Step 3 in LTE-M NTN: Acquisition of Additional SIBs

The LTE-M NTN UE learns about the SIBs being broadcast by the LTE-M NTN cell from the IE schedulingInfoList in the SIB1. Table 7.5 summarizes SI enhancements for the NTN (3GPP 2024n). The MIB and SIB1-BR enhancements have already been discussed earlier. The remaining enhancements are briefly summarized in Table 7.5.

---

13 In general, a UE may be in enhanced coverage when the normal coverage criterion S is not met but the enhanced coverage criterion corresponding to Coverage Enhanced (CE) RSRP threshold (and potentially RSRQ threshold) is met. There is a minor difference between LTE-M and NB-IoT cell selection; see (3GPP 2024k) for details.

Table 7.5 Key System Information Enhancements for the LTE-M NTN.

System Information (SI)	LTE-M NTN Related Changes
MIB	Cell barring information in MIB is ignored by the NR-NTN UE.
SIB1-BR	Contains cellBarredNTN to enable the NTN UE to determine if this is an allowed NTN cell or not. Includes the field "trackingAreaList-r17," containing a list of TAs to support multiple TAs per cell. The IE multiBandInfoList specifies a list of prioritized NTN frequency bands. A pre-Release 17 IE freqBandInfo can be used to define a list of additionalPmax and additionalSpectrumEmission values.
SIB2	The NTN-specific IE ntn-ConfigCommon-r17 provides NTN-specific parameters including ta-Report-r17, t318-r17, prach-TxDuration-r17, pucch-TxDuration-r17, and pusch-TxDuration-r17.
SIB3	Includes the parameter t-Service in support of quasi-Earth-fixed cells in an NTN.
SIB5	Provides a list of satellite IDs for a given carrier frequency.
SIB16	Contains time information to enable the LTE-M NTN UE to determine the UTC.
SIB31	Provides satellite assistance information for the serving cell.
SIB32	Satellite assistance information about the serving cell to enable the LTE-M NTN UE to predict discontinuous coverage.
SIB33	Satellite assistance information for neighboring cells.

SIB2 SIB2 provides radio resource configuration information that is common to all UEs. In particular, the IE RadioResourceConfigCommonSIB provides common radio resource configuration such as the random-access parameters and static physical layer parameters (3GPP 2024n). In support of the NTN, the IE RadioResourceConfigCommonSIB contains the IE ntn-ConfigCommon-r17 that provides NTN-specific parameters including ta-Report-r17, t318-r17, prach-TxDuration-r17, pucch-TxDuration-r17, and pusch-TxDuration-r17 (3GPP 2024n). The IE ta-Report-r17 indicates reporting of timing advance is enabled during Random Access during procedures, such as RRC connection establishment, RRC connection resume, or RRC connection reestablishment. The IE t318 specifies the value of timer T318. The IEs prach-TxDuration-r17, pucch-TxDuration-r17, and pusch-TxDuration-r17 are included to support segmented transmissions in an NTN. The IE prach-TxDuration specifies the duration of PRACH segment transmission in NTN transmission in the unit of the number of preamble transmissions (3GPP 2024i). The IE pucch-TxDuration specifies the duration of PUCCH segment transmission in NTN transmission in the unit of the number of subframes (3GPP 2024i). The IE pusch-TxDuration specifies the duration of PUSCH segment transmission in NTN transmission in the unit of the number of resource units (3GPP 2024i).

SIB3 The SystemInformationBlockType3 (SIB3) contains cell reselection information common for multiple types of cell reselection, such as intra-frequency, inter-frequency, and inter-RAT cell reselection (3GPP 2024n). The neighboring cell-related cell reselection information is not part of SIB3. In addition to containing IEs such as freqBandInfo and multiBandInfoList that are present in SIB2 and usable in an NTN, SIB3 also includes an NTN-specific parameter t-Service. The IE t-Service provides time information on when a quasi-Earth-fixed cell in an NTN will stop serving the currently served area (3GPP 2024k).

SIB5 In general, the SystemInformationBlockType5 (SIB5) contains information relevant to (i) inter-frequency cell reselection and (ii) E-UTRA and NR idle and inactive measurements (3GPP 2024n). Cell reselection parameters applicable to a frequency and cell-specific reselection parameters are included in SIB5. In addition to containing IEs such as freqBandInfo and multiBandInfoList that are present in SIB2 and SIB3 and usable in an NTN, SIB5 also includes an NTN-specific parameter satelliteAssistanceInfoList-r18, that provides a list of satellite IDs for a given carrier frequency. A set of carrier frequencies can be specified in the IE InterFreqCarrierFreqList-v1800, and satelliteAssistanceInfoList-r18 can be specified for each carrier frequency.

SIB16 The SystemInformationBlockType16 (SIB16) contains the IE timeInfoUTC that represents Coordinated Universal Time called UTC corresponding to the SFN boundary at or immediately after the ending boundary of the SI-window in which SIB16 is transmitted (3GPP 2024n). In an NTN cell, this indicated time is referenced at the uplink time synchronization Reference Point (RP). The LTE-M NTN UE considers the propagation delay between itself and the RP to determine the UTC time at the LTE-M NTN UE.

SIB31 As mentioned earlier, the 3GPP has introduced SystemInformationBlockType31 (SIB31) specifically for the NTN. SIB31 provides satellite assistance information for the serving cell. Table 7.6 lists key IEs of the IE ServingSatelliteInfo-r17 of SIB31 (3GPP 2024n). These IEs are briefly explained below.

ephemerisInfo-r17. Information about the satellite is specified in the form of stateVectors or orbitalParameters. The IE stateVectors specifies instantaneous values of the satellite state vectors, and the IE orbitalParameters provides instantaneous values of the satellite orbital parameters.

nta-CommonParameters-r17. Characteristics of the network-controlled common Timing Advance (TA) are specified in the form of three parameters: nta-Common-r17, nta-CommonDrift-r17, and nta-CommonDriftVariation-r17 (3GPP 2024i, 2024n). When one or more of these parameters are absent, the value of 0 is used. nta-Common-r17 is the network-controlled common TA expressed in the unit of μs and has a granularity of 32.55208×10^{-3}. The IE nta-CommonDrift-r17 is the drift rate of the common TA expressed in the unit of μs/s and has a granularity of 0.2×10^{-3} μs/s. The IE nta-CommonDriftVariation-r17 is the drift rate variation of the common TA expressed in the unit of μs/s^2 and has a granularity of 0.2×10^{-4} μs/s^2.

ul-SyncValidityDuration-r17. This is the validity duration of the satellite ephemeris data and common TA parameters. It is the maximum time duration from epochTime during which the UE can use the previously acquired satellite ephemeris without acquiring a new satellite ephemeris.

epochTime-r17. It is the epoch time or the satellite ephemeris data reference time and common TA parameters. It is the starting time of a DL subframe indicated by startSFN-r17 and startSubFrame-r17. The reference point for epoch time of the serving satellite ephemeris and Common TA parameters is the uplink time synchronization reference point for the NTN cell.

k-Offset-r17. It is the scheduling offset used in the timing relationships in the NTN and is expressed in the unit of ms.

Table 7.6 SIB 31 in LTE-M NTN: A Closer Look.

Information Element (IE)	Brief Description
ephemerisInfo-r17	Information about the satellite is specified in the form of stateVectors or orbitalParameters.
nta-Common Parameters-r17	Characteristics of the network-controlled common Timing Advance (TA) specified in the form of three parameters: nta-Common-r17, nta-CommonDrift-r17, and nta-CommonDriftVariation-r17.
ul-SyncValidity Duration-r17	Validity duration of the satellite ephemeris data and common TA parameters.
epochTime-r17	Epoch time of the satellite ephemeris data and common TA parameters expressed using startSFN-r17 and startSubFrame-r17.
k-Offset-r17	Scheduling offset is used in the timing relationships in the NTN.
k-Mac-r17	Scheduling offset is used when downlink and uplink frame timing are not aligned at the LTE-M NTN eNB.
satelliteId-r18	Identity of the satellite.
referenceLocation-r18	ReferenceLocation-r18 for fixedCell-r18 or movingCell-r18.
distanceThresh-r18	Distance from the serving cell reference location is used in location-based measurement initiation in RRC_IDLE (as specified in (3GPP 2024k)) and RRC_CONNECTED states. Each step represents 50 m.

k-Mac-r17. This scheduling offset parameter is used when downlink and uplink frame timing are not aligned at the LTE-M NTN eNB.

satelliteId-r18. It specifies the satellite identity.

referenceLocation-r18. This is the reference location of the NTN quasi-earth fixed cell (i.e., fixedCell-r18) or Earth-moving cell (i.e., movingCell-r18), used in location-based measurement initiation in the LTE-M NTN UE's RRC_IDLE and RRC_CONNECTED states. For the Earth-moving cell, the NTN UE derives the real-time reference location of the satellite based on (i) the broadcast reference location corresponding to the epoch time and (ii) the serving satellite ephemeris.

distanceThresh-r18. It is the distance of the LTE-M NTN UE from the serving cell reference location. This information is used in location-based measurement initiation in the LTE-M NTN UE's RRC_IDLE and RRC_CONNECTED states.

SIB32 As mentioned earlier, the 3GPP has introduced SystemInformationBlockType32 (SIB32) specifically for the NTN. SIB32 contains satellite assistance information to enable the LTE-M NTN UE to predict discontinuous coverage. Table 7.7 lists key IEs of SIB32 (3GPP 2024n). These IEs are briefly explained below.

satelliteInfoList-r17. This is a list of a maximum of maxSat-r17 entries of SatelliteInfo-r17. Each SatelliteInfo-r17 entry specifies (i) satelliteId-r17, (ii) serviceInfo-r17 containing tle-Ephemeris-Parameters-r17 and t-ServiceStart-r17, and (iii) footprintInfo-r17 containing referencePoint-r17 (longitude-r17 and latitude-r17), elevationAngles-r17 (elevationAngleRight-r17 and

Table 7.7 SIB 32 in LTE-M NTN: A Closer Look.

Information Element (IE)	Brief Description
satelliteInfoList-r17	A list of up to maxSat-r17 instances of the IE SatelliteInfo-r17. The IE SatelliteInfo-r17 includes (i) satelliteId-r17, (ii) serviceInfo-r17 containing tle-EphemerisParameters-r17 and t-ServiceStart-r17, and (iii) footprintInfo-r17 containing referencePoint-r17 (longitude-r17 and latitude-r17), elevationAngles-r17 (elevationAngleRight-r17 and elevationAngleLeft-r17), and radius-r17.
satelliteInfoList-v1800	A list of up to maxSat-r17 CarrierFreqList-v1800 with each CarrierFreqList-v1800 containing a list of up to maxSat-r17 carrier frequencies specified as ARFCN-ValueEUTRA.

elevationAngleLeft-r17), and radius-r17. The value of maxSat-r17 (which denotes the maximum number of satellites) is 4.

The IE satelliteId-r17 specifies the satellite identity.

The IE serviceInfo-r17 provides information on when the satellite will provide coverage. For example, the parameter tle-EphemerisParameters is specified for a satellite with Earth-moving cell(s), and the parameter t-ServiceStart is specified for a quasi-Earth-fixed cell. The IE title-EphemerisParameters include mean values of the satellite orbital parameters based on the TLE set format for estimating in-coverage and out-of-coverage periods for a satellite with Earth-moving cells (3GPP 2024k). The IE t-ServiceStart indicates when the incoming satellite will start serving the area for a quasi-Earth-fixed cell.

The IE footprintInfo-r17 contains (i) referencePoint-r17 (with longitude-r17 specifying the longitude of the reference point in degrees and latitude-r17 specifying the latitude of the reference point in degrees), (ii) elevationAngles-r17 (elevationAngleRight-r17 specifying rightmost elevation angle with reference to the satellite direction in degrees and elevationAngleLeft-r17 r17 specifying leftmost elevation angle with reference to the satellite direction in degrees), and (iii) radius-r17 (which is the distance between the reference point and the edge of the satellite or beam coverage in km).

The IE satelliteInfoList-v1800 consists of a list of up to maxSat-r17 CarrierFreqList-v1800 with each CarrierFreqList-v1800 containing a list of up to maxSat-r17 carrier frequencies specified as ARFCN-ValueEUTRA (which designates a carrier frequency).

SIB33 The SystemInformationBlockType33 (SIB33) is introduced for the NTN in Release 18 (3GPP 2024n). SIB33 provides satellite assistance information for neighbor cells, as specified in Table 7.8. These IEs are briefly explained below.

NeighSatelliteInfoList-r18. The IE neighSatelliteInfoList-r18 lists up to maxSat-r18 instances of the IE NeighSatelliteInfo-r18. maxSat-r18 is the maximum number of serving and neighbor satellites and is specified to be 4 in Release 18. The IE NeighSatelliteInfo-r18 includes the following IEs: (i) satelliteId-r18, (ii) ephemerisInfo-r18, (iii) nta-CommonParameters-r18, (iv) epochTime-r18, (v) k-Mac-r18, and (vi) t-ServiceStartNeigh-r18.

The IE satelliteId-r18 identifies the satellite.

The IE ephemerisInfo-r18 specifies stateVectors or orbitalParameters. The IE state vectors specify instantaneous values of the satellite state vectors. The IE orbitalParameters provides instantaneous values of the satellite orbital parameters.

Table 7.8 SIB 33 in LTE-M NTN: A Closer Look.

Information Element (IE)	Brief Description
neighSatelliteInfoList-r18	A list of up to maxSat-r18 instances of the IE NeighSatelliteInfo-r18. The IE NeighSatelliteInfo-r18 specifies (i) satelliteId-r18, (ii) ephemerisInfo-r18 (containing stateVectors or orbitalParameters), (iii) nta-CommonParameters-r18 (containing nta-Common-r18, nta-CommonDrift-r18, nta-CommonDriftVariation-r18), (iv) epochTime-r18 (containing startSFN-r18 and startSubFrame-r18), (v) k-Mac-r18, and (vi) t-ServiceStartNeigh-r18.
neighValidityDuration-r18	Validity duration of the neighbor satellite ephemeris data and common TA parameters.

The IE nta-CommonParameters-r18 specifies the common TA-related parameters, including nta-Common-r18, nta-CommonDrift-r18, and nta-CommonDriftVariation-r18, as discussed earlier for SIB32. nta-Common-r18 is the network-controlled common TA expressed in the unit of μs and has a granularity of 32.55208×10^{-3}. The IE nta-CommonDrift-r18 is the drift rate of the common TA expressed in the unit of μs/s and has a granularity of 0.2×10^{-3} μs/s. The IE nta-CommonDriftVariation-r18 is the drift rate variation of the common TA expressed in the unit of μs/s^2 and has a granularity of 0.2×10^{-4} μs/s^2.

The IE epochTime-r18 consists of startSFN-r18 and startSubFrame-r18. It is the epoch time of the neighbor satellite ephemeris data and common TA parameters (3GPP 2024i). epochTime-r18 is the starting time of a downlink subframe indicated by startSFN and startSubframe.

The IE k-Mac-r18 is the scheduling offset used when downlink and uplink frame timing are not aligned at the LTE-M NTN eNB and is expressed in ms.

The IE t-ServiceStartNeigh-r18 (TimeOffsetUTC-r17) indicates the earliest time when the area covered by the current serving cell is expected to be covered by the neighbor cell(s) served by the satellite indicated by satelliteId when quasi-Earth-fixed beams or cells are used.

neighValidityDuration-r18. It is the validity duration of the neighbor satellite ephemeris data and common TA parameters. It is the maximum time duration from epochTime during which the NTN UE can utilize previously acquired satellite ephemeris without acquiring a new satellite ephemeris (3GPP 2024i, 2024n). Its value ranges from 5 to 900 seconds.

7.3.2.7 Step 3 in NB-IoT NTN: Acquisition of Additional SIBs

The NB-IoT NTN UE learns about the SIBs being broadcast by the NB-IoT NTN cell from the IE schedulingInfoList in the SIB1-NB. Table 7.9 summarizes SI enhancements for the NB-IoT NTN (3GPP 2024n). The MIB and SIB1 enhancements have already been discussed earlier. The remaining enhancements are briefly summarized below.

SIB2-NB The NTN-specific IE ntn-ConfigCommon-r17 in the IE RadioResourceConfigCommon-SIB-NB provides NTN-specific parameters including ta-Report-r17, t318-r17, nprach-TxDurationFmt01-r17, nprach-TxDurationFmt2-r17, and npusch-TxDuration-r17 (3GPP 2024n). Furthermore, additional spectrum emission in SIB2-NB specifies spectrum emission-related requirements.

The IE ta-Report-r17 indicates reporting of timing advance is enabled during Random Access during procedures such as RRC connection establishment, RRC connection resume, or

Table 7.9 Key System Information Enhancements for the NB-IoT NTN.

System Information (SI)	NB-IoT NTN Related Changes
MIB-NB	earfcn-LSB specifies the two LSBs of the EARFCN for the NTN bands that are using a 100 kHz raster.
SIB1-NB	Contains cellBarredNTN to enable the NTN UE to determine if this is an allowed NTN cell or not. Includes the field "trackingAreaList-r17" containing a list of TAs to support multiple TAs per cell. The IE multiBandInfoList specifies a list of prioritized NTN frequency bands A pre-Release 17 IE freqBandInfo can be used to define a list of additionalPmax and additionalSpectrumEmission values.
SIB2-NB	The NTN-specific IE ntn-ConfigCommon-r17 in the IE RadioResourceConfigCommonSIB-NB provides NTN-specific parameters including ta-Report-r17, t318-r17, nprach-TxDurationFmt01-r17, nprach-TxDurationFmt2-r17, and npusch-TxDuration-r17. additionalSpectrumEmission specifies spectrum emission-related requirement.
SIB3-NB	Includes the parameter t-Service in support of quasi-Earth-fixed cells in an NTN. t-Service provides time information on when a quasi-Earth-fixed cell will stop serving the area it is currently serving. Includes multiBandInfoList containing a list of additionalPmax and additionalSpectrumEmission values applicable for the intra-frequency neighboring NB-IoT cells if the NB-IoT NTN UE selects the frequency band from freqBandIndicator in SystemInformationBlockType1-NB.
SIB5-NB	One or more instances of InterFreqCarrierFreqInfo-NB-r13with each instance specifying satelliteAssistanceInfo-r18 containing up to maxSat-r18 instances of SatelliteID-r18 (which is the identity of the satellite for neighbor cell measurements on a given carrier frequency).
SIB16-NB	Contains time information to enable the NB-IoT NTN UE to determine the UTC.
SIB31-NB	Provides satellite assistance information for the serving NB-IoT NTN cell.
SIB32-NB	Satellite assistance information to enable the NB-IoT NTN UE to predict discontinuous coverage.
SIB33-NB	Satellite assistance information for neighbor cells.

RRC connection reestablishment. The IE t318-r17 specifies the value of timer T318. The nprach-TxDurationFmt01-r17, nprach-TxDurationFmt2-r17, and npusch-TxDuration-r17 are included to support segmented transmissions in an NTN. The IE nprach-TxDurationFmt01-r17 specifies the duration of NPRACH segment transmission for NPRACH resource format 0 and format 1 in the NTN in the unit of preamble repetition unit, while the IE nprach-TxDurationFmt2-r17 specifies the duration of NPRACH segment transmission for NPRACH resource format 2 in the NTN in the unit of preamble repetition unit. The IE npusch-TxDuration-r17 specifies the duration of NPUSCH segment transmission in the NTN (3GPP 2024i).

SIB3-NB The SystemInformationBlockType3-NB (SIB3-NB) contains information common for multiple types of cell reselection, such as intra-frequency, inter-frequency, and inter-RAT cell reselection (3GPP 2024n). SIB3-NB includes the IE multiBandInfoList containing a list of additionalPmax and additionalSpectrumEmission values applicable for the intra-frequency neighboring NB-IoT cells if the NB-IoT NTN UE selects the frequency band from freqBandIndicator in SystemInformationBlock-Type1-NB. SIB3-NB also includes the parameter t-Service in support of quasi-Earth-fixed cells in an

NTN. t-Service provides time information on when a quasi-Earth-fixed cell will stop serving the area it is currently serving.

SIB5-NB In general, the SystemInformationBlockType5-NB (SIB5-NB) contains information relevant to inter-frequency cell reselection, such as information about other NB-IoT frequencies and inter-frequency neighboring cells (3GPP 2024n). Cell reselection parameters applicable to a frequency and cell-specific reselection parameters are included in SIB5-NB. SIB5-NB includes one or more instances of InterFreqCarrierFreqInfo-NB-r13with each instance specifying satelliteAssistanceInfo-r18 containing up to maxSat-r18 instances of SatelliteID-r18 (which is the identity of the satellite for neighbor cell measurements on a given carrier frequency).

SIB16-NB The SystemInformationBlockType16-NB (SIB16-NB) contains the IE timeInfo-r11, which includes timeInfoUTC, which specifies UTC (3GPP 2024n). In an NTN cell, timeInfoUTC is referenced at the uplink time synchronization RP. The NB-IoT NTN UE considers the propagation delay between itself and the RP to determine the UTC time at the NB-IoT NTN UE.

SIB31-NB As mentioned earlier, the 3GPP has introduced SystemInformationBlockType31-NB (SIB31-NB) specifically for the NTN (3GPP 2024n). SIB31-NB provides satellite assistance information for the serving cell. SIB31-NB contains the IE ServingSatelliteInfo-r17, which is also part of SIB31. See Table 7.6 and the following descriptions for details of the IEs of the IE ServingSatelliteInfo-r17 of SIB31-NB (3GPP 2024n).

SIB32-NB As mentioned earlier, the 3GPP introduced SystemInformationBlockType32-NB (SIB32-NB) specifically for the NTN (3GPP 2024n). SIB32-NB contains satellite assistance information to enable the NB-IoT NTN UE to predict discontinuous coverage. SIB32-NB contains the IE satelliteInfoList-r17. This IE is also used in SIB32; see Table 7.7 for details.

SIB33-NB The SystemInformationBlockType33-NB (SIB33-NB) is introduced for the NTN in Release 18 (3GPP 2024n). SIB33-NB provides satellite assistance information for neighbor cells through two IEs, neighSatelliteInfoList-r18 and neighValidityDuration-r18. These IEs in NB-IoT are the same as the IEs in SIB33 for LTE-M MTM. See Table 7.8 and details in the description following the table.

7.3.3 Random-Access Enhancements in IoT-NTN

The overall RA procedure is enhanced in IoT-NTN due to NTN-specific challenges such as long and variable propagation delays and high NTN cell mobility.

Figure 7.12 illustrates certain timing relationship parameters such as service link Round Trip Time (RTT), Feeder link RTT, Common TA, k_{mac}, and T_{TA} for an IoT-NTN using a transparent NTN payload. These parameters are the same as those discussed earlier for NR-NTN (3GPP 2024j).

Recall that the uplink time synchronization RP is the entity where the downlink frame and the uplink frame are aligned through the offset $N_{TA,offset}$ (3GPP 2024i, 2024j). Common TA is a configured timing offset that specifies the RTT between the RP and the NTN payload. K_{offset} is a configured scheduling offset that must be larger than or equal to the sum of the service link RTT and the Common TA. k_{mac} is a configured timing offset that specifies the RTT between the RP and the NTN

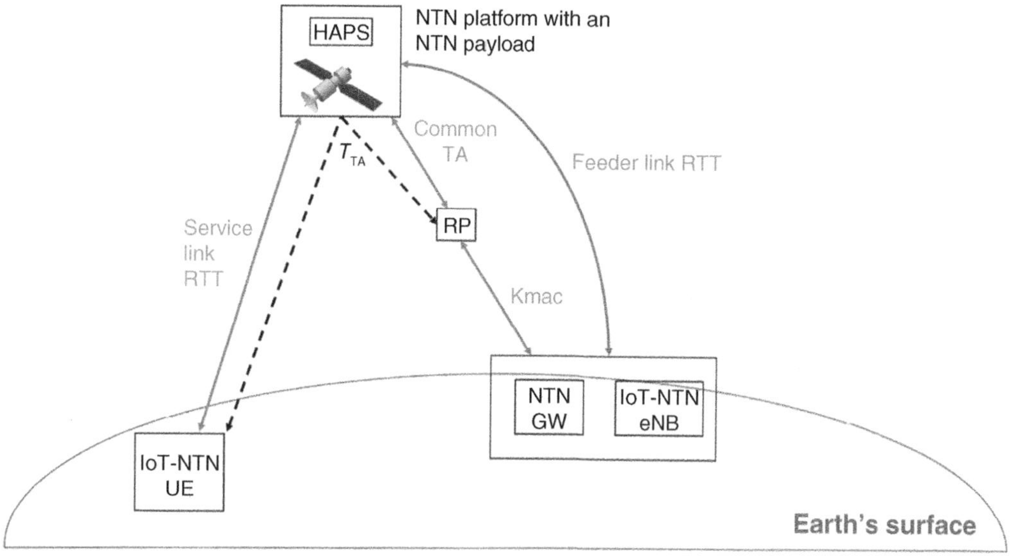

Figure 7.12 Timing Relationship in IoT-NTN.

eNB. k_{mac} is used to delay the application of a downlink configuration indicated by a MAC CE command on PDSCH/NPDSCH (3GPP 2024q) and to estimate the UE-eNB RTT (3GPP 2024u). The NTN eNB may provide k_{mac} when the downlink and the uplink frame timing are not aligned at the NTN eNB. k_{mac} is used to determine the start time of RAR window after an RA preamble transmission (3GPP 2024i). The scheduling offset K_{offset} provides the IoT-NTN UE sufficient processing time between a downlink reception and an uplink transmission (3GPP 2024i).

Figure 7.13 illustrates a simplified view of the 4-step RA procedure in an IoT-NTN. For a comprehensive view, see 3GPP (2024j) and 3GPP (2024m).

7.3.3.1 Step 0: Prior to RA Preamble Transmission

In *Step 0*, prior to the transmission of an RA preamble, the IoT-NTN UE determines a suitable Timing Advance to pre-compensate for the propagation delay so that its preamble arrives at the RP (approximately) at the target time and the IoT-NTN UE's signal does not cause timing misalignment-related interference in the uplink. To determine the timing of pre-compensation, the IoT-NTN UE utilizes its location (determined using the GNSS) and the NTN platform's ephemeris from System Information. Specifically, SIB31/ephemerisInfo-r17 in LTE-M or SIB31-NB/ephemerisInfo-r17 in NB-IoT, along with the GNSS-based IoT-NTN UE location, can be used to estimate the service link RTT. The IoT-NTN UE also obtains Common TA from SIB31/nta-CommonParameters-r17 in LTE-M or SIB31-NB/nta-CommonParameters-r17 in NB-IoT. The IoT-NTN UE computes the RTT between itself and the RP using the service link RTT and Common TA to pre-compensate for T_{TA} shown in Figure 7.12.

Furthermore, the IoT-NTN UE must pre-compensate for the Doppler shift so the IoT-NTN UE's signal does not cause frequency misalignment-related interference in the uplink. Hence, the IoT-NTN UE computes the Doppler shift of the service link based on its position and the ephemeris data from SIB31/ephemerisInfo-r17 in LTE-M or SIB31-NB/ephemerisInfo-r17 in NB-IoT and

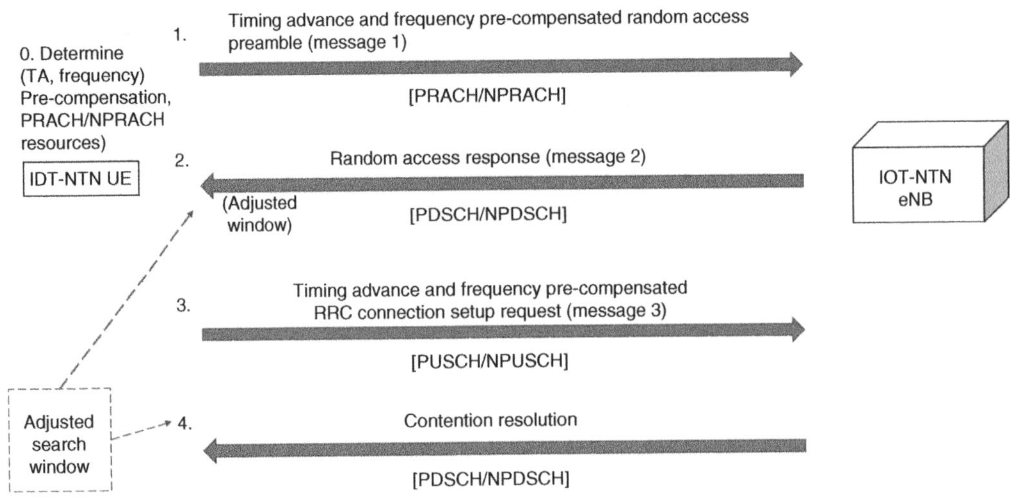

Figure 7.13 Simplified 4-Step Random-Access Procedure in IoT-NTN.

determines frequency pre-compensations for its RA preamble transmission so that the IoT-NTN UE's signal does not cause frequency misalignment-related interference in the uplink.

The IoT-NTN UE utilizes a suitable PRACH configuration based on the received SI and randomly selects an RA preamble. The IoT-NTN UE measures the RSRP of its serving IoT-NTN cell and determines the CE level. If the RSRP is too weak, the CE level corresponding to more repetitions is selected to increase reliability.

7.3.3.2 Step 1: RA Preamble Transmission

The LTE-M NTN UE sends a TA and Doppler shift pre-compensated RA preamble (also known as message 1 or Msg1) on the PRACH using the PRACH configuration specified in SIB2 and uses the correct number of preamble repetitions. The PRACH in LTE-M NTN uses the subcarrier spacing of 1.25 kHz.

The NB-IoT NTN UE sends a TA and Doppler shift pre-compensated RA preamble (also known as message 1 or Msg1) on the NPRACH with the appropriate number of repetitions using the NPRACH configuration specified in SIB2-NB. The NPRACH uses the subcarrier spacing of 3.75 kHz.

7.3.3.3 Step 2: Random-Access Response

In *Step 2*, the IoT-NTN eNB looks for the RA preambles from IoT-NTN UEs during certain periods. The IoT-NTN eNB responds with the RAR, considered to be message 2 or Msg2, that contains typical configuration parameters for the IoT-NTN UE, such as the Timing Advance Command to facilitate UL synchronization for the IoT-NTN UE, UL grant for message 3 transmission, and (Temporary) C-RNTI that identifies the IoT-NTN UE on the radio interface in the current IoT-NTN cell.

Note that the IoT-NTN UE looks for the Random-Access Response in the RA Response window, denoted by ra-ResponseWindow configured in the IE RACH-ConfigCommon in SIB2/RadioResourceConfigCommonSIB or RACH-ConfigCommon-NB-r13 in SIB2-NB/RadioResourceConfigCommonSIB-NB-r13, after sending the RA preamble. To accommodate long propagation delays in IoT-NTN, the start of the window is adjusted based on the propagation delay (3GPP

2024m). In the case of LTE-M NTN, the RA Response window starts at the subframe that contains the end of the last preamble repetition plus 3 subframes plus LTE-M NTN UE-eNB RTT and has length ra-ResponseWindowSize for the corresponding enhanced coverage level (3GPP 2024m). In the case of NB-IoT NTN, the RA Response window starts at the subframe that contains the end of the last preamble repetition plus X subframes plus UE-eNB RTT and has length ra-ResponseWindowSize for the corresponding enhanced coverage level, where X depends on the duplex mode (i.e., FDD vs. TDD), preamble format, and the number of NPRACH transmissions) (3GPP 2024m).

7.3.3.4 Step 3: Msg3 Transmission

In *Step 3*, using the allocated UL grant, the IoT-NTN UE sends Message 3/Msg 3 to initiate the establishment of an RRC connection between the UE and the eNB. Specifically, an LTE-M NTN UE sends the RRC Connection Setup Request message, and an NB-IoT NTN UE sends the RRC Connection Setup Request-NB message. This message contains the IoT-NTN UE's identity, such as S-TMSI, if it has been allocated a GUTI previously by an MME. S-TMSI is a short form of GUTI. If the IoT-NTN UE does not have S-TMSI, it generates a random number and includes this number in the RRC signaling message. Since more than one IoT-NTN UE might have used the same RA preamble ID, more than one IoT-NTN UEs may send the RRC signaling message using the UL grant specified by the IoT-NTN eNB in the RAR. The presence of the UE ID, such as S-TMSI or a random number, enables the IoT-NTN eNB to resolve the conflict among the NR-NTN UEs that might have used the same RA preamble at the same time.

Note that the IoT-NTN UE looks for the Contention Resolution message after sending Msg3 within a certain period. This period corresponds to the timer mac-ContentionResolutionTimer. To accommodate long propagation delays in IoT-NTN, the IoT-NTN UE starts mac-ContentionResolutionTimer at each HARQ retransmission of the bundle in the subframe containing the last repetition of the corresponding PUSCH/NPUSCH transmission plus IoT-NTN UE-eNB RTT (3GPP 2024m).

7.3.3.5 Step 4: Contention Resolution

The IoT-NTN eNB sends in the PDSCH/NPDSCH the Contention Resolution message containing the UE identity to the IoT-NTN UE that the IoT-NTN eNB has successfully detected. In Step 2, the location of this PDSCH/NPDSCH is identified by the PDCCH/NPDCCH associated with C-RNTI allocated to the IoT-NTN UE.

A successful RA procedure enables the RRC connection setup procedure to continue. In case the RA procedure fails, the IoT-NTN UE retries it by sending another RA preamble after a random wait time.

7.3.3.6 Random Access in IoT-NTN: Additional Considerations

The IoT-NTN UE reports Timing Advance during the RA procedure using Timing Advance Report MAC CE (3GPP 2024m) on the PUSCH/NPUSCH if SIB2/ta-Report-r17 in LTE-M or SIB2-NB/ta-Report-r17 in NB-IoT indicates that the TA report is enabled (3GPP 2024n). The Timing Advance reporting procedure is used in the NTN so that the IoT-NTN UE can provide the IoT-eNB with an estimate of the UE's Timing Advance value (i.e., T_{TA}).

The TA reporting procedure is also used in the NTN after the RA procedure (3GPP 2024m). For example, the TA reporting procedure is triggered if the IoT-NTN eNB configures the IoT-NTN UE with offsetThresholdTA. This procedure is also triggered if the IoT-NTN UE has not previously reported the Timing Advance value to the Timing Advance value of the current serving cell.

Furthermore, suppose the variation between the current estimate of the Timing Advance value and the last reported Timing Advance value is equal to or larger than offsetThresholdTA. In that case, the IoT-NTN UE sends a TA report to the IoT-NTN eNB.

Note that the IoT-NTN UE performs the pre-compensation of the Doppler shift on the service link. The management of (i) the Doppler shift on the feeder link and (ii) transponder frequency error is network implementation specific.

7.3.4 RRC Connection Setup in IoT-NTN

Figure 7.14 summarizes the main steps of a 3-way handshake between the IoT-NTN UE and the IoT-NTN eNB to set up an RRC connection in an IoT-NTN (3GPP 2024n). The RRC connection setup process is similar for a TN and an NTN. *The main difference between a TN RRC connection setup and an NTN RRC connection setup is that the IoT-NTN UE carries out Timing Advance and frequency compensation before uplink transmission of the first and the third message.*

In *Step 1*, like a TN, the IoT-NTN UE sends an RRC signaling message to set up an RRC connection. The LTE-M NTN UE sends the RRC Setup Request message, while an NB-IoT sends the RRC Setup Request-NB message. This RRC message includes the UE ID, such as S-TMSI or a random number. The IoT-NTN UE also indicates the purpose behind the message by specifying the establishment cause (e.g., mobile-originated signaling). The NB-IoT NTN UE also indicates its support for multitone transmission and multi-carrier transmission based on its capabilities. While not inside the RRC Setup Request message, a Timing Advance Report MAC CE can accompany the RRC Setup Request message if the NR-NTN gNB enables such reporting.

In *Step 2*, the IoT-NTN eNB responds with an RRC message to set up IoT-NTN UE-specific Signaling Radio Bearer 1 (SRB1). Specifically, the LTE-M NTN eNB sends the RRC Connection Setup containing LTE-M-specific configuration such as MPDCCH NarrowBand Indicator (NBI) and the number of MPDCCH repetitions to facilitate the MPDCCH monitoring by the LTE-M NTN UE. Similarly, the NB-IoT eNB sends the RRC Connection Setup-NB message to the NB-IoT NTN UE, which contains the number of RLC retransmissions, periodic BSR, the NPDCCH configuration, such as the number of repetitions, and the NPUSCH configuration, such as the number of ACK/ANCK repetitions.

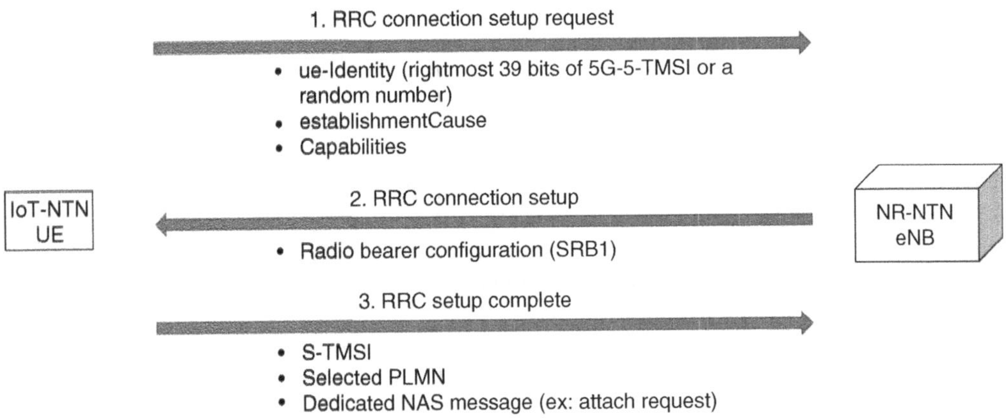

Figure 7.14 RRC Connection Setup in an IoT-NTN.

In *Step 3*, the LTE-M NTN UE uses SRB1 to send the RRC Setup Complete message to the IoT-NTN eNB, and the NB-IoT NTN UE uses SRB1 to send the RRC Setup Complete-NB to the IoT-NTN eNB. These messages contain S-TMSI, selected PLMN (and registered MME if available), and dedicated EPS NAS messages such as Attach Request (which in turn carries a PDN Connectivity Request message with a request to connect to a PDN). The LTE-M NTN UE also conveys certain capabilities useful for LTE-M operations, such as support for CE Mode B, attach without PDN Connectivity, and UP/CP Cellular IoT Optimizations. In contrast, the NB-IoT NTN UE conveys additional capabilities useful for NB-IoT operations, such as support for attach without PDN Connectivity and UP Cellular IoT Optimizations.

7.3.5 Initial Attach via IoT-NTN

Before an IoT-NTN UE can send or receive IoT traffic, prerequisites such as initial attach need to be met. Figure 7.15 shows the major steps of the initial attachment, including EPS bearer setup in IoT-NTN (3GPP 2024c).

7.3.5.1 Step 1: Network Acquisition

As described in Section 7.3.2, the IoT-NTN UE, upon power-up, detects the PSS and the SSS and determines the cell's PCI. The IoT-NTN UE then obtains System Information, including the MIB/MIB-NB and relevant SIBs/SIB-NBs. After obtaining random-access parameters, the IoT-NTN UE is ready to execute the RA procedure.

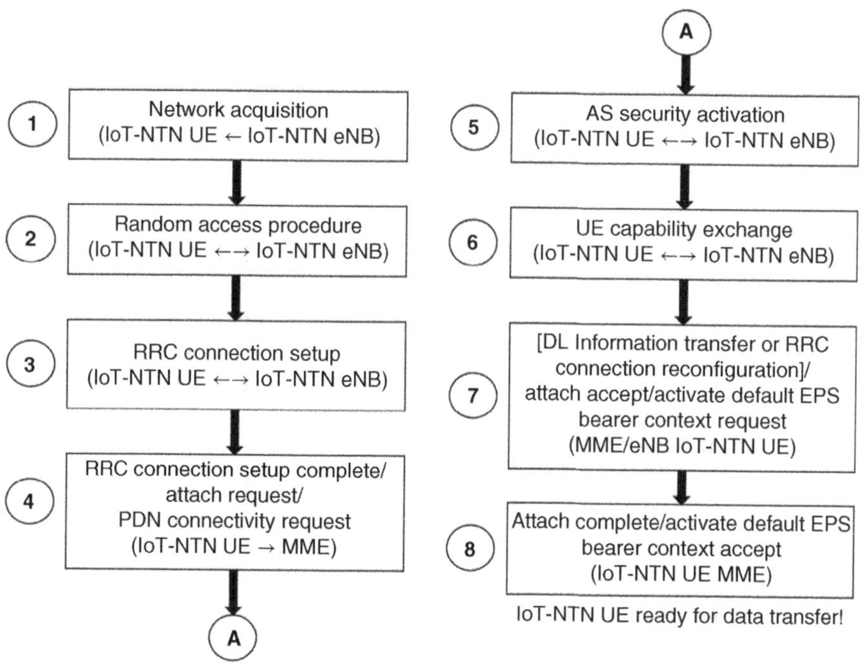

Figure 7.15 Initial Attach with EPS Bearer Setup in IoT-NTN.

7.3.5.2 Step 2: Random-Access Procedure

Section 7.3.3 explains the RA procedure in an IoT-NTN. The RA procedure has been enhanced for the IoT-NTN to accommodate much longer propagation delays in the NTN compared to a TN. The IoT-NTN UE sends a Timing Advance and frequency pre-compensated RA preamble to the selected IoT-NTN cell, which receives an RAR. Using the UL grant specified by the RAR, the IoT-NTN UE initiates the establishment of an RRC signaling connection.

7.3.5.3 Step 3: RRC Connection Setup

As discussed in Section 7.3.4, the IoT-NTN UE sends an RRC message to the IoT-NTN cell toward the end of the RA procedure to request the setup of the RRC connection. The IoT-NTN UE may also provide some IoT-related capabilities to the network. The IoT-NTN cell responds with an RRC message to set up the RRC connection and provides the radio interface configuration information to the IoT-NTN UE.

7.3.5.4 Step 4: Attach Request

The IoT-NTN UE sends the NAS Attach Request message to the MME via the IoT-NTN eNB. The IoT-NTN UE encapsulates this NAS message in the RRCConnectionSetupComplete/RRCConnectionSetupComplete-NB message. The IoT-NTN eNB selects a suitable MME for the IoT-NTN UE. The Attach Request message contains several IoT-NTN-specific parameters such as timer T3324 for PSM, timer T3412 (extended value) for periodic TAU, eDRX parameters, RoHC parameters for IP-based Control Plane for IoT traffic transport, and an indication for the support of SMS for the IoT traffic (3GPP 2024c). The NAS Attach Request message also contains a PDN Connectivity Request message for connecting to a PDN (e.g., Internet). The PDU Connectivity Request message includes typical Protocol Configuration Options (PCO) containing an IP address request and a DNS address request. The IoT-NTN UE can indicate its preference for non-IP data transfer by indicating the PDU type= "Non-IP."

Upon receiving the NAS Attach Request message, the MME may decide to verify the UE location based on the User Location Information (ULI) received from the IoT-NTN eNB to meet the regulatory requirements. Suppose the MME determines that the PLMN selected by the IoT-NTN UE is not allowed to operate at the present UE location. In that case, the MME rejects the IoT-NTN UE's attach request by replying with an Attach Reject message with a suitable Cause value. In such a case, the IoT-NTN UE attempts to select another PLMN.

Mutual authentication between the IoT-NTN UE and the MME occurs. The MME obtains authentication-related parameters from the HSS in support of mutual authentication. The NAS security between the IoT-NTN UE and the MME is also set up.

7.3.5.5 Step 5: Access Stratum Security Activation

The IoT-NTN UE and the IoT-NTN eNB exchange RRC messages to set up AS security. The RRC messages can now be both ciphered and integrity protected.

7.3.5.6 Step 6: UE Capability Transfer

The IoT-NTN eNB sends an RRC message to the IoT-NTN UE to obtain the capabilities of the IoT-NTN. The IoT-NTN UE specifies RAT Type as E-UTRA in its response message to the IoT-NTN eNB. The IoT-NTN UE also specifies its IoT-NTN capabilities. See Section 7.3.6 for more details.

7.3.5.7 Step 7: Attach Accept

The IoT-NTN eNB sends an RRC that encapsulates the NAS Attach Accept message received from the MME in the S1AP Initial Context Setup Request or DL NAS Transport message (3GPP 2024c). The LTE-M NTN eNB utilizes RRCConnectionReconfiguration or DLInformationTransfer, and the NB-IoT NTN eNB utilizes RRCConnectionReconfiguration-NB or DLInformationTransfer-NB. The NAS Attach Accept message specifies GUTI, PCO, eDRX parameters, T3412 extended value for periodic TAU, T3324 for PSM, "Control Plane Only Indicator" (if applicable), and RoHC for IP-based CP. The Attach Accept message contains the Activate Default EPS Bearer Context Request message to set up a default EPS bearer with specific QoS parameters for the IoT-NTN UE. The RRCConnectionReconfiguration from an LTE-M NTN eNB and RRCConnectionReconfiguration-NB from an NB-IoT NTN eNB specify the DRB portion of the EPS bearer.

If the MME receives an Attach Request indicating "Enhanced Discontinuous Coverage Support," it indicates to the IoT-NTN UE whether it supports Enhanced Discontinuous Coverage by including the "Enhanced Discontinuous Coverage Support" indication in the NAS Attach Accept message. The MME may also consider supporting discontinuous coverage while configuring the IoT-NTN UE with extended idle mode DRX parameters in the Attach Accept message.

Furthermore, the MME may include the "Return To Coverage Notification Not Required" IE to request the IoT-NTN UE in the ECM_IDLE state not to perform the TAU procedure when the IoT-NTN UE returns to coverage. Additionally, the MME may specify a maximum time offset. The MME also indicates in the Attach Accept message whether the IoT-NTN UE is not required to perform a TAU procedure when the unavailability period has ended.

The MME and the IoT-NTN UE use extended NAS timers to accommodate long propagation delays in the NTN (3GPP 2024f).

7.3.5.8 Step 8: Attach Complete

The IoT-NTN UE replies to the Attach Accept/Activate Default EPS Bearer Context Request message with the Attach Complete/Activate Default EPS Bearer Context Accept message by encapsulating such NAS message in (i) an RRC ULInformationTransfer message in LTE-M NTN and (ii) an RRC ULInformationTransfer-NB message in NB-IoT NTN (3GPP 2024c). The IoT-NTN eNB forward the Attach Complete message to the MME in an Uplink NAS Transport message. The network completes the S1-U/access bearer portion of the default EPS bearer so that all three components of the default EPS bearer are fully set up. Note that the EPS bearer consists of the following three components: (i) the DRB between the IoT-NTN UE and the IoT-NTN eNB in an IoT-NTN cell, (ii) the S1-U or access bearer between the IoT-NTN eNB and the S-GW, and (iii) the S5/S8 bearer between the S-GW and the P-GW.

7.3.5.9 Coarse UE Location Reporting in IoT-NTN

As in the case of NR-NTN, upon request by the network, an IoT-NTN UE reports its coarse UE location information[14] to the IoT-NTN eNB if such information is available at the IoT-NTN UE. Such coarse UE location reporting occurs after AS security is established for the IoT-NTN UE is in RRC_CONNECTED state (3GPP 2024j).

---

14 The coarse UE location is specified as the most significant bits of the GNSS coordinates to achieve location accuracy on the order of 2 km.

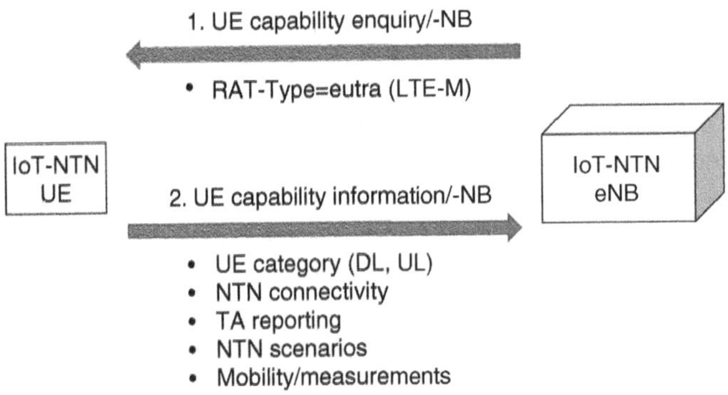

Figure 7.16 UE Capability Transfer in IoT-NTN.

7.3.6 UE Capability Transfer in IoT-NTN

The IoT-NTN eNB learns about the capabilities of the IoT-NTN UE through the UE capability transfer procedure (3GPP 2024n). Figure 7.16 summarizes the UE capability transfer-related RRC signaling message exchange between the IoT-NTN UE and the IoT-NTN eNB.

In *Step 1*, the IoT-NTN eNB sends an RRC message to the IoT-NTN UE to inquire about the IoT-NTN UE's radio access capabilities. Such messages are UECapabilityEnquiry in LTE-M and UECapabilityEnquiry-NB in NB-IoT. The LTE-M IoT eNB includes RAT-Type = eutra in the UECapabilityEnquiry message to request the transfer of the UE's LTE-M related radio access capabilities. The eNB may ask for the UE's radio access capabilities related to several non-LTE-M RATs, such as NR and UMTS, in the UECapabilityEnquiry message. In contrast, the UECapabilityEnquiry-NB message is specific to the UE's NB-IoT capabilities.

In *Step 2*, the IoT-NTN UE provides its capabilities to the IoT-NTN eNB. **Several NTN-related capabilities are common to both LTE-M and NB-IoT**.

7.3.6.1 LTE-M NTN UE Capabilities

The LTE-M NTN UE responds to the capability inquiry message from the LTE-M NTN eNB with a UECapabilityInformation message conveying its LTE-M radio access capabilities, including NTN capabilities. Specifically, this RRC message includes UE-CapabilityRAT-ContainerList, which is a list of containers (i.e., one or more instances of UE-CapabilityRAT-Container) associated with different RATs. For example, in the container for LTE-M, the LTE-M NTN UE includes RAT-Type=eutra and specifies numerous LTE-M capabilities as part of UECapabilityRAT-Container containing the IE UE-EUTRA-Capability. Note that LTE-M is considered part of LTE or E-UTRA. Table 7.10 specifies the LTE-M NTN UE's NTN capabilities defined in Release 17 and the UE category (3GPP 2024l, 2024n).

Table 7.11 specifies the LTE-M NTN UE's NTN capabilities defined in Release 18 (3GPP 2024l, 2024n). These capabilities are part of the IE UE-EUTRA-Capability-v1800-IEs/ntn-Parameters-v1800/NTN-Parameters-v1800.

7.3.6.2 NB-IoT NTN UE Capabilities

The NB-IoT NTN UE responds to the capability inquiry message from the NB-IoT eNB with a UECapabilityInformation-NB message conveying its NB-IoT radio access capabilities, including NTN capabilities in various IEs. Table 7.12 specifies the NB-IoT NTN UE's NTN capabilities defined in Release 17 and the UE category (3GPP 2024l, 2024n).

Table 7.10 IoT-NTN-Related Capabilities of an LTE-M NTN UE in Release 17.

Capability IE	Brief Explanation	Major Parent IE
UE-CategoryDL-v1310	Specifies Category M1 for the downlink.	UE-EUTRA-Capability-v1310-IEs
UE-CategoryUL-v1310	Specifies Category M1 for the uplink.	UE-EUTRA-Capability-v1310-IEs
ue-CategoryDL-v1430	Specifies Category M2 for the downlink.	UE-EUTRA-Capability-v1430-IEs
UE-CategoryUL-v1430	Specifies Category M2 for the uplink.	UE-EUTRA-Capability-v1430-IEs
ntn-Connectivity-EPC-r17	Indicates whether the UE supports NTN access.	UE-EUTRA-Capability-v1700-IEs/ ntn-Parameters-r17/ NTN-Parameters-r17
ntn-TA-Report-r17	Indicates whether the UE supports Timing advance reporting in an NTN cell.	UE-EUTRA-Capability-v1700-IEs/ ntn-Parameters-r17/ NTN-Parameters-r17
ntn-PUR-TimerDelay-r17	Indicates whether the UE supports delaying the start of the pur-ResponseWindowTimer in an NTN operation.	UE-EUTRA-Capability-v1700-IEs/ ntn-Parameters-r17/ NTN-Parameters-r17
ntn-OffsetTimingEnh-r17	Indicates whether the UE supports timing relationship enhancements using Differential Koffset.	UE-EUTRA-Capability-v1700-IEs/ ntn-Parameters-r17/ NTN-Parameters-r17
ntn-ScenarioSupport-r17	Indicates whether the UE supports NTN features in GSO or NGSO scenario.	UE-EUTRA-Capability-v1700-IEs/ ntn-Parameters-r17/ NTN-Parameters-r17
ntn-SegmentedPrecompensationGaps-r17	Specifies the supported gap length between segments for PUSCH and PUCCH (values: sym1 for 1 symbol, sl1 for 1 slot, sf1 for one subframe).	UE-EUTRA-Capability-v1720-IEs/ ntn-Parameters-v1720/ NTN-Parameters-v1720

Table 7.11 IoT-NTN-Related Capabilities of an LTE-M NTN UE in Release 18.

Capability IE	Brief Explanation
ntn-EventA4BasedCHO-r18	Indicates whether the UE supports Event A4-based conditional handover (called CondEvent A4).
ntn-LocationBasedCHO-EFC-r18	Indicates whether the UE supports location-based conditional handover for an Earth-Fixed Cell (EFC) (called CondEvent D1).
ntn-LocationBasedCHO-EMC-r18	Indicates whether the UE supports location-based conditional handover for an Earth-Moving Cell (EMC) (called CondEvent D1).
ntn-TimeBasedCHO-r18	Indicates whether the UE supports time-based conditional handover (called CondEvent T1).
ntn-LocationBasedMeasTrigger-EFC-r18	Indicates whether the UE supports location-based measurement trigger in RRC_CONNECTED in an Earth-Fixed Cell.
ntn-LocationBasedMeasTrigger-EMC-r18	Indicates whether the UE supports location-based measurement trigger in RRC_CONNECTED in an Earth-Moving Cell.
ntn-TimeBasedMeasTrigger-r18	Indicates whether the UE supports time-based measurement trigger in the RRC_CONNECTED state.

(*Continued*)

Table 7.11 (Continued)

Capability IE	Brief Explanation
ntn-RRC-HarqDisableSingleTB-CE-ModeA-r18	indicates whether the UE supports HARQ feedback disabling per HARQ process for downlink transmission by RRC configuration when operating in coverage enhancement mode A.
ntn-RRC-HarqDisableMultiTB-CE-ModeA-r18	Indicates whether the UE supports HARQ feedback disabling per HARQ process for downlink transmission by RRC configuration when operating in coverage enhancement mode A and when configured with ce-PDSCH-MultiTB-Config.
ntn-RRC-HarqDisableSingleTB-CE-ModeB-r18	Indicates whether the UE supports HARQ feedback disabling per HARQ process for downlink transmission by RRC configuration when operating in coverage enhancement mode B.
ntn-OverriddenHarqDisableSingleTB-CE-ModeB-r18	Indicates whether the UE supports DCI-based HARQ feedback disabling for downlink transmission by overriding the RRC configuration when operating in coverage enhancement mode B.
ntn-DCI-HarqDisableSingleTB-CE-ModeB-r18	Indicates whether the UE supports DCI-based HARQ feedback disabling for downlink transmission when HARQ feedback disabling per HARQ process for downlink transmission is not configured by RRC and operating in coverage enhancement mode B.
ntn-RRC-HarqDisableMultiTB-CE-ModeB-r18	Indicates whether the UE supports HARQ feedback disabling per HARQ process for downlink transmission by RRC configuration when operating in coverage enhancement mode B and when configured with ce-PDSCH-MultiTB-Config.
ntn-OverriddenHarqDisableMultiTB-CE-ModeB-r18	Indicates whether the UE supports DCI-based HARQ feedback disabling for downlink transmission by overriding the RRC configuration when operating in coverage enhancement mode B and when configured with ce-PDSCH-MultiTB-Config.
ntn-DCI-HarqDisableMultiTB-CE-ModeB-r18	Indicates whether the UE supports DCI-based HARQ feedback disabling for downlink transmission when HARQ feedback disabling per HARQ process for downlink transmission is not configured by RRC and operating in coverage enhancement mode B and when configured with ce-PDSCH-MultiTB-Config.
ntn-SemiStaticHarqDisableSPS-r18	Indicates whether the UE supports HARQ feedback transmission for the first SPS PDSCH transmission after activation when operating in coverage enhancement mode A.
ntn-UplinkHarq-ModeB-r18	Indicates whether the UE supports HARQ Mode B if the UE supports ce-ModeA-r13 or is an NB-IoT UE.
ntn-HarqEnhNGSO-Support-r18	Indicates whether the UL and DL HARQ process enhancements that are indicated as supported are applicable in NGSO scenarios for UE indicating support of GSO and NGSO scenarios.
ntn-Triggered-GNSS-Fix-r18	Indicates whether the UE supports network triggered GNSS position fix in the RRC_CONNECTED state.
ntn-Autonomous-GNSS-Fix-r18	Indicates whether the UE supports autonomous GNSS position fix in the RRC_CONNECTED state.
ntn-UplinkTxExtension-r18	Indicates whether the UE supports to perform UL transmission in a duration after original GNSS validity duration expires without GNSS re-acquisition.
ntn-GNSS-EnhNGSO-Support-r18	Indicates whether the GNSS measurement enhancements in the RRC_CONNECTED state that are indicated as supported are applicable in NGSO scenario for UE indicating support of GSO and NGSO scenarios.

Table 7.12 IoT-NTN-Related Capabilities of an NB-IoT NTN UE in Release 17.

Capability IE	Brief Explanation	Major Parent IE
ue-Category-NB-r13	Specifies the NB-IoT UE Category NB1.	UECapabilityInformation-NB-r13-IEs
ue-Category-NB-r14	Specifies the NB-IoT UE Category NB2.	UECapabilityInformation-NB-Ext-r14-IEs
ntn-Connectivity-EPC-r17	Indicates support for NTN access for connectivity with the EPC.	UE-Capability-NB-v1700-IEs/ntn-Parameters-r17/ NTN-Parameters-NB-r17
ntn-TA-Report-r17	Indicates whether the UE supports Timing advance reporting in an NTN cell.	UE-Capability-NB-v1700-IEs/ntn-Parameters-r17/ NTN-Parameters-NB-r17
ntn-PUR-TimerDelay-r17	Indicates whether the UE supports delaying the start of the pur-ResponseWindowTimer in an NTN operation.	UE-Capability-NB-v1700-IEs/ntn-Parameters-r17/ NTN-Parameters-NB-r17
ntn-OffsetTimingEnh-r17	Indicates whether the UE supports timing relationship enhancements using Differential Koffset.	UE-Capability-NB-v1700-IEs/ntn-Parameters-r17/ NTN-Parameters-NB-r17
ntn-ScenarioSupport-r17	Indicates what NTN scenario is supported-NGSO NTN platform or GSO NTN platform.	UE-Capability-NB-v1700-IEs/ntn-Parameters-r17/ NTN-Parameters-NB-r17
ntn-Segmented PrecompensationGaps-r17	Specifies the supported gap length between segments for PUSCH and PUCCH (values: sym1 for 1 symbol, sl1 for 1 slot, sf1 for one subframe).	UE-Capability-NB-v1720-IEs/ NTN-Parameters-NB-v1720/

Table 7.13 specifies the NB-IoT NTN UE's NTN capabilities defined in Release 18. These capabilities are part of the IE UE-Capability-NB-v1800-IEs/NTN-Parameters-NB-v1800 (3GPP 2024l, 2024n).

7.4 Major Takeaways

The 3GPP has two parallel work streams related to the NTN: NR-NTN and IoT-NTN. NR-NTN utilizes an NR-based air interface and the 5GC. The IoT-NTN supports LTE-M and NB-IoT air interfaces and utilizes the 4G EPC. Here are the major takeaways from this chapter.

- Before an NTN UE can exchange data with the radio and core network of the NTN, certain pre-data transfer operations occur between the NTN UE and the NTN radio and core networks. These operations include cell search, network acquisition, random access, RRC connection setup, registration or attach, and PDU session or EPS bearer setup. The frameworks for these procedures are reused, and suitable NTN-specific enhancements have been introduced.
- An NTN IE, cellBarredNTN, conveys accessibility to the 5GC/EPC via the NTN access.
- An NTN IE, trackingAreaList, supports broadcasting of multiple TAs per NTN cell.

Table 7.13 IoT-NTN-Related Capabilities of an NB-IoT NTN UE in Release 18.

Capability IE	Brief Explanation
ntn-LocationBasedMeasTrigger-EFC-r18	Indicates whether the UE supports location-based measurement trigger in RRC_CONNECTED in an Earth-Fixed Cell.
ntn-LocationBasedMeasTrigger-EMC-r18	Indicates whether the UE supports location-based measurement trigger in RRC_CONNECTED in an Earth-Moving Cell.
ntn-TimeBasedMeasTrigger-r18	Indicates whether the UE supports time-based measurement trigger in the RRC_CONNECTED state.
ntn-RRC-HarqDisableSingleTB-r18	Indicates whether the NB-IoT UE supports HARQ feedback disabling per HARQ process for downlink transmission by RRC configuration.
ntn-OverriddenHarqDisable SingleTB-r18	Indicates whether the NB-IoT UE supports DCI-based HARQ feedback disabling for DL transmission by overriding the RRC configuration.
ntn-DCI-HarqDisableSingleTB-r18	Indicates whether the NB-IoT UE supports DCI-based HARQ feedback disabling for DL transmission when HARQ feedback disabling per HARQ process for downlink transmission is not configured by RRC
ntn-RRC-HarqDisableMultiTB-r18	Indicates whether the NB-IoT UE supports HARQ feedback disabling per HARQ process for DL transmission by RRC configuration when configured with npdsch-MultiTB-Config.
ntn-OverriddenHarqDisable MultiTB-r18	Indicates whether the NB-IoT UE supports DCI-based HARQ feedback disabling for DL transmission by overriding the RRC configuration when configured with npdsch-MultiTB-Config.
ntn-DCI-HarqDisableMultiTB-r18	Indicates whether the NB-IoT UE supports DCI-based HARQ feedback disabling for DL transmission when HARQ feedback disabling per HARQ process for DL transmission is not configured by RRC and when configured with npdsch-MultiTB-Config.
ntn-UplinkHarq-ModeB-r18	Indicates whether the UE supports HARQ Mode B if the UE supports ce-ModeA-r13 or is an NB-IoT UE.
ntn-HarqEnhNGSO-Support-r18	Indicates whether the UL and DL HARQ process enhancements that are indicated as supported are applicable in NGSO scenarios for UE indicating support of GSO and NGSO scenarios.
ntn-Triggered-GNSS-Fix-r18	Indicates whether the UE supports network-triggered GNSS position fix in the RRC_CONNECTED state.
ntn-Autonomous-GNSS-Fix-r18	Indicates whether the UE supports autonomous GNSS position fix in the RRC_CONNECTED state.
ntn-UplinkTxExtension-r18	Indicates whether the UE supports to perform UL transmission in a duration after original GNSS validity duration expires without GNSS re-acquisition.
ntn-GNSS-EnhNGSO-Support-r18	Indicates whether the GNSS measurement enhancements in the RRC_CONNECTED state that are indicated as supported are applicable in NGSO scenario for a UE indicating support of GSO and NGSO scenarios.

- To address NTN-specific challenges such as long and variable propagation delays, specific timing relationship parameters, such as service link RTT, Feeder link RTT, Common TA, kmac, and T_{TA} are defined for an NTN.
- Enhancements to the RA procedure for the NTN include Timing Advanced and frequency pre-compensation by the NTN UE while transmitting in the uplink and adjusting search windows.
- An NTN UE may support reporting of Timing Advance during and after the RA procedure.
- An NTN UE conveys its regular and NTN-specific capabilities to the radio network through the UE capability transfer procedure.
- Certain AS and NAS timers are extended for the NTN to accommodate long propagation delays.
- The core network may verify that the NTN UE can use the selected NTN-related PLMN in its current location.
- In support of cell selection and SI acquisition, NTN-specific SIBs, including SIB19 and SIB25 have been introduced for NR-NTN in Releases 17 and 18. SIB19 contains numerous parameters that provide information on satellite assistance for NTN access. SIB25 contains TN coverage information to assist neighbor cell measurements for the UEs in an NTN cell.
- A new 5QI=10 is introduced for NR-NTN to accommodate long propagation delays for a GEO satellite-based NTN.
- In support of cell selection and SI acquisition, NTN-specific SIBs, including SIB31, SIB32, and SIB33, have been introduced for IoT-NTN in Releases 17 and 18. SIB31 provides satellite assistance information for the serving cell. SIB32 specifies satellite assistance information about the serving cell to enable the LTE-M NTN UE to predict discontinuous coverage. SIB33 provides satellite assistance information for neighbor cells.

References

3GPP (2024a). *Non-Access-Stratum (NAS) functions related to Mobile Station (MS) in idle mode*. TS23.122. https://portal.3gpp.org/desktopmodules/Specifications/SpecificationDetails.aspx?specificationId=789 (accessed 6 September 2024).

3GPP (2024b). *5G System (5GS) Location Services (LCS)*. TS23. 273. https://portal.3gpp.org/desktopmodules/Specifications/SpecificationDetails.aspx?specificationId=3577 (accessed 6 September 2024).

3GPP (2024c). *General Packet Radio Service (GPRS) enhancements for Evolved Universal Terrestrial Radio Access Network (E-UTRAN) access*. TS23.401. https://portal.3gpp.org/desktopmodules/Specifications/SpecificationDetails.aspx?specificationId=849 (accessed 6 September 2024).

3GPP (2024d). *System architecture for the 5G System (5GS)*. TS23.501. https://portal.3gpp.org/desktopmodules/Specifications/SpecificationDetails.aspx?specificationId=3144 (accessed 6 September 2024).

3GPP (2024e). *Procedures for the 5G System (5GS)*. TS23.502. https://portal.3gpp.org/desktopmodules/Specifications/SpecificationDetails.aspx?specificationId=3145 (accessed 6 September 2024).

3GPP (2024f). *Non-Access-Stratum (NAS) protocol for Evolved Packet System (EPS)*. TS24.301. https://portal.3gpp.org/desktopmodules/Specifications/SpecificationDetails.aspx?specificationId=1072 (accessed 6 September 2024).

3GPP (2024g). *Non-Access-Stratum (NAS) protocol for 5G System (5GS)*. TS24.501. https://portal.3gpp.org/desktopmodules/Specifications/SpecificationDetails.aspx?specificationId=3370 (accessed 6 September 2024).

3GPP (2024h). *Evolved Universal Terrestrial Radio Access (E-UTRA); User Equipment (UE) radio transmission and reception for satellite access*. TS36.102. https://portal.3gpp.org/desktopmodules/Specifications/SpecificationDetails.aspx?specificationId=4063 (accessed 6 September 2024).

3GPP (2024i). *Evolved Universal Terrestrial Radio Access (E-UTRA); Physical layer procedures*. TS36.213. https://portal.3gpp.org/desktopmodules/Specifications/SpecificationDetails.aspx?specificationId=2427 (accessed 6 September 2024).

3GPP (2024j). *Evolved Universal Terrestrial Radio Access (E-UTRA) and Evolved Universal Terrestrial Radio Access Network (E-UTRAN): Overall description*. TS36.300. https://portal.3gpp.org/desktopmodules/Specifications/SpecificationDetails.aspx?specificationId=2430 (accessed 6 September 2024).

3GPP (2024k). *User Equipment (UE) procedures in idle mode*. TS36.304. https://portal.3gpp.org/desktopmodules/Specifications/SpecificationDetails.aspx?specificationId=2432 (accessed 6 September 2024).

3GPP (2024l). *User Equipment (UE) radio access capabilities*. TS36.306. https://portal.3gpp.org/desktopmodules/Specifications/SpecificationDetails.aspx?specificationId=2434 (accessed 6 September 2024).

3GPP (2024m). *Evolved Universal Terrestrial Radio Access (E-UTRA); Medium Access Control (MAC) protocol specification*. TS36.321. https://portal.3gpp.org/desktopmodules/Specifications/SpecificationDetails.aspx?specificationId=2437 (accessed 6 September 2024).

3GPP (2024n). *Evolved Universal Terrestrial Radio Access (E-UTRA); Radio Resource Control (RRC); Protocol specification*. TS36.331. https://portal.3gpp.org/desktopmodules/Specifications/SpecificationDetails.aspx?specificationId=2440 (accessed 6 September 2024).

3GPP (2024o). *NR; User Equipment (UE) radio transmission and reception; Part 5: Satellite access Radio Frequency (RF) and performance requirements*. TS 38.101-5. https://portal.3gpp.org/desktopmodules/Specifications/SpecificationDetails.aspx?specificationId=3982 (accessed 6 September 2024).

3GPP (2024p). *NR; Requirements for support of radio resource management*. TS38.133. https://portal.3gpp.org/desktopmodules/Specifications/SpecificationDetails.aspx?specificationId=3204 (accessed 6 September 2024).

3GPP (2024q). *NR; Physical layer procedures for control*. TS38.213. https://portal.3gpp.org/desktopmodules/Specifications/SpecificationDetails.aspx?specificationId=3215 (accessed 6 September 2024).

3GPP (2024r). *NR; Physical layer measurements*. TS38.215. https://portal.3gpp.org/desktopmodules/Specifications/SpecificationDetails.aspx?specificationId=3217 (accessed 6 September 2024).

3GPP (2024s). *NR and NG-RAN Overall Description*. TS38.300. https://portal.3gpp.org/desktopmodules/Specifications/SpecificationDetails.aspx?specificationId=3191 (accessed 6 September 2024).

3GPP (2024t). *NR; User Equipment (UE) procedures in Idle mode and in RRC Inactive state*. TS38.304. https://portal.3gpp.org/desktopmodules/Specifications/SpecificationDetails.aspx?specificationId=3192 (accessed 6 September 2024).

3GPP (2024u). *NR; Medium Access Control (MAC) protocol specification*. TS38.321. https://portal.3gpp.org/desktopmodules/Specifications/SpecificationDetails.aspx?specificationId=3194 (accessed 6 September 2024).

3GPP (2024v). *NR; Radio Resource Control (RRC); Protocol specification*. TS38.331. https://portal.3gpp.org/desktopmodules/Specifications/SpecificationDetails.aspx?specificationId=3197 (accessed 6 September 2024).

3GPP (2024w). *NG-RAN; NG Application Protocol (NGAP)*. TS38. 413. https://portal.3gpp.org/desktopmodules/Specifications/SpecificationDetails.aspx?specificationId=3223 (accessed 6 September 2024).

Tripathi, N.D. and Reed, J.H. (2019). *5G Cellular Communications - Journey and Destination*. The Wireless University. https://thewirelessuniversity.com/ (accessed 6 September 2024).

8

Downlink and Uplink Data Transfer in an NTN

Data transfer in a Non-Terrestrial Network (NTN) must contend with unique factors of the radio environment, such as long and variable propagation delays. The data transfer in the NTN reuses the data transfer framework of the TN and makes enhancements to reflect NTN-specific radio channel challenges. This chapter discusses unique factors of the radio environment in an NTN. The chapter illustrates various timing relationships in the NTN. A new Timing Advance (TA) procedure is introduced for the NTN. Key prerequisites for data transfer, downlink data transfer, and uplink data transfer in New Radio-NTN (NR-NTN) are discussed. Similarly, key prerequisites for data transfer, downlink data transfer, and uplink data transfer in the Long-Term Evolution-Machine Type Communication (LTE-M) NTN and the Narrowband Internet of Things-NTN (NB-IoT-NTN) are described. Since an NTN User Equipment (UE) cannot perform data transfer without valid Global Navigation Satellite System (GNSS) measurements, the NTN supports a measurement gap for GNSS measurements. A procedure for managing GNSS measurement gaps is also summarized.

8.1 Characteristics of Data Transfer in an NTN: An Overview

Data transfer in an NTN must reflect the unique factors of the radio environment in an NTN. Figure 8.1 summarizes such factors. These factors are briefly discussed below.

8.1.1 Reuse of Data Transfer Frameworks

The NTN reuses the downlink data transfer and uplink data transfer frameworks defined for the TN and introduces incremental changes to reflect NTN-specific characteristics of the radio environment. As mentioned in Chapter 7, certain Access Stratum and Non-Access Stratum protocol times are extended to accommodate long propagation delays in an NTN. Depending on the achievable signal to interference plus noise ratio (SINR) in a given NTN deployment scenario (e.g., Geostationary Earth Orbit [GEO] satellite, Low Earth Orbit [LEO] satellite, or High Altitude Platform Station [HAPS]-based NTN), more aggressive MCSs corresponding to high Channel Quality Indicator (CQI) values may not be appropriate for an NTN. The type of the NTN payload (i.e., transparent or regenerative) and the antenna configuration at the NTN platform could restrict the scope of antenna techniques compared to a TN. For example, the TN UE can directly observe the propagation environment toward the serving TN gNB. However, when a transparent payload is used, the NTN UE can directly observe the radio environment toward the NTN payload only on the service

3GPP-based Non-Terrestrial Networks in 5G and 6G: Expanding the Frontiers of Wireless Communications, First Edition.
Nishith D. Tripathi and Jeffrey H. Reed.

Companion website: www.wiley.com/go/tripathi5g

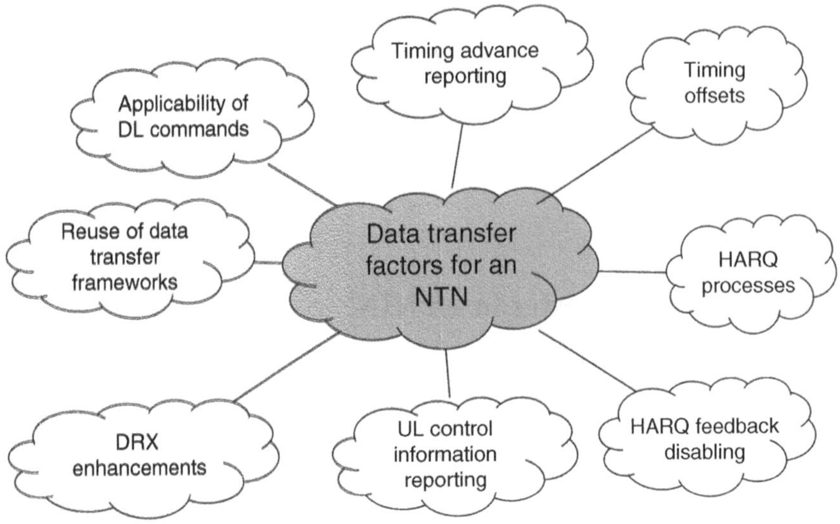

Figure 8.1 Important Factors for Data Transfer in an NTN.

link, although the overall propagation path between the NTN UE and the NTN gNB includes both the service link and the feeder link. The lack of visibility of the complete UE-gNB propagation path complicates spatial multiplexing when a transparent NTN payload is used. A regenerative NTN payload enables the NTN UE to directly observe the propagation environment toward the serving NTN gNB like a TN. The type of antenna available at the NTN gNB at the NTN platform and the service link radio environment dictate the degree of spatial multiplexing and the amount of beamforming gain. The IoT-NTN UE has a single receive antenna and, hence, it is much simpler than an NR-NTN UE from the data transfer perspective.

8.1.2 Timing Advance Reporting

Based on its capability, the NTN UE provides the gNB/eNB with a Timing Advance report. The gNB/eNB can use this report to determine any offsets and resource allocation.

8.1.3 Timing Offsets

To accommodate long round-trip delays in an NTN, a network-specific timing offset common for all NTN UEs in the NTN cell and an NTN UE-specific timing offset to reflect the service link delay are utilized, as described in Chapter 7.

8.1.4 HARQ Processes

Prior to the NTN, a maximum of 16 HARQ processes are supported for an NR UE in a cell in the downlink and the uplink. The NR NTN, based on UE and network capabilities, supports a maximum of 32 HARQ processes. The increased number of HARQ processes in the NTN increases the achievable throughput when 16 processes are inadequate to accommodate the round-trip propagation delay between the NTN UE and the NTN gNB.

8.1.5 HARQ Feedback Disabling

The NTN supports disabling HARQ feedback altogether. For example, the network may configure the NTN UE to disable reporting of HARQ feedback in response to a received transport block in the downlink. Such HARQ feedback disabling may be attractive when the round-trip propagation delay between the NTN UE and the gNB/eNB is too long. To improve the error rate or resiliency, simple blind repetition can be utilized. For example, the same transport block may be sent multiple times through mechanisms such as physical downlink shared channel (PDSCH) aggregation.

8.1.6 Uplink Control Information Reporting

Due to the long propagation delays, UL control information, such as CQI, reflects the radio environment observed by the NTN UE at an instant relatively further in the past compared to a TN UE. Hence, UL control information reporting by the UE is enhanced to accommodate long delays. Thus, the NTN gNB/eNB must consider long delays while assigning downlink and uplink radio resources to the NTN UE.

8.1.7 DRX Enhancements

Discontinuous Reception (DRX) is supported in the NTN. HARQ timers for DRX are applied after considering the round-trip propagation delay between the NTN UE and the NTN gNB/eNB.

8.2 Data Transfer in the NR-NTN

Data transfer in NR-NTN is based on NR-TN data transfer with NTN-specific enhancements. Section 8.2.1 discusses the prerequisites for data transfer in NR-NTN. Section 8.2.2 discusses downlink data transfer on the NR radio interface in the NR-NTN, and Section 8.2.3 discusses uplink data transfer on the NR radio interface in the NR-NTN.

8.2.1 Data Transfer Prerequisites in the NR-NTN

Figure 8.2 summarizes key prerequisites for data transfer in NR-NTN. These prerequisites must be fulfilled before data transfer can occur in the downlink and the uplink on the NR air interface in the NR-NTN.

Registration. As discussed in Chapter 7, the NR-NTN UE performs registration with the 5GC under a variety of circumstances, including upon power-up and for certain Tracking Area changes. Mutual authentication occurs during registration. The Access and Mobility Management Function (AMF) allocates a 5G-GUTI to the NR-NTN UE and provides the Registration Area consisting of a set of Earth-fixed Tracking Areas to the NR-NTN UE. The AMF also conveys allowed Network Slices to the NR-NTN UE.

PDU Session Establishment. The NR-NTN UE, any time after completion of registration, initiates the establishment of a protocol data unit (PDU) session for the selected Network Slice, such as an enhanced mobile broadband (eMBB) slice. The NR-NTN allocates an ID to the PDU session. The 5GC (e.g., the SMF) allocates an IP address to the NR-NTN UE for IP-based data transfer. A 5G QoS Flow is set up for a given PDU session. A 5G QoS Flow is associated with specific 5G QoS characterized by parameters including a 5QI. In the case of a GEO-based NTN, a new NTN-specific 5QI is more appropriate compared to other standard-defined 5QIs. A suitable Data

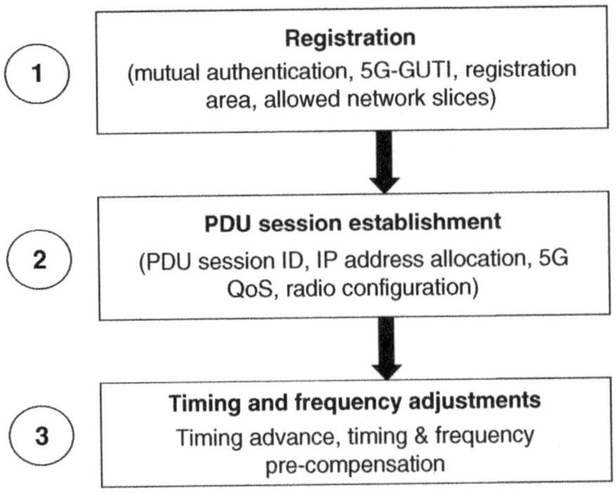

Figure 8.2 Key Data Transfer Prerequisites for the NR-NTN.

Radio Bearer (DRB) is set up between the NR-NTN UE and the NR-NTN gNB in an NTN cell for the related 5G QoS Flow. Multiple 5G QoS Flows and multiple PDU sessions toward one or more Data Networks may be set up. The NR-NTN gNB configures the NR-NTN UE by reporting (i) suitable measurements such as CSI and (ii) Timing Advance. The NR-gNB may configure the NR-NTN UE with up to 32 HARQ processes or may disable HARQ feedback.

Timing and Frequency Pre-compensation. The NR-NTN UE observes suitable SIBs, as discussed in Chapter 7, to obtain NTN-specific information such as satellite ephemeris. The NR-NTN UE utilizes its location along with the location of the satellite to determine the timing and frequency pre-compensation on an ongoing basis so that its transmissions minimize (or eliminate) intra-cell interference in the uplink. The NR-NTN gNB provides suitable timing adjustments to the NR-NTN UE via MAC signaling. Section 8.2.2 details the timing adjustments made in the NR-NTN.

8.2.2 Timing Adjustments in the NR-NTN

The NR-NTN introduces several enhancements to the existing 5G NR timing frameworks and a new framework for the Timing Advance reporting to reflect the propagation delays in the NTN (3GPP 2024l). Examples of the existing 5G NR timing frameworks include a framework for the timing relationship between a downlink slot carrying a certain type of information (e.g., a downlink control information [DCI], medium access control [MAC] control element [CE], or Transport Block) and an associated uplink slot and a framework for uplink transmission timing. Section 8.2.2.1 discusses an NTN-specific offset between the downlink (DL) and the uplink (UL), and Section 8.2.2.2 describes uplink transmission timing. Section 8.2.2.3 discusses the new framework for the Timing Advance reporting by the NR-NTN UE. The IE NTN-Config and the MAC CEs related to Timing Advance Command (which was defined originally in the 3GPP Release 15 for the NR-TN and reused in the NR-NTN), Differential Koffset, and TA Report support these timing frameworks.

As discussed in Section 7.2.3 in Chapter 7, DL and UL are frame-aligned at the uplink time synchronization reference point (RP) with an offset given by $N_{TA,offset}$. Furthermore, there are two scheduling offsets, K_{offset} and k_{mac}. K_{offset} is a *configured scheduling offset* that needs to be larger or equal to the sum of the service link round trip time (RTT) and the Common TA. The scheduling offset K_{offset} provides the UE sufficient processing time between a downlink reception and an uplink transmission (3GPP 2024j). k_{mac} is used to delay the application of a downlink configuration indicated by a MAC CE command on PDSCH (3GPP 2024j) and in the estimation of the UE-gNB RTT (3GPP 2024n). The NTN gNB may provide k_{mac} when the downlink and the uplink frame timing are not aligned at the NTN gNB.

The IE NTN-Config includes parameters for the NR-NTN access. This IE is part of the NTN-specific SIB19. The IE NTN-Config-r17 contains several parameters that govern timing adjustments: ephemerisInfo-r17, cellSpecificKoffset-r17, kmac-r17, and ta-Info-r17 (containing ta-Common-r17, ta-CommonDrift-r17, and ta-CommonDriftVariant-r17) (3GPP 2024o).

The IE EphemerisInfo specific satellite ephemeris is in the format of (i) a position and velocity state vector or (ii) orbital parameters. The NR-NTN UE utilizes its GNSS-based position and the satellite ephemeris to determine the service link RTT based on (i) the distance between itself and the NTN payload and (ii) the speed of light.

The IE cellSpecificKoffset is the scheduling offset used for the timing relationships in the NR-NTN. It is expressed in the unit of the number of slots for the subcarrier spacing of 15 kHz.

The IE kmac-r17 is the scheduling offset provided by the NR-NTRN gNB when the downlink and uplink frame timing are not aligned at the NR-NTN gNB and is expressed in the unit of the number of slots for the subcarrier spacing of 15 kHz.

The IE ta-Info-r17 provides Timing Advance (TA) parameters and includes ta-Common-r17, ta-CommonDrift-r17, and ta-CommonDriftVariant-r17. The IE ta-Common indicates network-controlled common TA value and has the granularity of 4.072×10^{-3} μs. The IE ta-CommonDrift indicates the drift rate of the common TA and has a granularity of 0.2×10^{-3} μs/s. The IE ta-CommonDriftVariant indicates the drift rate variation of the common TA and has a granularity of 0.2×10^{-4} μs/s^2.

The IE NTN-Config-r17 also contains other parameters that influence timing in the NR-NTN: epochTime-r17, ntn-UlSyncValidityDuration-r17, and ta-Report-r17.

The IE epochTime specifies the epoch (i.e., reference) time for the NTN assistance information and is the starting time of a DL subframe, indicated by a SFN and a subframe number signaled together with the assistance information.

The IE ntn-UlSyncValidityDuration-r17 is the validity duration (in seconds) configured by the network for assistance information. It indicates the maximum time duration from epochTime-r17, during which the NR-NTN UE can utilize the assistance information without acquiring new assistance information.

The IE ta-Report-r17 indicates that the reporting of TA is enabled during (i) the random-access procedure executed due to radio resource control (RRC) connection establishment or RRC connection resume and (ii) RRC connection reestablishment.

8.2.2.1 NTN-Specific DL-UL Offset

The timing relationship between (i) the downlink slot n containing information such as the DL command (e.g., DCI), configuration (e.g., MAC CE), or user traffic (e.g., a Transport Block) and (ii) the associated uplink slot m is expressed using the sum of (i) n and (ii) multiple offsets (i.e., $m = n +$ multiple offsets). These offsets are a function of factors such as the subcarrier spacing and the UE's processing capability in 5G NR for a TN and the type of DL information (3GPP 2024j). A term based on an NTN-specific offset, K_{offset}, is typically added to the sum of such offsets

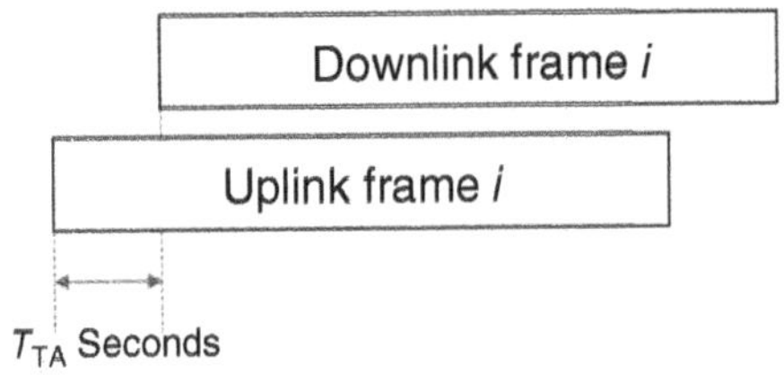

Figure 8.3 Timing Relationship Between the DL Frame and the UL Frame at the NR-NTN UE.

to reflect the propagation delays in the NTN (i.e., $m = n +$ multiple offsets + a term based on K_{offset})[1] (3GPP 2024j).

$$K_{offset} = K_{cell,offset} - K_{UE,offset}. \tag{8.1}$$

In Eq. (8.1), K_{offset} is given by cellSpecificKoffset in the IE NTN-Config as summarized earlier in this section, and $K_{UE,offset}$ is provided by a Differential Koffset MAC CE command (3GPP 2024n). Note that an offset in Eq. (8.1) is assumed to be 0 when the NR-NTN gNB has not provided such offset's value to the NR-NTN UE.

8.2.2.2 Uplink Transmission Timing

Figure 8.3 shows the relationship between the downlink frame[2] and the uplink frame at the NR NTN UE (3GPP 2024h).

The NR-NTN UE's transmission of an uplink frame number i starts T_{TA} before the start of the corresponding downlink frame i received at the NR-NTN UE. T_{TA} is given by

$$T_{TA} = \left(N_{TA} + N_{TA,offset} + N_{TA,adj}^{common} + N_{TA,adj}^{UE}\right) T_c. \tag{8.2}$$

In Eq. (8.2), N_{TA} is the Timing Advance between the downlink and the uplink, and $N_{TA,offset}$ is a fixed offset used to calculate the Timing Advance (3GPP 2024j). $N_{TA,adj}^{common}$ is network-controlled timing correction, and $N_{TA,adj}^{UE}$ is the UE-derived timing correction. $N_{TA,adj}^{common}$ and $N_{TA,adj}^{UE}$ are NTN-specific. T_c is the time unit used to express the size of various time domain structures[3] and is given by

$$T_c = \frac{1}{(\Delta f_{max} \cdot N_f)}. \tag{8.3}$$

In Eq (8.3), $\Delta f_{max} = 480 \times 10^3$ Hz and $N_f = 4{,}096$.

The NR-NTN gNB provides N_{TA} to the NR-NTN UE for different scenarios, such as in the Random Access Response during the random-access procedure and a Timing Advance MAC CE or the cell switch command, when the NR-NTN UE is in the RRC_CONNETED state.

The NR-NTN gNB may provide $N_{TA,offset}$ to the NR-NTN UE as a timing advance offset in the IE n-TimingAdvanceOffset as part of the serving cell radio configuration.

---

1 In case of the applicability of the Timing Advance command received by the NR-NTN UE from the NR-NTN gNB, such term is $2^{\mu} * K_{offset}$, where μ is the subcarrier spacing configuration (3GPP 2024j).

2 Recall that a frame in 5G NR is a 10 ms long time structure. A 5G NR frame consists of two half-frames with each half-frame containing five subframes. Thus, a frame consists of 10 subframes with each subframe having the length of 1 ms. A subframe has one or more slots based on the subcarrier spacing. A slot in 5G NR consists of 14 OFDM symbols in the case of normal Cyclic Prefix (CP) and 12 OFDM symbols in the case of extended CP. The length of the OFDM symbol is inversely proportional to the subcarrier spacing and includes CP.

3 T_c can be viewed as a reference time unit in seconds.

The NR-NTN UE calculates $N_{\mathrm{TA,adj}}^{\mathrm{common}}$ based on one-way propagation delay $\mathrm{Delay}_{\mathrm{common}}(t)$ as follows using ta-Common-r17, ta-CommonDrift-r17, and ta-CommonDriftVariant (3GPP 2024j).

$$\begin{aligned}\mathrm{Delay}_{\mathrm{common}}(t) = {} & \left[\frac{TA_{\mathrm{common}}}{2}\right] + \left[\left(\frac{TA_{\mathrm{CommonDrift}}}{2}\right)\left(t - t_{\mathrm{epoch}}\right)\right] + \\ & \left[\left(\frac{TA_{\mathrm{CommonDriftVariant}}}{2}\right)\left(t - t_{\mathrm{epoch}}\right)^2\right].\end{aligned} \tag{8.4}$$

In Eq. (8.4), TA_{common}, $TA_{\mathrm{CommonDrift}}$, and $TA_{\mathrm{CommonDriftVariant}}$ are specified by ta-Common-r17, ta-CommonDrift-r17, and ta-CommonDriftVariant, respectively. Furthermore, t_{epoch} corresponds to epochTime. $N_{\mathrm{TA,adj}}^{\mathrm{common}}$ is used by the NR-NTN UE to pre-compensate the two-way transmission delay between the uplink time synchronization reference point and the serving satellite. Note that the uplink time synchronization reference point is the point where the downlink frame and the uplink frame are frame-aligned with an offset given by $N_{\mathrm{TA,offset}}$.

The NR-NTN UE calculates $N_{\mathrm{TA,adj}}^{\mathrm{UE}}$ based on its position and serving satellite ephemeris (3GPP 2024j). The NR-NTN UE uses $N_{\mathrm{TA,adj}}^{\mathrm{UE}}$ to pre-compensate the two-way transmission delay on the service link.

In the network, as discussed in Section 7.2.3 in Chapter 7, DL and UL are frame-aligned at the uplink time synchronization RP with an offset given by $N_{\mathrm{TA,offset}}$.

8.2.2.3 Timing Advance Reporting

The NR-NTN gNB may configure the NR-NTN UE with a TA report-related configuration via RRC signaling. The NR-NTN UE's TA report provides the NR-NTN gNB with an estimate of the NR-NTN UE's TA value, denoted as TTA in Figure 8.3. In an implementation-specific manner, the NR-NTN gNB can utilize the NTN-UE's TA report to determine resource allocation and timing-related adjustments such as Timing Advance commands and Differential Koffset.

The NR-NTN gNB uses RRC signaling to control Timing Advance reporting by the NR-NTN UE by configuring ta-Report, offsetThresholdTA, and TimingAdvanceSR. An NR-NTN UE sends a Timing Advance Report (TAR) when any of the following events occur. Suppose ta-Report= "enabled" in SIB 19. In that case, the NR-NTN UE sends a TAR upon initiation of the Random-Access procedure due to initial access from RRC_IDLE, RRC Connection Resume procedure from RRC_INACTIVE, or RRC Connection Re-establishment procedure (3GPP 2024o). If ta-Report= "enabled" is indicated in the handover command, the NR-NTN UE sends a TAR upon initiation of the Random-Access procedure due to reconfiguration with sync. Suppose the parameter offset-ThresholdTA is configured or reconfigured and the NR-NTN UE has not previously reported the Timing Advance value to the current serving cell. In that case, the NR-NTN UE sends a TAR. Furthermore, if the variation between current information about Timing Advance and the last successfully reported information about Timing Advance is equal to or larger than offsetThresholdTA, the NR-NTN UE sends a TAR. If TimingAdvanceSR= "enabled," and if the NR-NTN UE does not have adequate resources to send a TAR, the NR-NTN UE sends a Scheduling Request (SR) to the NR-NTN gNB.

8.2.3 Downlink Data Transfer in the NR-NTN

As mentioned in Section 3.4.2 in Chapter 3, 5G NR provides a flexible framework for DL data transfer with various implementations possible. Figure 8.4 also illustrates the main steps of the DL data

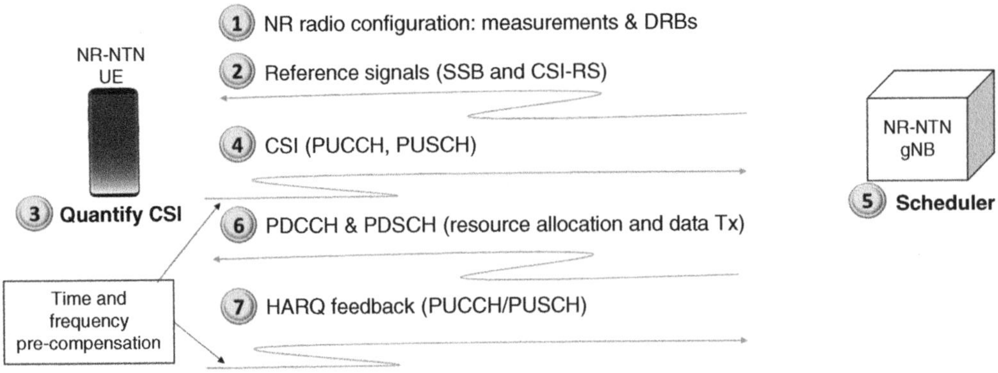

Figure 8.4 Downlink Data Transfer in the NR-NTN. *Source:* nadir_insan / Adobe Stock Photo.

transfer in 5G NR-NTN in an example implementation. Figure 8.4 summarizes dynamic scheduling-based scheduling downlink data transfer in NR-NTN.

In *Step 1*, the NR-NTN gNB configures the NR-NTN UE with one or more DRBs and measurements using the RRCReconfiguration message. The NR-NTN gNB may configure TA report-related parameters and longer protocol timers. For example, depending on the NTN type, the NR-NTN gNB may use t-ReassemblyExt, which defines an extended reassembly timer at the RLC layer instead of the regular t-Reassembly in the IE rlc-Config. The IE t-ReassemblyExt may be used in the Acknowledged Mode and Unacknowledged Mode of the RLC protocol in the IEs DL-AM-RLC-v1700 and dl-UM-RLC-v1700.

As in the case of a TN, the NR-NTN gNB may configure the NR-NTN UE to make measurements of synchronization signal/physical broadcast channel (SS/PBCH) only, Channel State Information-Reference Signal (CSI-RS) only, or both SS/PBCH and CSI-RS. The NR-NTN gNB may also configure the NR-NTN UE to make Channel State Information-Interference Measurement (CSI-IM) measurements. A given CSI-RS may be transmitted in (i) a specific portion of the NR-NTN cell in a CSI-RS beam or (ii) the entire NTN cell. The NR-NTN gNB may configure multiple UEs to observe the same CSI-RS. The NR-NTN gNB can specify the frequency band, the number of Physical Resource Blocks (PRBs), orthogonal frequency division multiplexing (OFDM) symbols, and subcarriers in a PRB where the NR-NTN UE can observe CSI-RS. The CSI-RS scrambling ID and periodicity of CSI-RS are also specified. The beam identities in the form of Transmission Configuration Indications (TCIs) are also specified by the NR-NTN gNB in support of multiple beams in an NTN cell associated with a given PCI.

The CSI in 5G NR includes CSI components such as CQI, PMI, RI, Layer 1-RSRP (L1-RSRP)[4], Layer Indicator (LI)[5], CSI-RS Resource Indicator (CRI), and SSB Resource Indicator (SSBRI). SSB is the Synchronization Signal Block, and it contains the primary and secondary synchronization signals and the PBCH. CRI refers to the CSI-RS beam, and SSBRI refers to the CSI-RS beam based on the configured TCIs. The CSI reporting can be periodic, semi-persistent, or aperiodic (e.g., based on a DCI command) (3GPP 2024i).

4 L1-RSRP is the RSRP measured on the CSI-RS or the secondary synchronization signal (SSS).

5 LI indicates the strongest layer among the spatial multiplexing layers.

Table 8.1 Characteristics of CQI Tables.

CQI Table Identity	CQI Table 1	CQI Table 2	CQI Table 3
BLER	0.1	0.1	0.00001
Modulation schemes	QPSK, 16QAM, and 64QAM	QPSK, 16QAM, 64QAM, and 256QAM	QPSK, 16QAM, and 64QAM
Least spectral efficiency (bits per modulation symbol)	0.1523	0.1523	0.0586
Highest spectral efficiency (bits per modulation symbol)	5.5547	7.4063	4.5234

Note that the NR-gNB needs to use the CSI report with caution. The CSI report may become stale in case of a long propagation delay, especially for a transparent NTN payload and a GEO satellite-based NTN. Short-term radio channel variations may not be reflected accurately by the CSI in such cases. The CSI reports likely reflect prevailing radio channel conditions for a HAPS-based NTNs and LEO satellite-based NTNs, and long-term variations for GEO satellite-based NTNs.

5G NR supports multiple CQI tables (3GPP 2024k). Table 8.1 summarizes selected characteristics of CQI tables.

CQI Table 1 may be more relevant to the NTN compared to CQI Table 2 and CQI Table 3. A challenging radio environment in the NTN limits the achievable SINR, preventing the use of an aggressive Modulation and Coding Scheme (MCS). Furthermore, long propagation delays make satellite-based NTNs less suitable for URLLC use cases.

5G NR supports much more comprehensive PMIs. Type-I codebooks suitable for SU-MIMO and Type-II codebooks suitable for MU-MIMO are supported. Single-panel and multi-panel antenna configurations are supported. The type of antenna array on the NTN platform with a regenerative payload dictates which codebooks can be utilized in an NTN. It is more challenging to support a variety of codebooks for a transparent payload due to the lack of visibility of the feeder link radio channel in a typical NTN with a transparent payload.

The NR-NTN gNB may configure CA for an NR-NTN UE. Up to four bandwidth parts (BWPs) may be defined on a given carrier frequency or component carrier, and one BWP on a carrier frequency is active at a time. The gNB allocates radio resources for the active BWP.

Compared to implementing advanced antenna techniques with high-order spatial multiplexing and high-order digital beamforming, Carrier Aggregation is much easier to utilize in an NR-NTN for both transparent and regenerative payloads.

In *Step 2*, the NR-NTN UE, based on the configuration by the serving NR-NTN gNB, observes the reference signals, such as the CSI-RS of the serving NR-NTN cell. Like a TN, the CSI-RS in the NR-NTN may be transmitted in a beam that is narrower than an SSB beam. The CSI-RS beams may cover the same region as the SSB beams.

In *Step 3*, the NR-NTN UE quantifies the prevailing DL radio channel conditions as CSI.

In *Step 4*, the NR-NTN UE sends the channel condition report to the NR-NTN gNB based on the NR-gNB-specified configuration of CSI reporting to enable the NR-NTN gNB to determine optimal PDSCH transmission parameters. The CSI report at a given instant contains one or more CSI components mentioned in Step 1 (i.e., CQI, PMI, RI, L1-RSRP, LI, CRI, and SSBRI). The channel condition report is sent on the physical uplink shared channel (PUCCH) or the physical uplink shared channel (PUSCH). Due to a potentially long delay in an NTN, an additional NTN-specific offset

Koffset is introduced, which affects the instant associated with the CSI reference resource used by the NR-NTN UE to quantify the CSI. Specifically, the CSI reference resource for a CSI reporting in the uplink slot n' is given by (3GPP 2024k)

$$n' = n - n_{\mathrm{CSI_{ref}}} - K_{\mathrm{offset}} \cdot \frac{2^{\mu_{\mathrm{DL}}}}{2^{\mu_{K_{\mathrm{offset}}}}} \tag{8.5}$$

In Eq. (8.5), n is the downlink slot number, $n_{\mathrm{CSI_{ref}}}$ is a reference value that depends on the factors such as the DL subcarrier spacing configuration and the type of CSI reporting (i.e., periodic, aperiodic, or semi-persistent reporting) (3GPP 2024k), K_{offset} is configured by the NR-NTN gNB for the NR-NTN UE and given by Eq. (8.1), $\mu_{K_{\mathrm{offset}}}$ is the subcarrier spacing configuration (with the value = 0 for FR1) associated with K_{offset}, and μ_{DL} is the subcarrier spacing configuration for the downlink.

In *Step 5*, the NR-NTN gNB executes an implementation-specific scheduling algorithm that considers the channel condition report along with numerous factors such as the NR-NTN UE capabilities, the number of PRBs in the UE's active BWP per carrier, the available resources on different carrier frequencies for CA, QoS needs of 5G QoS Flows of NR-NTN UEs, and NR-NTN gNB capabilities to allocate DL radio resources to the NR-NTN UE.

As in a TN, the NR-NTN gNB can convey allocated DL radio resources to the NR-NTN UE using DCI Format 1_0 or 1_1 on the PDCCH in *Step 6*. DCI Format 1_0 is simpler and hence more compact, while DCI Format 1_1 is more comprehensive.

Both DCI Formats can convey parameters such as (i) frequency domain assignment (e.g., PRBs), (ii) time domain assignment (i.e., starting OFDM symbol number and length of the allocation), (iii) MCS from the configured CQI table, (iv) New Data Indicator (NDI), (v) Redundancy Version (RV), (vi) HARQ process number (i.e., one of sixteen processes), (vii) PDSCH-to-HARQ feedback timing indicator, and (viii) PUCCH Resource Indicator for UL feedback to the DL transmission. 5G NR utilizes LDPC coding on the PDSCH to increase the effectiveness of channel coding and reduce the error rate while reducing energy consumption. Additionally, DCI Format 1_1 can indicate TCI for future communications and Code Block Group (CBG) transmission information. See Section 3.4.2 in Chapter 3 for details of NDI, RV, and CBG.

In an NR-NTN, the HARQ feedback for a given HARQ process may be disabled by the NR-gNB for a given NR-NTN UE using the IE downlinkHARQ-FeedbackDisabled. For a unicast transmission, in one option, PDSCH-to-HARQ_feedback timing indicator field values can be provided in pucch-Config/pucch-ConfigurationList in the IE such as dl-DataToUL-ACK-v1700 (3GPP 2024j). The IE dl-DataToUL-ACK-v1700 is specifically introduced for the NTN (3GPP 2024o).

The NR-NTN UE retrieves the packet from the PDSCH transmitted by the NR-NTN gNB based on the PDCCH-specified parameters in *Step 6*.

The NR-NTN UE attempts to decode the packet on the PDSCH. In *Step 7*, the NR-NTN UE conveys the HARQ ACK feedback to the NR-NTN gNB depending on whether it has successfully decoded the packet or not, based on the "PDSCH-to-HARQ feedback timing indicator" specified by the DCI (e.g., one of the values specified in dl-DataToUL-ACK-v1700). The NR-NTN UE sends this HARQ ACK feedback on the PUCCH or the PUSCH.

The NR-NTN gNB sends a new packet if it has received HARQ ACK, or retransmits a previously sent packet if it has received a HARQ NACK provided the maximum number of HARQ retransmissions has not been reached yet.

As in a TN, the NR-NTN gNB may utilize dynamic resource allocation by specifying radio resources dynamically using a DCI. Furthermore, like a TN, the NR-NTN gNB may also utilize semi-persistent scheduling (SPS), where the resources are specified through RRC signaling and activated or deactivated based on DCI signaling. The time domain allocation may use the slot (other than the OFDM symbols occupied by PDCCHs), a mini slot (i.e., a few symbols of the slot), or multiple slots. The frequency domain allocation may use PRBs or Resource Block Groups (RBGs), where an RBG consists of multiple PRBs.

8.2.4 Uplink Data Transfer in the NR-NTN

Like DL data transfer, NR-NTN reuses the flexible UL data transfer framework defined for the TN (Tripathi and Reed 2019). Figure 8.5 (Tripathi and Reed 2019) illustrates the main steps of the UL data transfer in NR-NTN in an example implementation. Specifically, Figure 8.5 summarizes dynamic scheduling-based uplink data transfer in the NR-NTN.

The NR-NTN gNB configures the NR-NTN UE to facilitate UL data transfer in *Step 1* by providing configuration related to SR, sounding reference signal (SRS), buffer status report (BSR), and power headroom report (PHR). The NR-NTN gNB provides configurations for different PUCCH formats that can be used by the NR-NTN UE depending on the content to be transmitted on the UL. As part of the SR configuration, a set of PUCCH format 0/1 resources is specified. The periodicity, offset, and number of symbols for the SR-related PUCCH are specified. In the absence of SR configuration or SR failure, the NR-NTN UE can use the RA procedure to make a scheduling request to the NR-NTN gNB. The NR-NTN gNB may configure TA report-related parameters to facilitate TA reporting by the NR-NTN UE.

The NR-NTN gNB specifies SRS configuration to the NR-NTN UE, such as the number of OFDM symbols for the SRS transmission (e.g., last 1, 2, or 4 OFDM symbols), the SRS transmission bandwidth with two transmission combs (e.g., SRS on every 2nd or 4th subcarrier), and the number of ports (e.g., 1, 2, or 4).

As in a TN, the NR-NTN gNB may configure the NR-NTN UE with one or more SRS resource sets using SRS-ResourceSet or SRS-PosResourceSet parameters. For each SRS resource set, the NR-NTN UE may be configured with K SRS resources via a parameter SRS-Resource. The maximum value of K depends on the UE capability (3GPP 2024m). The SRS transmission may be aperiodic, semi-persistent, or periodic.

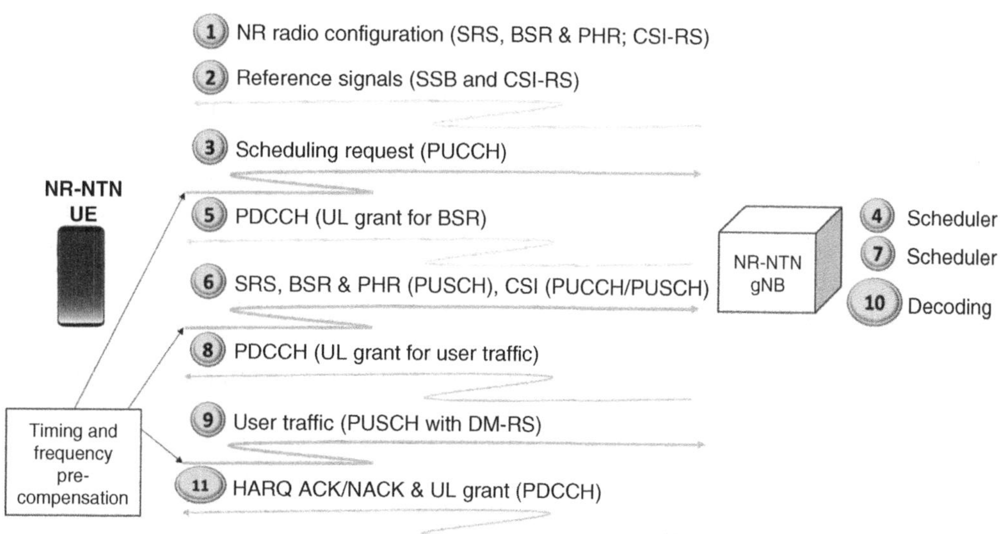

Figure 8.5 Uplink Data Transfer in the NR-NTN. *Source:* nadir_insan / Adobe Stock Photo.

In the case of aperiodic SRS, when the NR-NTN UE receives the DCI triggering aperiodic SRS in the downlink slot n, the NR-NTN UE transmits the aperiodic SRS in the slot K_s given by (3GPP 2024k)

$$K_s = \left\lfloor n \frac{2^{\mu_{SRS}}}{2^{\mu_{PDCCH}}} \right\rfloor + k + K_{offset} \cdot \frac{2^{\mu_{SRS}}}{2^{\mu_{K_{offset}}}} \tag{8.6}$$

In Eq. (8.6), n is the downlink slot number, $\mu_{SRS}$is the subcarrier spacing configuration for the SRS, μ_{PDCCH} is the subcarrier spacing configuration for the PDCCH, k is configured by the parameter slotOffset for each aperiodic SRS resource in each triggered SRS resources set and is based on the subcarrier spacing of the triggered SRS transmission, K_{offset} is configured by the NR-NTN gNB for the NR-NTN UE and given by Eq. (8.1), and $\mu_{K_{offset}}$ is the subcarrier spacing configuration (with the value = 0 for FR1) associated with K_{offset}.

The NR-NTN gNB can utilize the received SRS to quantify the UL radio channel conditions to guide the UL radio resource allocation. However, the NR-gNB must use such UL channel estimation with caution. The quantified UL channel conditions may become stale in case of a long propagation delay, especially for a transparent NTN payload and a GEO satellite-based NTN. Short-term radio channel variations may not be reflected accurately by the quantified UL channel conditions in such cases. The quantified UL channel conditions are likely to reflect prevailing radio channel conditions for a HAPS-based NTN and a LEO satellite-based NTN, and long-term variations in the radio channel for a GEO satellite-based NTN.

The NR-NTN gNB also specifies BSR configurations and periodic BSR timers. A short BSR provides UL buffer information (e.g., amount of data) about one Logical Channel Group (LCG), while a long BSR provides UL buffer information about multiple LCGs. The gNB specifies PHR-related parameters, such as a periodic timer for the PHR Type 1 PHR reports the difference between the NR-NTN UE's maximum transmit power and the estimated PUSCH power. Type 2 PHR reports the difference between the UE's maximum transmit power and the E-UTRA power. Type 3 PHR reports the difference between the UE's maximum transmit power and the estimated SRS power. Depending upon the type of the NR-NTN UE (e.g., a smartphone or a very small aperture terminal [VSAT]), the type of NTN platform (e.g., a GEO satellite or a LEO satellite), and the receive antenna gain of the NTN payload, the amount of transmit power utilized by the NR-NTN UE may be pretty close to the maximum transmit power capability of the NR-NTN UE. In such a case, the PHR is likely to report relatively little PHR value consistently.

In *Step 2*, the NR-NTN UE observes the SSB and the CSI-RS transmitted by the NR-NTN gNB to quantify the radio channel conditions in the form of CSI. While the CSI is directly helpful for the DL data transfer, it is also beneficial for the UL data transfer, especially for beam selection by the NR-NTN gNB to receive signals from the NR-NTN UE. Again, beams in an NTN cell are expected to be wider in an NTN compared to a TN due to the long distances between the Earth's surface and the NTN platform and larger NTN cell sizes.

In *Step 3*, when the NR-NTN UE has data such as a signaling message or traffic to send, it sends an SR on the configured PUCCH.

The NR-NTN gNB executes a scheduling algorithm in *Step 4* to allocate adequate UL resources to the NR-NTN UE so that the NR-NTN UE can send the BSR, which occurs in *Step 6*.

The NR-NTN gNB specifies the UL grant on the PDCCH for the NR-NTN UE's BSR in *Step 5*.

In *Step 6*, the NR-NTN UE sends the SRS, BSR, and PHR on the PUSCH per the configuration specified by the NR-NTN gNB. The NR-NTN UE also sends the CSI on the PUCCH or the PUSCH.

The NR-NTN gNB executes a scheduling algorithm in *Step 7*. The NR-NTN gNB learns about the UL resource requirements of the UL data waiting at the NR-NTN UE by processing the BSR.

The NR-NTN gNB evaluates the UL channel conditions by observing the SRS and determines the power availability at the NR-NTN UE by considering the PHR. The NR-NTN gNB considers a variety of factors such as the NR-NTN UE capabilities, UL channel conditions, available transmit power at the NR-NTN UE, available PRBs in the active BWP in the uplink, the number of carrier frequencies, the type and the amount of data in the NR-NTN UE's UL buffer, and QoS needs of NR-NTN UEs to allocate UL radio resources. As mentioned earlier in Step 1, the UL channel estimation needs to be used by the NR-gNB with caution due to a potentially long propagation delay. For example, a more conservative MCS may be used by the NR-NTN gNB. Furthermore, based on the scheduler strategy, PUSCH aggregation may be used to improve reliability.

In *Step 8*, the NR-NTN gNB conveys allocated resources to the NR-NTN UE using the DCI format 0_0 or 0_1 on the PDCCH. Both DCI Formats can convey parameters such as (i) frequency domain assignment (e.g., PRBs) with or without hopping, (ii) time domain assignment (i.e., starting OFDM symbol number and length of the allocation), (iii) MCS from the configured CQI table up to 64QAM or 256QAM, (iv) NDI, (v) RV, (vi) HARQ process number (i.e., one of thirty-two processes per NR-NTN UE capability), (vii) TPC command for scheduled PUSCH, and (viii) UL/Supplementary Uplink (SUL) Indicator (for UEs configured with SUL). 5G NR utilizes LDPC coding on the PUSCH to increase the effectiveness of channel coding and reduce the error rate while reducing energy consumption. Additionally, DCI Format 0_1 can indicate SRS Resource Indicator (SRI) and precoding information.

The NR-NTN UE in *Step 9* transmits a packet on the PUSCH based on the received UL grant in Step 8. The PUSCH can be transmitted with or without transform precoding. Multiple spatial multiplexing layers can be used on the PUSCH. The PUSCH in 5G NR can use quadrature phase shift keying (QPSK), 16QAM, 64QAM, and optionally 256QAM. Frequency hopping may or may not be used on the PUSCH; both intra-slot and inter-slot frequency hopping are supported.

Compared to high-order spatial multiplexing and high-order digital beamforming, Carrier Aggregation is much easier to utilize in an NR-NTN for transparent and regenerative payloads. Furthermore, depending upon the type of the NR-NTN UE (e.g., a smartphone or a VSAT), the type of NTN platform (e.g., a GEO satellite or a LEO satellite), and the receive antenna gain of the NTN payload, the amount of transmit power utilized by the NR-NTN UE may be very close to the maximum transmit power capability of the NR-NTN UE, regardless of the location of the NR-NTN UE in the NTN cell. In contrast, the transmit power of a UE in a TN cell will be quite different in the cell center and near the cell edge due to different path losses, especially in medium or large-sized TN cells.

As in the case of a DCI in a TN, when the NR-NTN UE is scheduled to transmit a Transport Block with or without a CSI report, the time domain resource assignment field value m in the NR-NTN provides a row index $(m + 1)$ to a resource allocation table (3GPP 2024k). The indexed row defines the slot offset $K2$, the start and length indicator SLIV or the start symbol S and the allocation length L, the PUSCH mapping type, the number of slots used for Transport Block Size (TBS) determination, and the number of PUSCH repetitions. The slot Ks in which the NR-NTN UE transmits the PUSCH (when K_{offset} is configured) is given by (3GPP 2024k)

$$K_S = \left\lfloor n \frac{2^{\mu_{\text{PUSCH}}}}{2^{\mu_{\text{PDCCH}}}} \right\rfloor + K_2 + K_{\text{offset}} \cdot \frac{2^{\mu_{\text{PUSCH}}}}{2^{\mu_{K_{\text{offset}}}}} \tag{8.7}$$

In Eq. (8.7), n is the downlink slot number, μ_{PUSCH} is the subcarrier spacing configuration for the PUSCH, μ_{PDCCH} is the subcarrier spacing configuration for the PDCCH, K_2 is given by the index row to a table based on the time domain resource assignment field value, K_{offset} is configured by the NR-NTN gNB for the NR-NTN UE and given by Eq. (8.1), and $\mu_{K_{\text{offset}}}$ is the subcarrier spacing configuration (with the value = 0 for FR1) associated with K_{offset}.

In *Step 10*, the NR-NTN gNB attempts to decode the packet on the PUSCH in a suitable beam through receive beamforming in an implementation-specific manner and conveys the HARQ ACK or NACK feedback to the NR-NTN UE depending on whether it has successfully decoded the packet or not. The NR-NTN gNB sends this HARQ feedback on the PDCCH. The NR-NTN gNB also allocates UL resources for new packet transmission or retransmission on the PUSCH depending upon the HARQ feedback it sends to the NR-NTN UE.

As in a TN, the NR-NTN gNB may utilize dynamic scheduling, SPS, or Configured Scheduling (CS). Dynamic scheduling is illustrated in Figure 8.4, where resource allocation dynamically occurs via a DCI. In the case of SPS, the resources are specified through RRC signaling and activated or deactivated via DCI signaling and SPS C-RNTI. In CS, RRC signaling specifies the resource allocation.

When the PUSCH resource allocation is semi-statically configured via RRC signaling, the parameter configuredGrantConfig in BWP-UplinkDedicated IE is used. In the case of Type 1 PUSCH transmissions with a configured grant, the NR-NTN UE utilizes the parameters in configuredGrant-Config as in a TN. In the case of Type 2 PUSCH transmissions with a configured grant, the resource allocation is based on the MAC layer-specified configuration (3GPP 2024n) and the DCI-specified UL grant, and K_{offset} is applied by the NR-NTN UE while determining the first transmission opportunity.

8.3 Data Transfer in the LTE-M NTN

Data transfer in LTE-M NTN is based on LTE-M data transfer in a TN and NTN-specific enhancements. Section 8.3.1 discusses prerequisites for data transfer in LTE-M NTN. Section 8.3.2 discusses downlink data transfer on the radio interface in the LTE-M NTN, and Section 8.3.3 discusses uplink data transfer on the radio interface in the LTE-M NTN.

8.3.1 Data Transfer Prerequisites in the LTE-M NTN

Figure 8.6 summarizes the key prerequisites for data transfer in the LTE-M NTN.

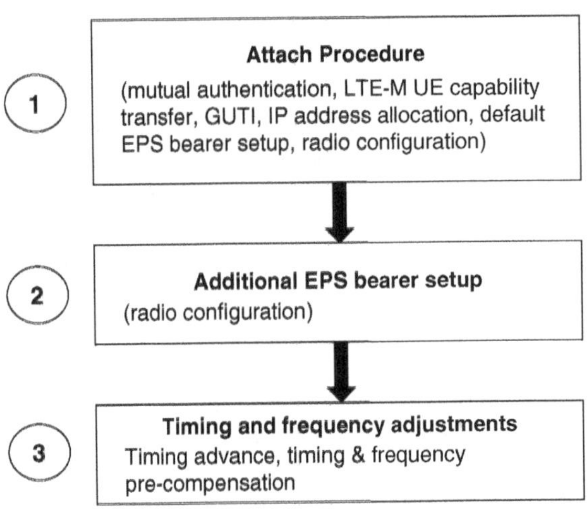

Figure 8.6 Key Data Transfer Prerequisites for the LTE-M NTN.

In *Step 1*, the LTE Attach procedure is carried out with the initiation of the procedure by the LTE-M NTN UE as described in Section 7.3.1 in Chapter 7. During the Attach procedure, the LTE-M NTN UE and the Evolved Packet Core (EPC) authenticate each other. The capabilities of the LTE-M NTN UE are transferred to the LTE-M NTN eNB via the UE capability transfer procedure. The Mobility Management Entity (MME) registers the LTE-M NTN UE in a set of Earth-fixed Tracking Areas. The EPC also assigns an IP address to the LTE-M NTN UE for IP-based data transfer. The MME assigns a globally unique temporary identity (GUTI) to the LTE-M NTN UE. During the Attach procedure, a default EPS bearer is set up. As part of this EPS bearer setup, radio configuration for a suitable DRB is also carried out to support the QoS associated with the default EPS Bearer. LTE-M NTN-specific radio configuration (e.g., RLC timer configuration, HARQ configuration, scheduling offsets, and Timing Advance reporting) is carried out in support of LTE-M NTN.

In the optional *Step 2*, additional default or dedicated EPS bearers may be established to meet different QoS requirements or to connect the LTE-M NTN UE with additional Packet Data Networks associated with different Access Point Names (APNs).

In *Step 3*, the LTE-M NTN UE carries out Timing Advance and frequency pre-compensation for all of its UL transmissions on an ongoing basis to eliminate or minimize intra-NTN cell interference.

Note that the LTE-M NTN UE in the RRC_CONNECTED mode continuously updates the Timing Advance and frequency pre-compensation. The LTE-M NTN UE enters the RRC_IDLE mode if the GNSS position that it has acquired is outdated and uplink transmission extension is not active unless GNSS acquisition was triggered, or uplink transmission extension is active (3GPP 2024d). The LTE-M NTN performs reporting of the "remaining GNSS validity duration" after completing the GNSS acquisition (3GPP 2024f).

Note that the Control Plane-based data transfer may be used as discussed in Section 3.1.1 in Chapter 3, where the IoT data is transported via NAS signaling between the IoT-NTN UE and the EPC. Such NAS signaling is encapsulated in RRC signaling on the radio interface. Furthermore, a non-IP-based data transfer may also be used in IoT-NTN as in a TN.

8.3.2 Timing Relationships in the LTE-M NTN

The LTE-M NTN introduces several enhancements to the existing LTE-M timing frameworks and a new framework for the Timing Advance reporting to reflect the propagation delays in the NTN. Examples of the existing LTE-M timing frameworks include a framework for the timing relationship between a downlink subframe carrying a specific type of information (e.g., a DCI, MAC CE, or Transport Block) and an associated uplink subframe and a framework for uplink transmission timing. Section 8.3.2.1 describes uplink transmission timing in the LTE-M NTN. Section 8.3.2.2 discusses an NTN-specific offset between the DL and the UL in the LTE-M NTN. Note that Section 8.2.2.3 discusses the framework for the Timing Advance reporting by the NR-NTN UE; this framework is similar to the framework followed by an LTE-M NTN UE. Section 8.3.2.3 summarizes the Timing Advance reporting by an LTE-M NTN UE.

8.3.2.1 Uplink Transmission Timing

Figure 8.7 shows the relationship between the downlink frame and the uplink frame at the NR-NTN UE (3GPP 2024a).

The LTE-M NTN UE's transmission of an uplink frame number *i* starts T_{TA} seconds before the start of the corresponding downlink frame *i* received at the LTE-M NTN UE. T_{TA} is given by

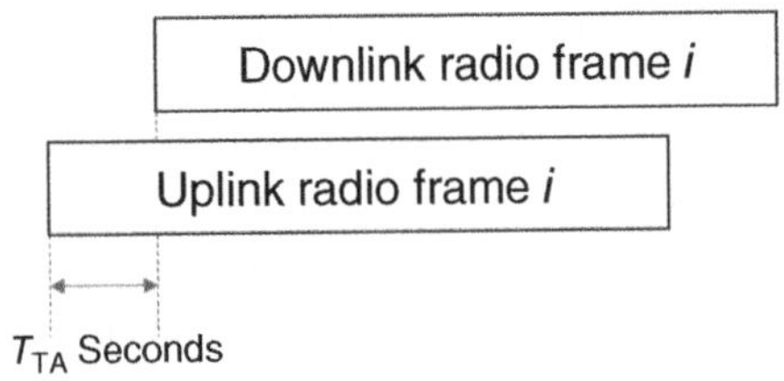

Figure 8.7 Timing Relationship Between the DL Frame and the UL Frame at the LTE-M NTN UE.

$$T_{TA} = \left(N_{TA} + N_{TA,offset} + N_{TA,adj}^{common} + N_{TA,adj}^{UE}\right) T_s. \tag{8.8}$$

In Eq. (8.2), N_{TA} is the Timing advance between the downlink and the uplink, and $N_{TA,offset}$ is a fixed offset used to calculate the Timing Advance (3GPP 2024a,c). N_{TA} has different ranges based on the LTE-M NTN UE configuration (3GPP 2024a). In general, $N_{TA,offset}$ is 0 for Frame Structure Type 1 and 624 for Frame Structure Type 2[6]. $N_{TA,adj}^{common}$ is network-controlled timing correction, and $N_{TA,adj}^{UE}$ is the UE-derived timing correction. $N_{TA,adj}^{common}$ and $N_{TA,adj}^{UE}$ are NTN-specific. T_c is the basic time unit in seconds and is given by

$$T_s = \frac{1}{(15,000^{*}\ 2,048)} s. \tag{8.9}$$

The LTE-M NTN eNB provides N_{TA} to the LTE-M NTN UE in different scenarios, such as in the Random-Access Response during the random-access procedure and a Timing Advance MAC CE when the LTE-M NTN UE is in the RRC_CONNETED state.

The LTE-M NTN UE calculates $N_{TA,adj}^{common}$ based on one-way propagation delay $\text{Delay}_{common}(t)$ as follows using N_{TA}^{common}, $N_{TA}^{commonDrift}$, and $N_{TA}^{commonDriftVariant}$ (3GPP 2024c).

$$\begin{aligned}\text{Delay}_{common}(t) &= \frac{1}{2}\left[N_{TA}^{common}\right] + \frac{1}{2}\left[N_{TA}^{commonDrift}\left(t - t_{epoch}\right)\right] + \\ &\frac{1}{2}\left[N_{TA}^{commonDriftVariant}\left(t - t_{epoch}\right)^{2}\right].\end{aligned} \tag{8.10}$$

In Eq. (8.10), N_{TA}^{common}, $N_{TA}^{commonDrift}$, and $N_{TA}^{commonDriftVariant}$ are specified by the LTE-M NTN eNB in SIB31 in the IE nta-CommonParameters-r17 as nta-Common-r17, nta-CommonDrift-r17, and nta-CommonDriftVariant, respectively (3GPP 2024c,g). Furthermore, t_{epoch} corresponds to epochTime. $N_{TA,adj}^{common}$ is used by the LTE-M NTN UE to pre-compensate the two-way transmission delay between the uplink time synchronization reference point and the serving satellite. Note that the uplink time synchronization reference point is the point where the downlink frame and the uplink frame are frame-aligned with an offset given by $N_{TA,offset}$.

The LTE-M NTN UE calculates $N_{TA,adj}^{UE}$ based on its position and serving satellite ephemeris (3GPP 2024c). The LTE-M NTN UE uses $N_{TA,adj}^{UE}$ to pre-compensate the two-way transmission delay on the service link.

In the radio network, as discussed in Section 7.3.3 in Chapter 7, DL and UL are frame-aligned at the uplink time synchronization RP with an offset T_{TA}, which is determined by $N_{TA,offset}$.

6 Frame Structure Type 1 is defined for FDD, and Frame Structure Type 1 is defined for TDD. Frame Structure Type 3 is defined for the Licensed Assisted Access (LAA) secondary cell.

8.3.2.2 NTN-Specific DL-UL Offset

An NTN-specific offset (3GPP 2024c), Koffset (3GPP 2024c), influences the timing relationship between the downlink command (e.g., a DCI or a MAC CE) and the associated uplink transmission.

$$K_{\text{offset}} = K_{\text{cell,offset}} - K_{\text{UE,offset}}. \tag{8.11}$$

In Eq. (8.11), K_{offset} is given by k-Offset-r17 in SIB31 as discussed in Section 7.3 in Chapter 7, and $K_{\text{UE,offset}}$ is provided by a Differential Koffset MAC CE command (3GPP 2024f). Note that an offset in Eq. (8.11) is assumed to be 0 when the LTE-M NTN eNB has not provided such offset's value to the LTE-M NTN UE.

8.3.2.3 Timing Advanced Reporting by an LTE-M NTN UE

The LTE-M NTN eNB may configure the LTE-M NTN UE with a TA report-related configuration via RRC signaling. The LTE-M NTN UE's TA report provides the LTE-M NTN eNB with an estimate of the LTE-M NTN UE's TA value called TTA in Figure 8.7. In an implementation-specific manner, the LTE-M NTN eNB can utilize the LTE-M NTN UE's TA report to determine resource allocation and timing-related adjustments such as Timing Advance commands and Differential Koffset.

The LTE-M NTN eNB uses RRC signaling to control how the LTE-M NTN UE reports Timing Advance by configuring parameters such as ta-Report and offsetThresholdTA (3GPP 2024f) (3GPP 2024g). An LTE-M NTN UE sends a Timing Advance Report (TAR) when a suitable event occurs. For example, assume that ta-Report = "enabled" in SystemInformationBlockType2. In that case, reporting of timing advance is enabled during Random Access due to RRC connection establishment, RRC connection resume, or RRC connection reestablishment (3GPP 2024g). Similarly, when ta-Report = "enabled" is included in MobilityControlInfo, TA reporting is enabled during Random Access due to handover. In another example, suppose the parameter offsetThresholdTA is configured or reconfigured and the LTE-M NTN UE has not previously reported the Timing Advance value to the current serving cell. In that case, the LTE-M NTN UE sends a TAR. Furthermore, if the variation between current information about Timing Advance and the last successfully reported information about Timing Advance is equal to or larger than offsetThresholdTA, the LTE-M NTN UE sends a TAR.

8.3.3 Downlink Data Transfer in the LTE-M NTN

After one or more EPS bearers are set up, the LTE-M NTN UE can exchange traffic with the LTE-M NTN eNB. The main steps of the DL data transfer in the LTE-M NTN are illustrated in an example implemented as in Figure 8.8. Figure 8.8 also summarizes dynamic scheduling-based downlink data transfer in the LTE-M NTN. The steps of downlink data transfer are similar for the TN and the NTN, with timing-related enhancements made for the NTN.

In *Step 1*, the LTE-M NTN UE observes the Cell-specific Reference Signal (CRS) of the serving LTE-M NTN cell to quantify the prevailing DL radio channel conditions. The CRS in LTE-M NTN reuses the same structure as the CRS in LTE.

In *Step 2*, the LTE-M NTN UE sends the channel condition report to the LTE-M NTN eNB to enable the eNB to determine optimal PDSCH transmission parameters. This report contains information such as the CQI, Precoding Matrix Indicator (PMI), and Rank Indicator (RI). Since the LTE-M NTN UE, like an LTE-M TN UE, has a single receive antenna, RI cannot be greater than one in LTE-M NTN. LTE-M supports QPSK and 16QAM to reduce complexity. Additionally, LTE-M supports simpler convolutional coding for the PDSCH. The channel condition report is sent on the PUCCH or the PUSCH. If the LTE-M NTN UE has been allocated a PUSCH in the current

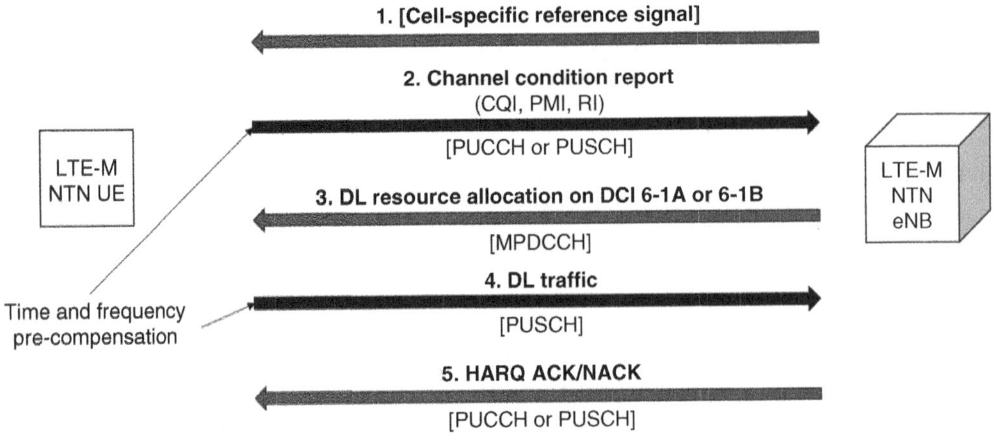

Figure 8.8 Downlink Data Transfer in the LTE-M NTN.

subframe, the channel condition report is sent on the PUSCH along with other PUSCH payloads, such as IoT traffic. If no PUSCH is allocated to the LTE-M NTN UE for the current subframe, the LTE-M NTN UE utilizes the PUCCH to convey the channel condition report. For increased reliability, the channel condition report may be repeated. The CSI reference resource[7] that the LTE-M NTN UE reports in subframe n is the subframe $\left(n - n_{\mathrm{CQI}_{\mathrm{ref}}} - K_{\mathrm{offset}}\right)$ where $n_{CQI_{\mathrm{ref}}}$ is $> = 4$, and K_{offset} is given by Eq. (8.11).

The LTE-M NTN eNB executes a scheduling algorithm that considers the channel condition report along with numerous factors such as the LTE-M NTN UE capabilities (e.g., half-duplex FDD) (3GPP 2024e), the number of PRBs in the narrowband, the number of narrowbands, and QoS needs of LTE-M NTN UEs to allocate DL radio resources. As mentioned earlier, in the case of NR-NTN, the radio channel condition report needs to be used by the LTE-M NTN eNB with caution due to a potentially long propagation delay. For example, a more conservative MCS may be used by the LTE-M NTN eNB.

The LTE-M NTN eNB then conveys allocated resources to the LTE-M NTN UE using Downlink Control Information (DCI) Format 6-1A or 6-1B on the MPDCCH in *Step 3*. The MPDCCH itself may use repetitions for increased reliability. The MPDCCH may also use frequency hopping. The DCI format conveys PDSCH resource allocation parameters such as the allocated narrowband index, the number of PRBs assigned to the LTE-M NTN UE in the allocated narrowband index, MCS, and the number of PDSCH repetitions. The LTE-M NTN UE, like a TN LTE-M UE, begins to receive the associated PDSCH subframe $(n + 2)$ if the DCI carried by the MPDCCH has indicated the resource allocation in subframe n. If the MPDCCH has been repeated, subframe n contains the last repetition of the MPDCCH.

In *Step 4*, the LTE-M NTN UE retrieves the packet from the PDSCH. Note that the PDSCH may be repeated as specified in the DCI Format in Step 3. The PDSCH utilizes QPSK or 16QAM and a single-layer transmission. Frequency hopping may be used on the PDSCH.

The LTE-M NTN UE attempts to decode the packet on the PDSCH. The LTE-M NTN UE conveys the HARQ ACK or NACK feedback to the LTE-M NTN eNB, depending on whether it has successfully decoded the packet or not in *Step 5*. The LTE-M NTN UE sends this HARQ feedback on the

7 The CSI reference resource is a group of downlink PRBs that the UE observes to quantify the CQI. The CRS typically resides in these PRBs. A subframe where the CSI reference resource exists is a valid downlink or special subframe.

PUCCH or the PUSCH. If the PDSCH (or the last repetition of the PDSCH if repetitions are used) were received in subframe m by an LTE-M UE in a TN, such UE would send the HARQ ACK feedback in subframe (m + 4). It is also possible to add additional delay to allow the LTE-M UE in a TN to send the HARQ feedback in a subframe after subframe (m + 4). However, in the case of LTE-M NTN, an additional NTN-specific delay of K_{offset} is added. In general, the LTE-M NTN UE sends the HARQ-ACK in subframe n in response to the PDSCH transmission received in subframe $(n - k - K_{\text{offset}})$. The value of k depends on factors such as HARQ-ACK delay field in the DCI as well as tables specified in 3GPP (2024j) based on various conditions related to half-duplex operation, CE Mode A, the use of fourteen HARQ processes, and HARQ ACK bundling. The HARQ ACK/NACK may be repeated for increased reliability. In the case of HARQ-ACK bundling, the LTE-M NTN UE, like an LTE-M TN UE, transmits a single HARQ-ACK feedback for multiple DL transport blocks (referred to as packets here).

Note that the LTE-M NTN UE may be configured by the LTE-M NTN eNB to disable HARQ feedback for a given HARQ process through the parameter downlinkHARQ-FeedbackDisabled-Bitmap or downlinkHARQ-FeedbackDisabled-DCI. In such a case, the LTE-M NTN UE does not send HARQ feedback to the LTE-M NTN eNB in response to a Transport Block received on the PDSCH in subframe n. Furthermore, for such HARQ process, the LTE-M NTN UE does not expect to receive (i) an MPDCCH or (ii) a PDSCH without a corresponding MPDCCH in the downlink subframes $n + 1$, $n + 2$, and $n + 3$.

If the LTE-M NTN eNB has configured (i) downlinkHARQ-FeedbackDisabled-DCI or (ii) downlinkHARQ-FeedbackDisabled-Bitmap and downlinkHARQ-FeedbackDisabled-DCI for the LTE-M NTN UE, the value of "3" in the DCI field HARQ-ACK resource offset functions as a HARQ feedback disabled indicator.

When the HARQ feedback is not disabled, the LTE-M NTN eNB sends a new packet (i.e., a Transport Block) if it has received HARQ ACK from the LTE-M NTN UE or retransmits a previously sent packet if it has received a HARQ NACK provided a maximum number of HARQ retransmissions has not been reached.

8.3.4 Uplink Data Transfer in the LTE-M NTN

Figure 8.9 summarizes dynamic scheduling-based uplink data transfer in the LTE-M NTN in an example implementation. The steps of uplink data transfer are similar for the TN and the NTN, with timing-related enhancements made for the NTN. Note that all the uplink transmissions involve Timing Advance and frequency pre-compensation.

In *Step 1*, when the LTE-M NTN UE has data (e.g., a signaling message or IoT traffic) to send, it sends a SR on the configured PUCCH. For increased reliability, the PUCCH may use repetitions. Furthermore, frequency hopping is inherently supported on the PUCCH for enhanced reliability in the UL.

The LTE-M NTN eNB executes a scheduling algorithm to allocate adequate UL resources to the LTE-M NTN UE so that the LTE-M NTN UE can send the BSR. The LTE-M NTN eNB specifies the UL grant for the LTE-M NTN UE's BSR using the DCI 6-0A/6-0B on the MPDCCH (3GPP 2024c) in *Step 2*. Notre that the DCI Format 6-0A/6-0B is used for (i) the scheduling of PUSCH in an uplink cell, (ii) the indication of HARQ-ACK feedback, and (iii) operation on preconfigured UL resources. The MPDCCH itself may use repetitions for increased reliability. The MPDCCH may also use frequency hopping.

In *Step 3*, the LTE-M NTN UE sends the BSR on the PUSCH to the LTE-M NTN eNB using the allocated UL grant. Note that the LTE-M NTN UE typically performs PUSCH transmission in the

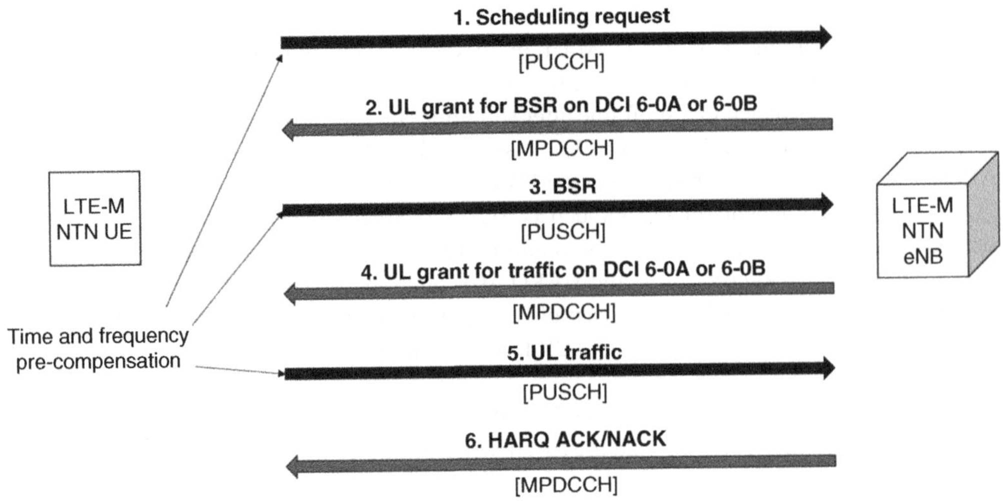

Figure 8.9 Uplink Data Transfer in the LTE-M NTN.

subframe(s) $(n + k_i + K_{offset})$ (3GPP 2024c), where n is the last subframe in which the MPDCCH is transmitted, k_i reflects (i) the number of subframes in which repetitions are carried out and (ii) multiple transport blocks (if any) as specified in suitable tables in 3GPP (2024c), and K_{offset} is defined in Eq. (8.11). Note that $(n + k_i)$ is used in a TN, and K_{offset} is NTN-specific.

The LTE-M NTN eNB learns about the UL resource requirements of the UL data waiting at the LTE-M NTN UE by processing the BSR. The LTE-M NTN eNB then executes a scheduling algorithm that considers a variety of factors, such as the LTE-M NTN UE capabilities (e.g., half-duplex FDD), UL channel conditions (estimated by the LTE-M NTN eNB based on the observation of the SRS), the number of PRBs in the narrowband, the number of narrowbands, the type and the amounts of data in the LTE-M NTN UE's UL buffer, and QoS needs of LTE-M NTN UEs to allocate UL radio resources. As mentioned earlier, in the case of DL data transfer, the LTE-M NTN eNB needs to use the radio channel estimate cautiously due to a potentially long propagation delay. For example, a more conservative MCS may be used by the LTE-M NTN eNB.

The LTE-M NTN eNB conveys allocated resources to the LTE-M NTN UE using the DCI format 6-0A (for CE Mode A) or 6-0B (for CE Mode B) on the MPDCCH in *Step 4*. As mentioned in Step 2, the MPDCCH may use repetitions and frequency hopping. The DCI formats 6-0A and 6-0B include the resource allocation information, such as the narrowband index specifying the narrowband, the allocated PRBs in the narrowband, MCS, and the number of PUSCH repetitions.

The LTE-M NTN UE transmits a packet on the PUSCH based on the received UL grant in *Step 5*. LTE-M NTN utilizes cross-subframe scheduling, where the LTE-M NTN UE typically utilizes the allocated PUSCH resources to transmit a packet in the subframe(s) $(n + k_i + K_{offset})$, as explained in Step 3, where n is the last subframe in which the MPDCCH is transmitted. If the MPDCCH has been repeated, subframe n contains the last repetition of the MPDCCH. The LTE-M NTN UE transmits on a single antenna. The PUSCH in LTE-M NTN can make use of QPSK or 16QAM. Frequency hopping is also supported on the PUSCH in LTE-M NTN.

In *Step 6*, the LTE-M NTN eNB attempts to decode the packet on the PUSCH and conveys the HARQ ACK feedback to the LTE-M NTN UE, depending on whether it has successfully decoded the packet or not. The LTE-M NTN eNB sends this HARQ feedback on the MPDCCH in a DCI. The

LTE-M NTN eNB also allocates UL resources for new packet transmission or retransmission on the PUSCH, depending upon the HARQ feedback it sends to the LTE-M NTN UE. As in a TN, the LTE-M NTN eNB may configure mpdcch-UL-HARQ-ACK-FeedbackConfig and use the DCI Format 6-0A to indicate ACK feedback. The NDI field in the DCI conveys to the UE if the resource allocation for the HARQ process is for new transmission or retransmission; this field is absent if the LTE-M eNB has enabled ce-PUSCH-MultiTB-Config.

For both a TN and an NTN, LTE-M supports a maximum of eight uplink HARQ processes per serving cell in CE Mode A in frequency division duplex/duplexing (FDD). For both a TN and an NTN, for CE Mode B, LTE-M supports a maximum of four uplink HARQ processes per serving cell if the UE is configured with ce-PUSCH-MultiTB-Config; otherwise, a maximum of two uplink HARQ processes per serving cell are supported. If preconfigured uplink resources (PURs) are used on a PUSCH, one uplink HARQ process per serving cell is supported.

The LTE-M NTN eNB may trigger SRS transmission by the LTE-M NTN UE using (i) type 0 trigger (which is via RRC signaling) or (ii) type 1 or type 2 trigger[8] (which is by DCI signaling). with the SRS so that the eNB can estimate UL channel conditions. When the LTE-M NTN eNB requests an LTE-M NTN UE for SRS transmission in subframe n using the DCI format 6-0A/6-1A, such UE begins SRS transmission in the first subframe satisfying the condition $(n + k + K_{\text{offset}})$, with $k \geq 4$, and a certain additional condition is satisfied based on the duplex type (e.g., time division duplex/duplexing (TDD) or FDD) and other SRS parameters such as $T_{\text{SRS},1}$ and k_{SRS}; see 3GPP (2024c) for more details.

The PUR method may be used in LTE-M for UL resource allocation in a TN or an NTN. If the LTE-M NTN UE's PUR-based PUSCH transmission ends in subframe n, the UE looks for its MPDCCH in a search space window starting in subframe $(n + 4 + K_{\text{mac}})$ with window duration specified by the parameter pur-MPDCCH-SS-window-duration. K_{mac} is specified in SIB31 as k-Mac-r17. If the LTE-M NTN UE detects an MPDCCH carrying the DCI format 6-0A/6-0B with the cyclic redundancy check (CRC) scrambled by the UE's PUR-RNTI, and if such DCI corresponds to PUR ACK/fallback indication, the UE does not need to monitor an MPDCCH in the remaining search space window duration.

8.3.5 GNSS Measure Gap Procedure in the LTE-M NTN

An LTE-M NTN UE cannot utilize the NTN for data transfer if it does not have valid GNSS measurements. To facilitate the UE's GNSS measurements in an NTN FDD serving cell, a GNSS measurement gap needs to be created. The LTE-M NTN eNB can send a GNSS Measurement Command MAC CE in a PDSCH to initiate the GNSS measurement process at the UE. Assume that the LTE-M NTN UE receives the MAC CE in subframe n. If the UE does not provide HARQ-ACK information for the HARQ process associated with the transport block carrying the GNSS Measurement Command MAC CE, the UE considers the subframe $(n + 6)$ to be the start of the GNSS measurement gap. Otherwise, the UE considers the subframe $(k + 2)$ to be the start of the GNSS measurement gap, where k is the first DL subframe after the end of the transmission of the PUSCH carrying ACK/NACK response for the HARQ process associated with the transport block in the PDSCH (3GPP 2024c). Furthermore, the LTE-M NTN UE is not required to monitor an MPDCCH during the GNSS measurement gap until it reacquires the GNSS position and performs a contention-based Random Access (3GPP 2024f).

8 Type 1 trigger is applicable to one set of DCI formats, and type 2 trigger is applicable to other DCI formats (3GPP 2024c).

8.4 Data Transfer in the NB-IoT NTN

Data transfer in NB-IoT NTN is based on NB-IoT data transfer in TN and NTN-specific enhancements. There are many similarities between the data transfer in LTE-M and NB-IoT for both a TN and an NTN. Section 8.4.1 discusses the prerequisites for data transfer in NB-IoT NTN. Timing relations in the NB-IoT NTN are similar to those in the LTE-M NTN and are summarized in Section 8.4.2. Section 8.4.3 discusses downlink data transfer on the radio interface in the NB-IoT NTN, and Section 8.4.4 discusses uplink data transfer on the radio interface in the NB-IoT NTN.

8.4.1 Data Transfer Prerequisites in the NB-IoT NTN

Figure 8.10 summarizes key prerequisites for data transfer in NB-IoT NTN.

In *Step 1*, the LTE Attach procedure is carried out with the initiation of the procedure by the NB-IoT NTN UE, as described in Section 7.3.1 in Chapter 7. During the Attach procedure, the NB-IoT NTN UE and the EPC authenticate each other. The capabilities of the NB-IoT NTN UE are transferred to the NB-IoT eNB via the UE capability transfer procedure. The MME registers the NB-IoT NTN UE in a set of Earth-fixed Tracking Areas. The EPC also assigns an IP address to the NB-IoT NTN UE for IP-based data transfer. The MME assigns a GUTI to the NB-IoT NTN UE. During the Attach procedure, a default EPS bearer is set up. As part of this EPS bearer setup, radio configuration for a suitable DRB is also carried out to support the QoS associated with the default EPS Bearer. NB-IoT NTN-specific radio configuration (e.g., RLC timer configuration, HARQ configuration, scheduling offsets, and Timing Advance reporting) is carried out in support of NB-IoT NTN.

In the optional *Step 2*, additional default or dedicated EPS bearers may be established to meet different QoS requirements or to connect the NB-IoT NTN UE with additional Packet Data Networks associated with different APNs.

In *Step 3*, the NB-IoT NTN UE carries out Timing Advance and frequency pre-compensation for all of its UL transmissions on an ongoing basis to eliminate or minimize intra-NTN cell interference.

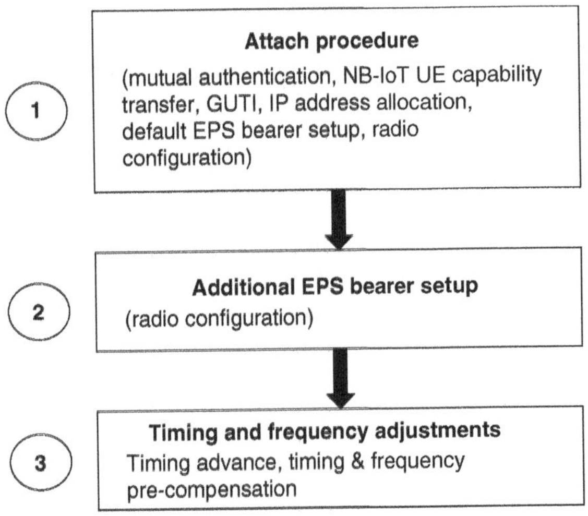

Figure 8.10 Key Data Transfer Prerequisites for the NB-IoT NTN.

As mentioned in Section 8.2.1 for the case of LTE-M NTN, Control plane-based data transfer and non-IP-based data transfer may also be used for the NB-IoT NTN.

8.4.2 Timing Relationships in the NB-IoT NTN

The NB-IoT NTN and the LTE-M NTN share similar timing relationships. Like the LTE-M NTN, the NB-IoT NTN introduces several enhancements to the existing NB-IoT timing frameworks. Examples of the existing NB-IoT timing frameworks include a framework for the timing relationship between a downlink subframe carrying a certain type of information (e.g., a DCI, MAC CE, or Transport Block) and an associated uplink subframe and a framework for uplink transmission timing. Section 8.4.2.1 describes uplink transmission timing in the NB-IoT NTN. Section 8.4.2.2 discusses an NTN-specific offset between the DL and the UL in the NB-IoT NTN.

8.4.2.1 Uplink Transmission Timing in the NB-IoT NTN

Figure 8.11 shows the relationship between the downlink frame and the uplink frame at the NB-IoT UE (3GPP 2024a), *which is identical to that in the LTE-M NTN.*

The NB-IoT NTN UE's transmission of an uplink frame number i starts T_{TA} seconds before the start of the corresponding downlink frame i received at the NB-IoT NTN UE. T_{TA} is given by

$$T_{\text{TA}} = \left(N_{\text{TA}} + N_{\text{TA,offset}} + N_{\text{TA,adj}}^{\text{common}} + N_{\text{TA,adj}}^{\text{UE}}\right) T_s. \tag{8.12}$$

In Eq. (8.12), N_{TA} is the Timing advance between the downlink and the uplink, and $N_{\text{TA,offset}}$ is a fixed offset used to calculate the Timing Advance (3GPP 2024a,c). N_{TA} has different ranges based on the NB-IoT NTN UE configuration (3GPP 2024a). In general, $N_{\text{TA,offset}}$ is 0 for Frame Structure Type 1 and 624 for Frame Structure Type 2. $N_{\text{TA,adj}}^{\text{common}}$ is network-controlled timing correction, and $N_{\text{TA,adj}}^{\text{UE}}$ is the UE-derived timing correction. $N_{\text{TA,adj}}^{\text{common}}$ and $N_{\text{TA,adj}}^{\text{UE}}$ are NTN specific. T_c is the basic time unit in seconds and is given by

$$T_s = \frac{1}{(15{,}000 * 2{,}048)} s. \tag{8.13}$$

The NB-IoT NTN eNB provides N_{TA} to the NB-IoT NTN UE in different scenarios, such as in the Random-Access Response during the random-access procedure and a Timing Advance MAC CE when the LTE-M NTN UE is in the RRC_CONNETED state.

The NB-IoT NTN UE calculates $N_{\text{TA,adj}}^{\text{common}}$ based on one-way propagation delay $\text{Delay}_{\text{common}}(t)$ as follows using $N_{\text{TA}}^{\text{common}}$, $N_{\text{TA}}^{\text{commonDrift}}$, and $N_{\text{TA}}^{\text{commonDriftVariant}}$ (3GPP 2024c).

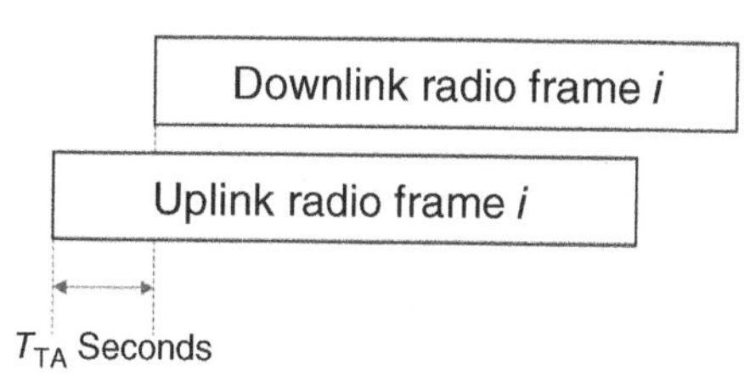

Figure 8.11 Timing Relationship Between the DL Frame and the UL Frame at the NB-IoT NTN UE.

$$\text{Delay}_{\text{common}}(t) = \frac{1}{2}\left[N_{\text{TA}}^{\text{common}}\right] + \frac{1}{2}\left[N_{\text{TA}}^{\text{commonDrift}}\left(t - t_{\text{epoch}}\right)\right] + \frac{1}{2}\left[N_{\text{TA}}^{\text{commonDriftVariant}}\left(t - t_{\text{epoch}}\right)^2\right]. \tag{8.14}$$

In Eq. (8.14), $N_{\text{TA}}^{\text{common}}$, $N_{\text{TA}}^{\text{commonDrift}}$, and $N_{\text{TA}}^{\text{commonDriftVariant}}$ are specified by the NB-IoT NTN eNB in SIB31-NB in the IE nta-CommonParameters-r17 as nta-Common-r17, nta-CommonDrift-r17, and nta-CommonDriftVariant, respectively (3GPP 2024c,g). Furthermore, t_{epoch} corresponds to epochTime. $N_{\text{TA,adj}}^{\text{common}}$ is used by the NB-IoT NTN UE to pre-compensate for the two-way transmission delay between the uplink time synchronization reference point and the serving satellite. Note that the uplink time synchronization reference point is the point where the downlink frame and the uplink frame are frame-aligned with an offset given by $N_{\text{TA,offset}}$.

The NB-IoT NTN UE calculates $N_{\text{TA,adj}}^{\text{UE}}$ based on its position and serving satellite ephemeris (3GPP 2024c). The NB-IoT NTN UE uses $N_{\text{TA,adj}}^{\text{UE}}$ to pre-compensate the two-way transmission delay on the service link.

In the radio network, as discussed in Section 7.3.3 in Chapter 7, DL and UL are frame-aligned at the uplink time synchronization RP with an offset T_{TA}, which is determined by $N_{\text{TA,offset}}$.

When the NB-IoT NTN UE is transmitting information in multiple uplink segments, time and frequency pre-compensation are adjusted per uplink segment with a transmission duration of $N_{\text{segment}}^{\text{precompensation}}$ time units (3GPP 2024c,g).

8.4.2.2 NTN-Specific DL-UL Offset in the NB-IoT NTN

The timing relationship between the downlink command (e.g., a DCI or a MAC CE) and the associated uplink transmission is influenced by an NTN-specific offset, K_{offset} (3GPP 2024c). *This relationship in the NB-IoT NTN is identical to that in the LTE-M NTN.*

$$K_{\text{offset}} = K_{\text{cell,offset}} - K_{\text{UE,offset}}. \tag{8.15}$$

In Eq. (8.11), K_{offset} is given by k-Offset-r17 in SIB31-NB as discussed in Section 7.3 in Chapter 7, and $K_{\text{UE,offset}}$ is provided by a Differential Koffset MAC CE command (3GPP 2024f). Note that an offset in Eq. (8.11) is assumed to be 0 when the NB-IoT eNB has not provided such an offset value to the NB-IoT UE.

8.4.3 Downlink Data Transfer in the NB-IoT NTN

Figure 8.12 summarizes dynamic scheduling-based downlink data transfer in NB-IoT NTN.

In *Step 1*, the NB-IoT NTN eNB executes an implementation-specific scheduling algorithm that considers a variety of factors such as historical HARQ feedback, the NB-IoT NTN UE capabilities, such as Cellular IoT optimization, the available radio resources on the anchor or non-anchor carrier, the amount of data in the buffer, and QoS requirements to allocate the DL radio resources to the NB-IoT NTN UE. Note that no explicit channel condition report contains information such as CQI in NB-IoT and the NB-IoT NTN to reduce the processing requirements on the NB-IoT NTN UE.

The NB-IoT NTN eNB, in *Step 2*, conveys resource allocation to the NB-IoT NTN UE using DCI Format N1 on the NPDCCH. Examples of the allocation information include the number of subframes allocated to the NB-IoT NTN UE for transmission, MCS, the number of narrowband physical downlink shared channel (NPDSCH) repetitions, HARQ Resource Information for the NPUSCH, DCI Subframe Repetition Number, and NDI. The NPDCCH itself may be repeated for reliability. In general, the DCI format N1 is used for (i) scheduling of one NPDSCH codeword

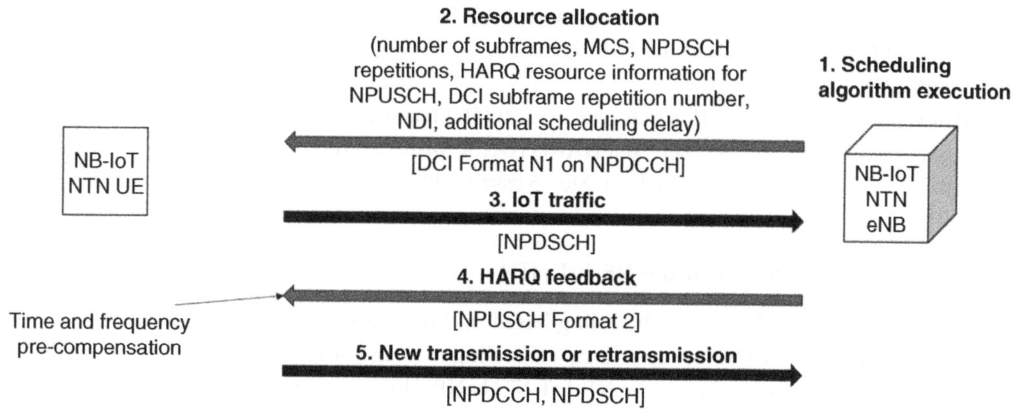

Figure 8.12 Downlink Data Transfer in the NB-IoT NTN.

per TTI in one cell, (ii) random-access procedure initiated by an NPDCCH order, (iii) notifying SC-MCCH change, and (iv) the operation on PUR (3GPP 2024b).

NB-IoT utilizes cross-subframe scheduling in a TN and an NTN. The specified resource allocation usually applies to the NPDSCH transmission starting in subframe $(n + 5)$ if the associated NPDCCH has occurred in subframe n. If the NPDCCH has been repeated, subframe n contains the last repetition of the NPDCCH. Thus, the typical gap or the scheduling delay between the DL resource allocation by the NPDCCH and the beginning of the associated NPDSCH transmission is four subframes in NB-IoT. However, the NB-IoT NTN eNB can also increase such scheduling delay beyond four subframes by specifying an additional scheduling delay k_0 (3GPP 2024c). The timing association of the NPDCCH with the NPDSCH is the same in both a TN and an NTN.

The NB-IoT NTN UE retrieves the packet from the NPDSCH in *Step 3*. Note that the NPDSCH may be repeated as specified in the DCI Format in Step 2. The NPDSCH utilizes QPSK and a single-layer transmission, which may be single-antenna or space frequency block code (SFBC) transmission. One or two HARQ processes may be used. The NPDSCH may be repeated up to a maximum of 2,048 times.

The NB-IoT NTN UE attempts to decode the packet on the NPDSCH. The NB-IoT NTN UE conveys the HARQ ACK feedback to the NB-IoT NTN eNB, depending on whether it has successfully decoded the packet or not in *Step 4*. The NB-IoT NTN UE sends this HARQ feedback using the NPUSCH Format 2. If the NPDSCH (or the last repetition of the NPDSCH if repetitions are used) is received in subframe n, the NB-IoT UE sends the HARQ ACK/NACK in subframe $n + k_0' + K_{\text{offset}} - 1$ in the case of FDD, where k_0' is k_0. and K_{offset} is defined by Eq. (8.11). Furthermore, the HARQ ACK/NACK on the NPUSCH Format 2 may be repeated for increased reliability. The subcarriers allocated for ACK/NACK and k_0 are determined by the ACK/NACK resource field in the DCI format and a subcarrier spacing-based table in 3GPP (2024b).

An NB-IoT UE in a TN or an NTN may support two HARQ processes for the NPDSCH. Such UE capability can be specified by the parameter two HARQ processes-r14. The NB-IoT eNB may configure an NB-IoT UE with npdsch-MultiTB-Config to use two HARQ processes.

Note that the NB-IoT NTN UE may be configured by the NB-IoT NTN eNB to disable HARQ feedback through the configured parameter downlinkHARQ-FeedbackDisabled-DCI-NB. Furthermore, when the parameters downlinkHARQ-FeedbackDisabled-Bitmap-NB and downlinkHARQ-FeedbackDisabled-DCI-NB are configured, the value of HARQ-ACK resource= "15" serves as a

HARQ feedback disabled indicator. When the HARQ feedback is disabled, the NB-IoT NTN UE does not send HARQ feedback to the NB-IoT NTN eNB in response to a Transport Block received on the NPDSCH in subframe n.

In *Step 5*, the NB-IoT NTN eNB sends a new packet if it has received HARQ ACK or retransmits a previously sent packet if it has received a HARQ NACK, provided the maximum number of HARQ retransmissions has not been reached yet. Step 5 essentially repeats Steps 2 and 3 discussed earlier.

8.4.4 Uplink Data Transfer in the NB-IoT NTN

Figure 8.13 summarizes dynamic scheduling-based uplink data transfer in the NB-IoT NTN.

When the NB-IoT NTN UE has data (e.g., a signaling message or IoT traffic) to send, it needs to convey the need for UL resources to the NB-IoT NTN eNB. However, there is no PUCCH in NB-IoT for the UE to send a Scheduling request. NB-IoT relies on the random-access procedure to convey the UL resource needs. Hence, in *Step 1*, the NB-IoT NTN UE executes the random-access procedure.

In *Step 2*, the NB-IoT NTN eNB executes a scheduling algorithm to allocate adequate UL resources to the UE so that the UE can send the BSR. The NB-IoT NTN eNB specifies the UL grant for the NB-IoT NTN UE's BSR using the DCI Format N0 NPDCCH. The DCI Format N0 is used for the scheduling of the NPUSCH as well as the operation of the PUR in one UL cell.

In *Step 3*, the NB-IoT NTN UE sends the BSR on the NPUSCH using the NPUSCH Format using the allocated UL grant. If the NPDCCH (or the last repetition of the NPDCCH if repetitions are used) is received in subframe n, the NB-IoT UE transmits the NPUSCH beginning in subframe $n + k_0 + K_{\text{offset}}$ in the case of FDD, where k_0 is determined by the scheduling delay field in the DCI Format N0 along with a duplex-based table defined in 3GPP (2024c), and K_{offset} is defined by Eq. (8.11). The NPUSCH may be repeated for increased reliability.

The NB-IoT NTN eNB learns about the UL resource requirements of the UL data waiting at the NB-IoT NTN UE by processing the BSR. The NB-IoT NTN eNB then executes a scheduling

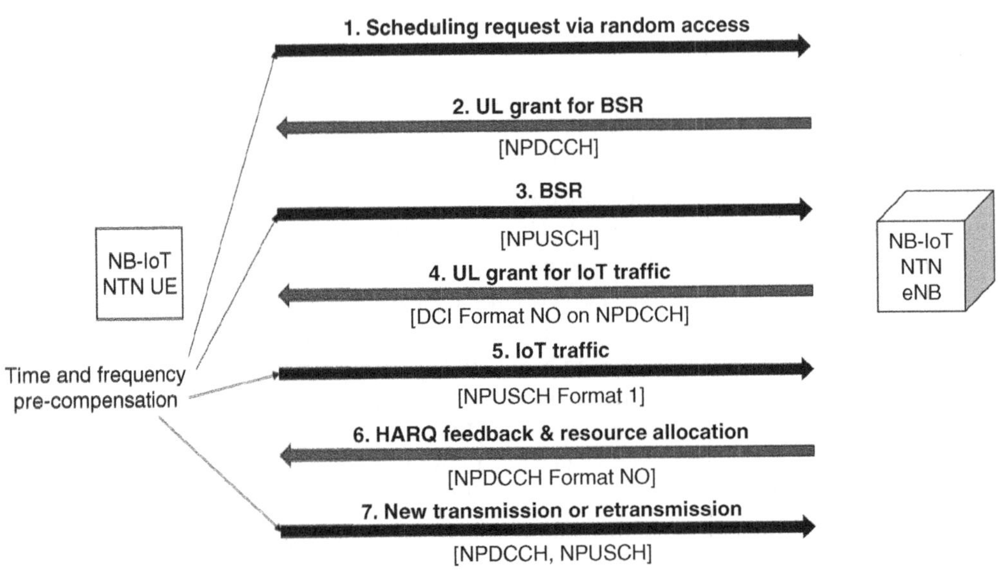

Figure 8.13 Uplink Data Transfer in the NB-IoT NTN.

algorithm that considers a variety of factors such as the NB-IoT NTN UE capabilities (e.g., single-tone transmission and multi-carrier transmission), UL channel conditions (estimated by the eNB based on the observation of the DMRS accompanying the NPUSCH; there is no SRS in NB-IoT), available radio resources, the amount of data in the NB-IoT NTN UE's UL buffer, and QoS needs of NB-IoT NTN UEs to allocate UL radio resources. The NB-IoT NTN eNB conveys allocated resources to the NB-IoT NTN UE using the DCI Format N0 on the NPDCCH in *Step 4*. As mentioned in Step 2, the NPDCCH may use repetitions. The DCI Format N0 includes the resource allocation information such as the allocated Resource Units (RUs), MCS, scheduling delay, and NDI. An RU is associated with a specific subcarrier spacing (SCS), a certain number of subcarriers, the TTI, and the supported modulation scheme. For example, an RU may use 1 subcarrier and a 32 ms TTI with the SCS of 3.75 kHz and may allow binary phase shift keying (BPSK) or quadrature phase shift keying (QPSK). Another RU may use 12 subcarriers, the TTI of 1 ms for the SCS of 15 kHz, and QPSK. The NPUSCH supports the SCS of 3.75 and 15 kHz. The NPUSCH supports 1, 3, 6, or 12 subcarriers. Specific combinations of the SCS and the number of subcarriers are used to define RUs, and the NB-IoT allocates RUs to the NB-IoT NTN UE for the NPUSCH transmission.

The NB-IoT NTN UE transmits a packet on the NPUSCH Format 1 along with the associated DMRS based on the received UL grant in *Step 5*. NB-IoT utilizes cross-subframe scheduling for the UL as well. Specifically, the NB-IoT NTN UE sends a transport block on the NPUSCH using the NPUSCH Format 1 that uses the allocated UL grant. If the NPDCCH (or the last repetition of the NPDCCH if repetitions are used) is received in subframe n, the NB-IoT UE transmits the NPUSCH beginning in subframe $n + k_0 + K_{\text{offset}}$ in the case of FDD, where k_0 is determined by the scheduling delay field in the DCI Format N0 along with a duplex-based table defined in 3GPP (2024c), and K_{offset} is defined by Eq. (8.11). The NPUSCH may be repeated for increased reliability. As in a TN, the NB-IOT NTN UE transmits on a single antenna. The NPUSCH in NB-IoT typically uses BPSK or QPSK as needed. Based on the NB-IoT UE capability, a higher-order modulation scheme such as 16QAM may be supported (but potentially difficult to use in an NTN that has low SINR in the uplink). Frequency hopping is supported on the NPUSCH in NB-IoT in a TN and an NTN.

In *Step 6*, the NB-IoT NTN eNB attempts to decode the packet on the NPUSCH and conveys the HARQ ACK or NACK feedback to the NB-IoT NTN UE depending on whether it has successfully decoded the packet or not. The NB-IoT eNB sends this HARQ feedback using the DCI Format N0 on the NPDCCH. The NB-IoT NTN eNB also allocates UL resources for new packet transmission or retransmission on the NPUSCH depending upon the HARQ feedback it sends to the NB-IoT NTN UE.

In *Step 7*, the NB-IoT NTN UE transmits a new packet or retransmits the previously transmitted packet on the NPUSCH, like Step 5.

Like the LTE-M NTN, the NB-IoT NTN may utilize the PUR method for UL resource allocation. If the NTN UE's PUR-based NPUSCH transmission ends in subframe n, the NB-IoT NTN UE looks for its NPDCCH in a search space window starting in subframe $(n + 4 + K_{\text{mac}})$ with a window duration NB-IoT specified by the parameter pur-SS-window-duration. K_{mac} is specified in SIB31-NB as k-Mac-r17. If the NB-IoT NTN UE detects an NPDCCH carrying the DCI format N0 with the CRC scrambled by the NB-IoT NTN UE's PUR-RNTI and the value of "modulation and coding scheme" field in the corresponding DCI Format is set to "14," the UE does not need to monitor an NPDCCH in the remaining search space window duration.

Note that NPUSCH Format 1 transmission can be scheduled by an NPDCCH with DCI format N0, as shown in Figure 8.12 for a TN and an NTN. Additionally, NPUSCH Format 1 transmission can use a PUR configured by RRC signaling. Retransmission of a transport block transmitted using the PUR is scheduled by an NPDCCH using the DCI Format N0.

8.4.5 GNSS Measure Gap Procedure in the NB-IoT NTN

An NB-IoT NTN UE cannot utilize the NTN for data transfer if it does not have valid GNSS measurements. To facilitate the UE's GNSS measurements in an NTN FDD serving cell, a GNSS measurement gap needs to be created. The NB-IoT NTN eNB can send a GNSS Measurement Command MAC CE in a NPDSCH to initiate the GNSS measurement process at the UE. Assume that the NB-IoT NTN UE receives such a MAC CE in subframe n. If the UE does not provide HARQ-ACK information for the HARQ process associated with the transport block carrying the GNSS Measurement Command MAC CE, the UE considers the subframe $(n + 13)$ to be the start of the GNSS measurement gap. Otherwise, the UE considers the subframe $(k + 2)$ to be the start of the GNSS measurement gap, where k is the first DL subframe after the end of the transmission of the NPUSCH carrying ACK/NACK response for the HARQ process associated with the transport block in the NPDSCH (3GPP 2024c). Furthermore, the NB-IoT NTN UE is not required to monitor an NPDCCH during the GNSS measurement gap until it reacquires the GNSS position and performs a contention-based Random Access (3GPP 2024f).

8.5 Major Takeaways

Data transfer in an NTN must reflect unique factors of the radio environment, such as long propagation delays and challenging radio environments. The data transfer framework of the TN is not only reused in the NTN but also enhanced to reflect NTN-specific radio channel challenges. Here are the major takeaways from this chapter.

- Data transfer in an NTN needs to reflect unique factors of the radio environment in an NTN, such as long propagation delays and a challenging radio environment.
- Key aspects of data transfer in an NTN include reuse of downlink and uplink data transfer frameworks defined in a TN, Timing Advance reporting, timing offsets, increased number of HARQ processes, HARQ feedback disabling, uplink control information reporting, and DRX enhancements.
- Key prerequisites for data transfer in NR-NTN include registration, PDU session establishment, and timing and frequency pre-compensation.
- K_{offset} is a *configured scheduling offset* that needs to be larger than or equal to the sum of the service link RTT and the Common TA. K_{offset} provides the UE with sufficient processing time between a downlink reception and an uplink transmission.
- k_{mac} is used to delay the application of a downlink configuration indicated by a MAC CE command on PDSCH/NPDSCH and to estimate the UE-gNB/eNB RTT (3GPP 2024n). The NTN gNB/eNB may provide k_{mac} when the downlink and the uplink frame timing are not aligned at the NTN gNB/eNB.
- The NR-NTN UE's transmission of an uplink frame number i starts T_{TA} before the start of the corresponding downlink frame i received at the NR-NTN UE. The DL and the UL are frame-aligned at the uplink time synchronization reference point (RP) with an offset given by $N_{\text{TA,offset}}$.

- The NR-NTN gNB may configure the NR-NTN UE with a TA report-related configuration via RRC signaling. The NR-gNB uses RRC signaling to control Timing Advance reporting by the NR-NTN UE by configuring ta-Report, offsetThresholdTA, and TimingAdvanceSR.
- The NR-NTN downlink data transfer can utilize more (up to 32) HARQ processes, HARQ feedback disabling, conservative MCS on the PDSCH, increased time between the CSI reference resource and CSI reporting, and increased time between the PDSCH reception and HARQ feedback reporting.
- The NR-NTN uplink data transfer can utilize more (up to 32) HARQ processes, conservative MCS on the PUSCH, increased time between the DCI triggering the SRS and the SRS transmission by the UE, increased time between the PDCCH reception and the PUSCH transmission, and Timing Advance reporting.
- Key prerequisites for data transfer in the LTE-M NTN and the NB-IoT NTN are the Attach procedure with the default EPS bearer setup, (optional) setup of additional EPS bearers, and timing and frequency pre-compensation.
- The DL and UL data transfer enhancements in the LTE-M NTN and the NB-IoT NTN are similar to those in NR-NTN, where timing relationships are adjusted to compensate for potentially long propagation delays, and timing and frequency pre-compensation are carried out by the UE for uplink transmissions.
- Since an NTN UE cannot perform data transfer in the absence of valid GNSS measurements, a measurement gap for GNSS measurements is supported in the NTN.

References

3GPP (2024a). *E-UTRA; Physical channels and modulation*. TS36.211. https://portal.3gpp.org/desktopmodules/Specifications/SpecificationDetails.aspx?specificationId=2425 (accessed 6 September 2024).

3GPP (2024b). *E-UTRA; Multiplexing and channel coding*. TS36.212. https://portal.3gpp.org/desktopmodules/Specifications/SpecificationDetails.aspx?specificationId=2426 (accessed 6 September 2024).

3GPP (2024c). *Evolved Universal Terrestrial Radio Access (E-UTRA); Physical layer procedures*. TS36.213. https://portal.3gpp.org/desktopmodules/Specifications/SpecificationDetails.aspx?specificationId=2427 (accessed 6 September 2024).

3GPP (2024d). *Evolved Universal Terrestrial Radio Access (E-UTRA) and Evolved Universal Terrestrial Radio Access Network (E-UTRAN): Overall description*. TS36.300. https://portal.3gpp.org/desktopmodules/Specifications/SpecificationDetails.aspx?specificationId=2430 (accessed 6 September 2024).

3GPP (2024e). *User Equipment (UE) radio access capabilities*. TS36.306. https://portal.3gpp.org/desktopmodules/Specifications/SpecificationDetails.aspx?specificationId=2434 (accessed 6 September 2024).

3GPP (2024f). *Evolved Universal Terrestrial Radio Access (E-UTRA); Medium Access Control (MAC) protocol specification*. TS36.321. https://portal.3gpp.org/desktopmodules/Specifications/SpecificationDetails.aspx?specificationId=2437 (accessed 6 September 2024).

3GPP (2024g). *Evolved Universal Terrestrial Radio Access (E-UTRA); Radio Resource Control (RRC); Protocol specification*. TS36.331. https://portal.3gpp.org/desktopmodules/Specifications/SpecificationDetails.aspx?specificationId=2440 (accessed 6 September 2024).

3GPP (2024h). *NR; Physical channels and modulation.* TS38.211. https://portal.3gpp.org/desktopmodules/Specifications/SpecificationDetails.aspx?specificationId=3213 (accessed 6 September 2024).

3GPP (2024i). *NR; Multiplexing and channel coding.* TS38.212. https://portal.3gpp.org/desktopmodules/Specifications/SpecificationDetails.aspx?specificationId=3214 (accessed 6 September 2024).

3GPP (2024j). *NR; Physical layer procedures for control.* TS38.213. https://portal.3gpp.org/desktopmodules/Specifications/SpecificationDetails.aspx?specificationId=3215 (accessed 6 September 2024).

3GPP (2024k). *NR; Physical layer procedures for data.* TS38.214. https://portal.3gpp.org/desktopmodules/Specifications/SpecificationDetails.aspx?specificationId=3216 (accessed 6 September 2024).

3GPP (2024l). *NR and NG-RAN overall description.* TS38.300. https://portal.3gpp.org/desktopmodules/Specifications/SpecificationDetails.aspx?specificationId=3191 (accessed 6 September 2024).

3GPP (2024m). *NR; User Equipment (UE) radio access capabilities.* TS38.306. https://portal.3gpp.org/desktopmodules/Specifications/SpecificationDetails.aspx?specificationId=3193 (accessed 6 September 2024).

3GPP (2024n). *NR; Medium Access Control (MAC) protocol specification.* TS38.321. https://portal.3gpp.org/desktopmodules/Specifications/SpecificationDetails.aspx?specificationId=3194 (accessed 6 September 2024).

3GPP (2024o). *NR; Radio Resource Control (RRC) protocol specification.* TS38.331. https://portal.3gpp.org/desktopmodules/Specifications/SpecificationDetails.aspx?specificationId=3197 (accessed 6 September 2024).

Tripathi, N.D. and Reed, J.H. (2019). *5G Cellular Communications: Journey and Destination.* The Wireless University. https://thewirelessuniversity.com/ (accessed 6 September 2024).

9

Mobility Management in an NTN

Mobility management principles for a TN are reused in a Non-Terrestrial Network (NTN) along with NTN-specific enhancements. The NTN UE's Radio Resource Control (RRC) state influences mobility management. This chapter provides a brief overview of the NTN UE's RRC states, followed by a summary of major NTN mobility challenges and candidate solutions. Mobility management in the NR-NTN is addressed by discussing measurements and handover triggers, types of NTN handover, and activities of the NR-NTN UE in the RRC_IDLE and RRC_INACTIVE states. A detailed signaling flow for the Conditional Handover (CHO) in the NR-NTN is illustrated. Similarly, mobility management in the LTE-M NTN is addressed by discussing measurements and handover triggers, a detailed signaling flow for the CHO, and activities of the LTE-M NTN UE in the RRC_IDLE state. A detailed signaling flow for the CHO is illustrated. Mobility management in the NB-IoT NTN is also described. Feeder link switchover, a unique aspect of the NTN, is explained. The issue of discontinuous coverage is also summarized.

9.1 RRC States and Mobility Management in a TN and an NTN

A 5G NR UE may be in one of the three RRC states—RRC_CONNECTED, RRC_IDLE, and RRC_INACTIVE (Tripathi and Reed 2019; 3GPP 2024g). The 5G NR UE's RRC state influences mobility management, as summarized in Figure 9.1.

When a 5G NR UE is in the RRC_CONNECTED state, it has a dedicated RRC signaling connection with a serving cell. Such a cell has a certain Physical layer Cell Identity (PCI) and a global cell identifier such as an NR Cell Global Identity (NCGI). Such UE can exchange data traffic with the network. Mobility management for a UE in the RRC_CONNECTED state is quite comprehensive. It requires significant processing, especially at the UE and radio network, and sometimes the core network. Specifically, the gNB configures the UE with suitable measurement reporting events, and the UE sends a measurement report to the serving cell (or equivalently, serving gNB). The serving gNB makes a handover decision, selects a new serving cell for the UE (if needed), and sends a handover command to the UE (if needed). When needed, the UE performs registration update with the Access and Mobility Management Function (AMF) in the 5GC. Section 9.3.1 discusses handover in the NR-NTN.

When a 5G NR UE is in the RRC_IDLE state, it does not have a dedicated RRC signaling connection with any cell. Mobility management for a UE in the RRC_IDLE state is significantly simpler

3GPP-based Non-Terrestrial Networks in 5G and 6G: Expanding the Frontiers of Wireless Communications, First Edition.
Nishith D. Tripathi and Jeffrey H. Reed.

Companion website: www.wiley.com/go/tripathi5g

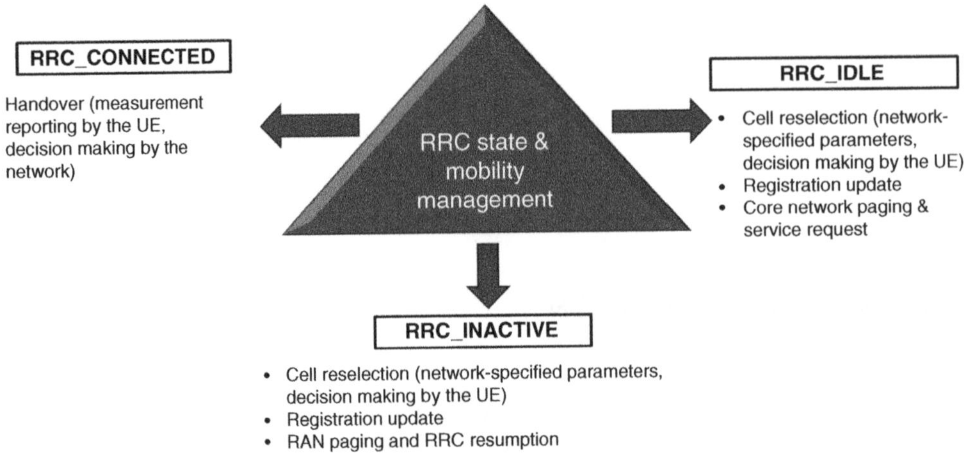

Figure 9.1 Impact of the RRC State on Mobility Management in 5G NR.

than mobility management in the UE's RRC_CONNECTED state. The UE in the RRC_IDLE state monitors the best cell from the radio perspective and performs cell reselection if it encounters a better cell. Cell reselection-related parameters are broadcast by the gNBs. The UE in the RRC_IDLE state does not send any measurement reports to the radio network even when the UE selects a new cell and possibly a new gNB. When needed, the UE performs a registration update with the core AMF in the 5GC. The UE in the RRC_IDLE state must enter the RRC_CONNECTED state before exchanging data traffic with the network. Section 9.3.2 discusses mobility management for RRC_IDLE UEs in the NR-NTN.

When a 5G NR UE is in the RRC_INACTIVE state, it does not have a dedicated RRC signaling connection with any cell, just like in the RRC_IDLE state. The last serving gNB of the UE is the gNB that controls the cell in which the UE enters the RRC_INACTIVE state. Such last serving gNB preserves the UE's access stratum context. The UE also preserves its AS context. Mobility management for a UE in the RRC_INACTIVE state is similar to that in RRC_IDLE state. The UE in the RRC_INACTIVE state monitors the best cell from the radio perspective and performs cell reselection if it encounters a better cell. Cell reselection-related parameters are broadcast by the gNBs. The UE in the RRC_INACTIVE state does not send any measurement reports to the radio network even when the UE selects a new cell and possibly a new gNB. To send a signaling message or data traffic, the UE enters the RRC_CONNECTED state by resuming its RRC connection based on the Access Stratum (AS) context. If data traffic comes from the core network to the last serving gNB, RAN paging is carried out, where a page message is sent from all the cells of an RAN Notification Area (RNA). If the UE crosses from one RNA to another RNA, an RNA update is carried out, and a new gNB obtains the UE's AS context for the last serving gNB (Tripathi and Reed 2019). Section 9.3.2 discusses mobility management for RRC_INACTIVE UEs in the NR-NTN. When needed, like the RRC_IDLE UE, the UE in the RRC_INACTIVE state performs a registration update with the core AMF in the 5GC. It is up to the gNB to ask the UE to enter the RRC_INACTIVE or RRC_IDLE state upon the lack of activity for a certain period (e.g., few seconds).

An LTE-M UE typically uses the two RRC states originally defined in LTE, RRC_CONNECTED and RRC_IDLE. Handover in LTE-M is similar to handover in LTE and 5G NR. Section 9.4.1 explains handover in the LTE-M NTN. The LTE-M UE's activities in the RRC_IDLE state are similar to those of the NR UE in the RRC_IDLE state. The LTE-M UE performs a tracking area update

(TAU) with the MME in the EPC when needed. Section 9.4.2 describes the LTE-M NTN UE's activities in the RRC_IDLE state.

An NB-IoT UE, like an LTE-M UE, typically uses the RRC_CONNECTED and RRC_IDLE states. However, the NB-IoT UE performs cell reselection on both RRC states. The NB-IoT UE does not send any measurement reports to the radio network even when the NB-IoT UE selects a new cell and possibly a new gNB. When needed, the NB-IoT UE performs a TAU with the MME in the EPC. Section 9.5 describes the mobility management of the NB-IoT NTN UE.

9.2 Mobility Challenges in the NTN and Associated Solutions

The NTN involves unique challenges associated with mobility. Hence, NTN-specific solutions are needed to address these challenges. Figure 9.2 summarizes major NTN mobility challenges and potential mechanisms that can be utilized to address these challenges. Details of these mechanisms are discussed in Sections 9.3 through Section 9.8.

Moving NTN Cells. Moving NTN cells pose unique challenges in an NTN. For example, the quasi-Earth-fixed cells involve a change in the serving cell for NTN-UEs such that traditional handover triggers are not practical. Hence, a time-based trigger has been introduced for the NTN. Such a trigger provides adequate time for the NTN UEs to detect an incoming cell and enables the network to move NTN UEs from an outgoing NTN cell to an incoming NTN cell. Information about serving and neighboring satellites is provided in NTN-specific system information. Dedicated RRC signaling can be used to convey information about incoming NTN cells. To avoid massive mobility-based registration updates, Earth-fixed tracking areas (TAs) are used, and each NTN cell can broadcast multiple TAs. When a change in the serving NTN cell leads to a change in the serving satellite, a satellite switch with resync is supported for a more efficient handover. Furthermore, due to long propagation delays and different propagation delays between a serving satellite and a neighboring satellite, SMTC-related enhancements are made to improve the detection of neighboring satellites.

Signal Strength Variations. As discussed in Section 4.9 in Chapter 4, in the case of the NTN, the difference in the signal strength near the cell center and the cell edge may not be significant. Hence, Reference Signal Received Power (RSRP) as the sole handover criterion is often inadequate, requiring a different mobility management approach in the NTN. Location-based handover triggered has

Challenge	Candidate solutions
Moving NTN cell	1. Time-based handover trigger 2. Enhanced SI & RRC signaling 3. Earth-Fixed TAs & multiple TAs per NTN cell 4. Satellite switch with resync
Signal strength variations	Location-based handover trigger
Feeder link switchover	Enhanced NTN-gNB implementation and handover
Massive handover	RACH-less handover
TN-NTN mobility	SI and RRC signaling enhancements
Discontinuous coverage	Signaling enhancements

Figure 9.2 NTN Mobility: Challenges and Candidate Solutions.

been introduced so that the distance between the NTN UE and the reference point (i.e., the cell center) can be used to determine whether the NTN UE is near the NTN cell edge.

Feeder Link Switch Over. A moving NTN payload (e.g., a Low Earth Orbit [LEO] satellite) may not have the visibility of the same NTN-GW all the time. Hence, a feeder link switch occurs when an NTN payload needs to connect to a different NTN-GW. The NTN-gNB implementation needs to support such feeder link switchover. NTN-based handover mechanism can move NTN UEs from one NTN-GW/NTN-gNB to another NTN-GW/NTN-gNB.

Massive Handover. Reducing the amount of random-access resources and the random-access A scenario (e.g., quasi-Earth-fixed cells) could require transferring many NTN UEs from one NTN cell to another. RACH-less handover, initially defined for a TN, can be applied to an NTN to reduce the amount of random-access resources and signaling.

TN-NTN Mobility. There are various scenarios requiring mobility between a TN and an NTN. For example, an NTN-UE communicating with an NTN may enter a TN coverage area. Handover from an NTN to a TN in such a scenario can provide better performance to the UE. Similarly, when a UE communicating with a TN may leave a TN coverage area. Handover from a TN to an NTN in such a scenario can provide seamless communications experience to the UE.

Discontinuous Coverage. An NTN may not have continuous coverage. For example, an LEO satellite-based NTN may not have an adequate number of LEO satellites to provide continuous coverage to an NTN UE. Hence, signaling enhancements are made in the NTN to reduce the processing requirements for an NTN UE and improve the NTN UE's battery life.

9.3 Mobility Management in the NR-NTN

Mobility management in the NR-NTN reuses the principles of mobility management used in the NR-based TN. As in a TN, mobility management is based on the RRC state of the NR-NTN UE. Section 9.3.1 discusses mobility management for an NR-NTN UE in the RRC_CONNECTED state, while Section 9.3.2 describes mobility management for an NR-NTN UE in the RRC_IDLE and RRC_INACTIVE states.

9.3.1 Handover in the NR-NTN

Handover is carried out for an NR-NTN UE in the RRC_CONNECTED state. Key handover principles of the NR-based TN, such as neighbor cell search, measurements, measurement or handover triggers, and inter-gNB signaling for handover, are reused in the NR-NTN. However, NTN-specific enhancements, such as location and time-based triggers, are defined for more reliable handover. Section 9.3.1.1 discusses measurements, measurement events, and handover triggers for the NR-NTN. Different types of intra-NTN handover in the NR-NTN are explained in Section 9.3.1.2. Types of handovers unique to an NTN, such as NTN-NTN handover, NTN-TN handover, and satellite switching, are briefly summarized in Section 9.3.1.3. The overall signaling for the CHO in the NR-NTN is illustrated in Section 9.3.1.4.

9.3.1.1 Measurements, Measurement Events, and Triggers for Handover

A location-based handover trigger has been introduced to address the NTN-specific challenge of the lack of sufficient signal strength variations between the cell center and the cell edge in an NTN. Furthermore, a time-based handover trigger has been introduced to address the challenge of ensuring reliable handover of NR-NTN UEs from the outgoing NR-NTN cell to the incoming NR-NTN cell in the case of quasi-Earth-fixed beams or cells.

Location-Based Handover Trigger for the NTN The NTN-UE's location relative to the serving cell and a neighbor cell can be used to reliably estimate the need for handover from the serving cell to a neighbor target cell. When the NR-NTN UE is farther from the currently serving NR-NTN cell and closer to the neighbor NR-NTN cell, RSRPs of the serving and neighbor NTN cells measured at the NTN-UE may not be widely different. However, distances between the NR-NTN UE and the serving and neighbor cells can be used to determine the need for handover more reliably. Figure 9.3 illustrates how the location-based trigger can be used for handover.

The distance-based trigger can be implemented using the criteria defined by Eqs. (9.1) and (9.2) (3GPP 2024g).

$$\text{Ml1} - \text{Hys} > \text{Thresh1} \tag{9.1}$$

$$\text{Ml2} + \text{Hys} < \text{Thresh2} \tag{9.2}$$

In Eqs. (9.1) and (9.2), Ml1 is the distance between the NR-NTN UE and a reference location specified as referenceLocation1 in the IE reportConfigNR (e.g., the center of the serving NTN cell), Ml2 is the distance between the NR-NTN UE and a reference location specified as referenceLocation2 in reportConfigNR (e.g., the center of the neighbor NTN cell), Hys is the hysteresis parameter specified as hysteresisLocation in reportConfigNR, Thresh1 is the threshold represented by the parameter distanceThreshFromReference1 specifying the distance from a reference location configured with the parameter referenceLocation1 in reportConfigNR, and Thresh2 is the threshold represented by the parameter distanceThreshFromReference2 specifying the distance from a reference location configured with the parameter referenceLocation2 in reportConfigNR (3GPP 2024g). Ml1, Ml2, Hys, Thresh1, and Thresh2 are expressed using meters.

Equations (9.1) and (9.2) represent the entering conditions 1 and 2, respectively. The related leaving conditions are specified in Eqs. (9.3) and (9.4) (3GPP 2024g).

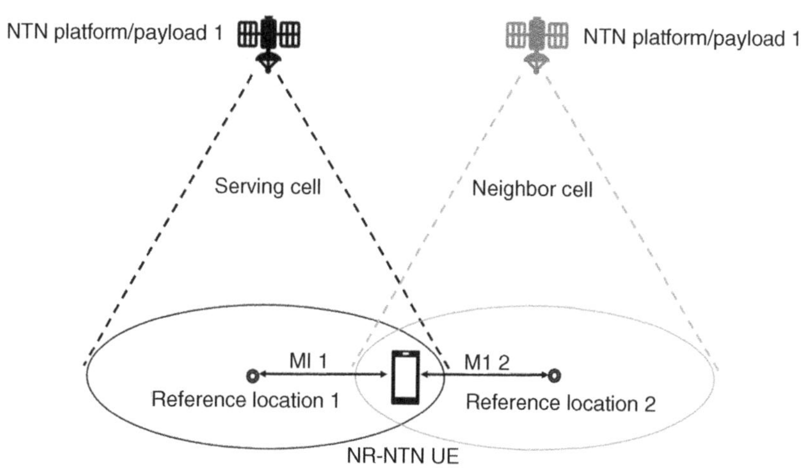

Figure 9.3 Location-Based Trigger in the NR-NTN.

$$\text{Ml1} + \text{Hys} < \text{Thresh1} \tag{9.3}$$

$$\text{Ml2} - \text{Hys} > \text{Thresh2} \tag{9.4}$$

The location-based trigger is used in NTN-specific events such as regular Event D1 and conditional event condEvent D2 (3GPP 2024g). In essence, Event D1 corresponds to the case where the distance between the UE and referenceLocation1 is above threshold1, and the distance between the UE and referenceLocation2 is below threshold2.

Time-Based Handover Trigger for the NTN Figure 9.4 illustrates a scenario where a time-based handover trigger can be utilized. A quasi-Earth-fixed NTN cell 1 covers a given geographic area X during one period and a different geographic area Y during another period. Hence, another quasi-Earth-fixed NTN cell 2 serves the geographic area X while quasi-Earth-fixed NTN cell 1 serves the geographic Area Y. Since a given geographic area is served by a quasi-Earth-fixed NTN cell based on time, a time-based handover trigger is helpful so that the NTN UEs can be moved from an outgoing NTN cell to an incoming NTN cell via handover. A certain time overlap between the outgoing NTN cell and the incoming cell is useful so that many NTN UEs can be successfully moved to the incoming cell before the outgoing cell ceases to serve the target coverage area.

The key idea is that the NR-NTN UE needs to be moved to the incoming NTN cell before the outgoing NTN cell stops serving the currently served geographic area. Hence, the NR-NTN UE compares its time with time parameters defined for the time-based event. Specifically, the conditional event condEvent T1 makes use of the time-based parameters. The entering and leaving conditions are specified in Eqs. (9.5) and (9.6) (3GPP 2024g).

$$\text{Mt} > \text{Thresh1} \tag{9.5}$$

$$\text{Mt} > \text{Thresh1} + \text{Duration} \tag{9.6}$$

In Eqs. (9.5) and (9.6), Mt is the time measured at the NR-NTN UE, Thresh1 is the threshold specified as t_1-Threshold in reportConfigNR, and Duration is the duration specified in reportConfigNR (3GPP 2024g). Mt, Thresh1, and Duration are expressed in ms. The serving NR-NTN gNB can

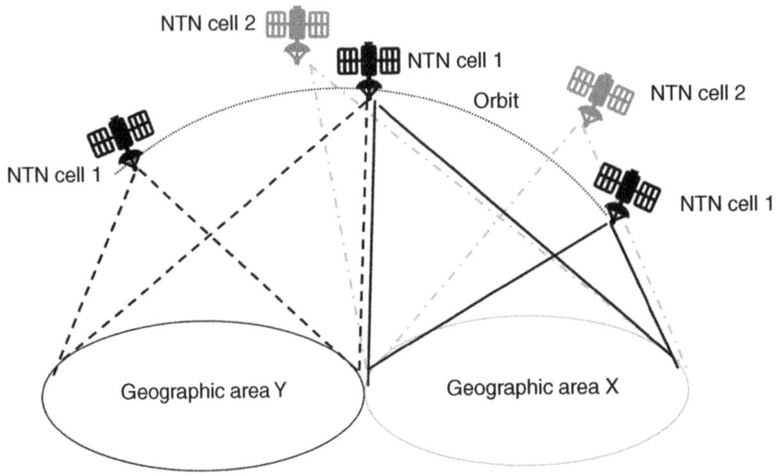

Figure 9.4 Time-Based Trigger in the NR-NTN.

provide suitable handover parameters to provide NR-NTN UEs adequate time to transition to a different NTN cell.

SMTC Enhancements for Neighbor Cell Measurements An NR UE utilizes SS/PBCH block Measurement Timing Configuration (SMTC) to measure neighbor cell reference signals. Propagation delay differences for the signals of the serving cell and neighbor cells received at a UE are relatively small in a TN. However, propagation delay differences for the currently serving NTN and neighbor satellites' signals can be quite large. Hence, the SMTC windows defined for a TN could be inadequate in an NTN, potentially preventing an NTN UE from detecting reference signals of neighbor NTN cells. Hence, SMTC windows are adjusted in an NTN to enable the NTN UE to reliably measure neighbor NTN cells.

9.3.1.2 Types of Intra-NTN Handover in an NR-NTN

Different types of intra-NTN handover may occur between NTN cells. Examples of handover include traditional handover, Dual-Active Protocol Stack (DAPS) handover, RACH-less handover, and CHO. These handover types are briefly introduced later.

Traditional Handover Traditional handover is supported in a TN as well as an NTN. In traditional handover in the NR-NTN like a TN handover, the NTN UE makes measurements based on the configuration by the NR-NTN gNB, the NR-NTN UE provides a measurement report to the serving NR-NTN gNB when a configured measurement event occurs, the serving NR-NTN gNB makes a handover request to the target NR-NTN gNB, the target NR-NTN gNB reserves radio resources for the NR-NTN UE in the target cell, the serving NR-NTN gNB sends a handover command to the NR-NTN UE containing the information specified by the target NR-NTN gNB, and the NR-NTN UE performs random access in the target cell and establishes a dedicated radio connection with the target cell.

The measurement events (e.g., A3) defined for a TN are likely to be inadequate or ineffective by themselves in various NTN scenarios, such as quasi-Earth-fixed cells. The location-based trigger for Event D1 can be used for traditional handover and enhanced NTN handover. Both time and location-based triggers are available for use in conjunction with CHO in Release 17 and Release 18. See below for a discussion on NTN enhancements in the CHO section. Furthermore, CHO signaling flow is also summarized in Section 9.3.1.4.

DAPS Handover DAPS handover is not supported for NTN. Note that DAPS handover involves simultaneous communication of the UE with both the source and target cells. The absence of DAPS handover in an NTN simplifies handover for an NTN.

RACH-Less Handover RACH-less handover, originally defined for a TN, can also be utilized in an NTN (3GPP 2024d,f). When the source and target cells have similar propagation delays toward the NTN UE, and the UE can be considered uplink synchronized with the target cell when it is synchronized with the source cell, RACH-less handover can be used. RACH-less handover can reduce the handover delay, save RACH resources on the radio interface, and reduce the processing load in the radio network. The network may configure the UE with RACH-less handover by configuring the parameter rach-lessHO. The initial uplink transmission for RACH-less handover can make use of a dynamic uplink grant or a configured uplink grant Type 1 pre-allocated by RRC signaling using the parameters, such as cg-RACH-less-Configuration (containing configured grant information), tci-StateID (containing the TCI state information associated with a beam), and dg-beam (containing the SSB index).

Conditional Handover (CHO) A CHO is a handover executed by the UE when one or more handover execution conditions are met (3GPP 2024d). The CHO was originally defined for a TN and enhanced for an NTN. In support of the CHO, the UE starts evaluating the execution condition(s) upon receiving the CHO configuration from the serving gNB and stops evaluating the execution condition(s) after a handover is executed. The CHO configuration is an important element of the CHO. The CHO configuration contains (i) the configuration of CHO candidate cells generated by the candidate gNB(s) and (ii) execution condition(s) generated by the source gNB. An execution condition for the CHO may include one or two trigger conditions. A gNB may configure up to two different trigger quantities, such as (i) RSRP and RSRQ or (ii) RSRP and SINR simultaneously evaluating the CHO execution condition of a given CHO candidate cell.

In support of the NTN, a time-based trigger condition and a location-based trigger condition are introduced for CHO. The RRM measurement-based event A4 plays an important role in CHO for an NTN. An NR-NTN gNB may configure time-based or location-based trigger conditions independently from the measurement condition for CHO in scenarios, such as a hard satellite switch case, where the service discontinuity gap time length is assumed to be zero or negligible. In general, the NR-NTN gNB configures the NR-NTN UE with a time-based or a location-based trigger condition along with one of the measurement-based trigger conditions, such as CHO events A3, A4, and A5 (3GPP 2024d,g).

To facilitate time trigger-based CHO, the source NR-NTN gNB may specify relevant parameters to a single target NR-NTN gNB using the Source NG-RAN Node to Target NG-RAN Node Transparent Container in an NG-C based handover (3GPP 2024d; 3GPP TS23.502-2024). The source NR-NTN gNB specifies the relevant CHO configuration for the NR-NTN UE in the RRC reconfiguration message during handover execution. The source NR-NTN gNB can consider the time indicated to the NR-NTN UE to decide when to start forwarding the early data to the target NR-NTN gNB. Note that time-based CHO can also make use of RACH-less handover. See Section 9.3.1.4 for an example of signaling flow for CHO in the NR-NTN.

9.3.1.3 NTN-Specific Handover in the NR-NTN

In addition to handover within an NTN, additional scenarios such as NTN-NTN handover, TN-NTN handover, and satellite switch with re-sync are also supported. These scenarios are briefly summarized later.

NTN-NTN Handover UE may support mobility between gNBs operating with NTN payloads in different orbits (e.g., Geo Synchronous Orbit [GSO] and Non Geo Synchronous Orbit [NGSO] satellites at different altitudes).

TN-NTN Handover During mobility between NTN and terrestrial network (TN), a UE is not required to connect to both NTN and TN simultaneously.

NOTE: The NTN TN handover refers to mobility in both directions, i.e., from NTN to TN (hand-in) and from TN to NTN (hand-out).

Satellite Switch with Re-sync Upon both hard and soft satellite switchover in the quasi-Earth-fixed scenario with the same sychronization signal block (SSB) frequency and the same gNB, the satellite switch with re-sync procedure is supported. The satellite switch with re-sync avoids L3 mobility for UEs in the cell by maintaining the same PCI on the geographical area covered by the quasi-Earth-fixed beam. CHO can be configured simultaneously with the satellite switch with a re-sync procedure.

For soft satellite switchover, the UE can start synchronizing with the target satellite before the source satellite stops serving the cell. The UE does not need to be connected to the source satellite when it switches to the target satellite.

9.3.1.4 NR-NTN Handover Signaling Flow

As discussed earlier in Section 9.3.1.2, both location-based and time-based handover triggers are supported in the CHO for the NTN. Figure 9.5 summarizes the key steps of the CHO in the NR-NTN when the AMF and the UPF remain the same. See 3GPP (2024d) for details on these steps.

The NR-NTN gNB configures the UE with suitable measurements using the RRC reconfiguration message in *Step I*. The NR-NTN gNB specifies one or more carrier frequencies that the NR-NTN UE should measure. The NR-NTN gNB can also specify whitelisted cells that the NR-NTN UE considers for higher priority measurements. Additionally, the NR-NTN gNB specifies what measurements to make, such as RSRP, RSRQ, and signal-to-interference plus noise ratio (SINR). Furthermore, the NR-NTN gNB configures the measurement reporting or handover events for which the UE needs to send a measurement report message. The NR-NTN gNB may specify the periodicity of measurement reports in case of periodic reporting. While the NR-NTN can use measurement reporting or handover events, NTN-specific events, such as location-based Event D1 for traditional handover, may be configured. Time- and location-based events are available for configuration for the CHO in the NR-NTN. The NR-NTN gNB can configure the NR-NTN UE with the SMTC configuration that has been enhanced for the NTN.

The 3GPP has defined six intra-RAT handover events and two inter-RAT handover events originally for a TN, which can be reused in an NTN. A1 to A6 are intra-RAT handover events, while B1

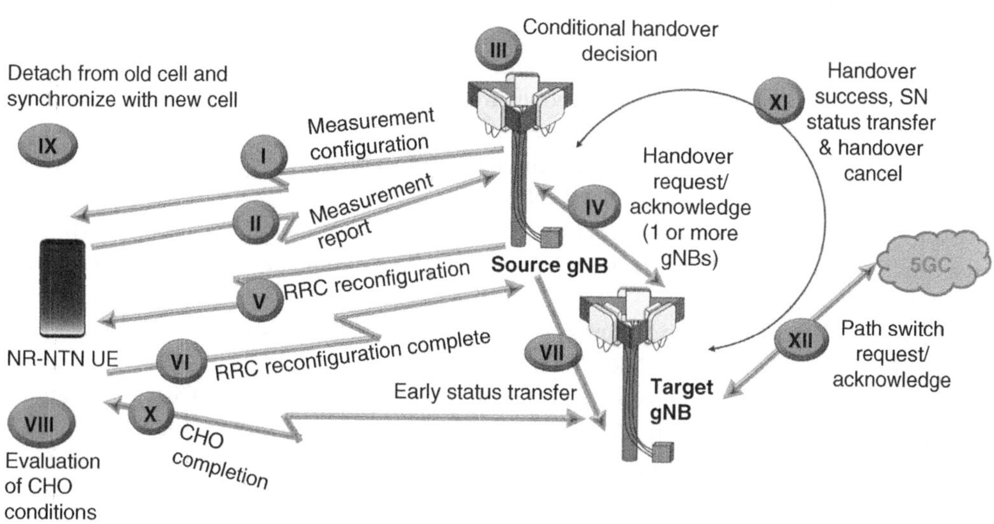

Figure 9.5 Conditional Handover in the NR-NTN. *Source:* nadir_insan / Adobe Stock Photo.

and B2 are inter-RAT handover events. See Section 3.2.2 in Chapter 3 for brief descriptions on these events; details can be found in 3GPP (2024d).

In *Step II*, the NR-NTN UE makes measurements of the serving cell and neighboring cells. The specific periods during which the UE can make measurements are influenced by the NR-NTN gNB-specified SSB Measurement Timing Configuration (SMTC). On a given carrier frequency, the received signal contains signals from the serving cell and the neighboring cells, and the 5G NR UE can use the PCIs of cells to determine the RSRP and RSRQ of cells. When the reporting criterion for the measurement reporting is met (e.g., the occurrence of an event such as A3), the NR-NTN UE sends the Measurement Report message containing the cell identity (i.e., PCI) and RSRP and RSRQ of the serving and relevant neighbor cell(s).

In *Step III*, the serving NR-NTN gNB processes the NR-NTN UE's measurement report and decides to use the CHO instead of traditional handover.

In *Step IV*, the serving NR-NTN gNB sends a Handover Request message to one or more target NR-NTN gNBs to request the CHO for the identified CHO candidate cells. The Handover Request message contains information such as the target PCI and 5G QoS parameters (e.g., 5G QoS flows and UE-AMBR) for currently established 5G QoS flows. The target NR-NTN gNB executes an implementation-specific admission control algorithm. It evaluates the QoS needs of the current 5G QoS flows of the NR-NTN UE against the availability of radio resources. If the target NR-NTN gNB decides to accept the CHO request, it responds to the serving NR-NTN gNB with the Handover Request Acknowledge containing the CHO configuration for the CHO candidate cell. The Handover Request Acknowledge message may contain parameters such as C-RNTI allocated to the NR-NTN UE in the CHO candidate cell, a dedicated random-access preamble (to avoid contention with other UEs using the PRACH in the new CHO candidate cell), and system information. This message includes the list of 5G QoS flows admitted by the target NR-NTN gNB. If packet forwarding is intended for the NR-NTN UE, the Xn-U tunnel information is also provided in this message. Note that the serving NR-NTN gNB may receive positive responses to its CHO request from multiple target NR-NTN gNBs. Furthermore, the CHO response message is sent by the target NR-NTN gNB for each candidate cell.

In Step V, the currently serving NR-NTN gNB sends the RRCReconfiguration message to the NR-NTN UE. This message contains the configuration of CHO candidate cells provided by target NR-NTN gNBs and CHO execution conditions determined by the source NR-NTN gNB.

After receiving the RRCReconfiguration message containing CHO-specific information, the NR-NTN UE in *Step VI* replies with the RRCReconfiguration message to the serving NR-NTN gNB to acknowledge its reception.

In *Step VII*, if early data forwarding is used, the source NR-NTN gNB sends the EARLY STATUS TRANSFER message to the target NR-NTN gNBs.

In *Step VIII*, the NR-NTN UE evaluates the CHO execution conditions. If one or more CHP execution conditions are satisfied, the NR-NTN UE selects such CHP candidate cell as the new serving cell and proceeds to Step IX.

In *Step IX*, the NR-NTN UE discontinues communications with the serving NR-NTN cell and synchronizes with the new serving NR-NTN cell.

In *Step X*, the NR-NTN UE completes the CHO procedure by sending the RRCReconfigurationComplete message to the target NR-NTN gNB associated with the new serving NR-NTN cell. The NR-NTN UE releases previously stored CHO configurations after completing the CHO procedure.

In *Step XI*, the target NR-NTN gNB (which is now the new serving NR-NTN gNB for the NR-NTN UE) sends the HANDOVER SUCCESS message to the source NR-NTN gNB (which is the previously serving gNB for the UE) on the Xn interface to convey to the source NR-NTN gNB that the NR-NTN UE has successfully accessed the target NR-NTN cell. The source NR-NTN gNB replies with the SN STATUS TRANSFER message. Note that late data forwarding may be initiated as soon as the source NR-NTN gNB receives the HANDOVER SUCCESS message from the target (i.e., new serving) NR-NTN gNB. The source NR-NTN gNB sends the HANDOVER CANCEL message to the other candidate target NR-NTN gNBs (if any) to cancel CHO for the NR-NTN UE so that any reserved resources can be released.

In *Step XII*, the NR-NTN gNB-core network connectivity is modified. For example, the NR-NTN UE-specific N2 connection is changed so the AMF can exchange UE-specific signaling messages with the new serving NR-NTN gNB. Furthermore, the N3 connection is modified to enable the UPF to exchange user traffic with the new serving NR-NTN gNB. To modify N2 and N3 connections, the target NR-NTN gNB (i.e., new serving NR-NTN gNB) sends the Path Switch Request to the AMF. The AMF exchanges PDU session update messages with the SMF to provide the SMF with information about the target gNB side of the N3 tunnel and to obtain information about the UPF side of the N3 tunnel from the SMF. Next, the AMF responds to the new serving NR-NTN gNB with the Path Switch Request Acknowledge message to inform the NR-NTN gNB about the change in the network's UE-specific N2 and N3 connections (3GPP TS23.502-2024).

L1/L2-Triggered Mobility in the NR-NTN Handover is typically carried out via RRC signaling. However, to reduce handover delay, the 3GPP has defined L1/L2-Triggered Mobility (LTM) for a TN. The LTM is a procedure in which a gNB receives L1 measurement report(s) from a UE, and the gNB changes the UE's serving cell by a cell switch command signaled via a MAC CE (3GPP 2024d). The cell switch command indicates an LTM candidate configuration that the serving gNB has previously prepared and provided to the UE through RRC signaling. The UE then performs handover to the target cell. While the LTM was originally defined for a TN, it can also be applied to an NTN.

Based on the availability of a valid TA value, RACH-less LTM cell switch or RACH-based LTM cell switch is performed. The gNB may configure the UE to initiate UL TA acquisition, which is called early TA, for one or multiple target cells. If the target cell has the same N_{TA} as the current serving cell or N_{TA} is 0, the early TA acquisition procedure is unnecessary. The network may request the UE to perform early TA acquisition of a candidate cell before a cell switch. The early TA acquisition procedure can be (i) triggered by a PDCCH order or (ii) realized through UE-based TA measurements configured via RRC signaling. In the PDCCH order-based approach, the gNB of the candidate cell calculates the TA value and sends it to the gNB of the serving cell. The serving cell sends the TA value in the LTM cell switch command MAC CE while triggering the LTM-based cell switch or handover. In the RRC signaling-based approach, the UE performs TA measurement for the candidate cells, applies the TA value measured, and performs RACH-less LTM upon receiving the cell switch command. Note that the gNB may also include a TA value in the LTM cell switch command MAC CE without early TA acquisition.

9.3.2 Mobility Management of the NR-NTN UE in the RRC_IDLE and RRC_INACTIVE States

9.3.2.1 NTN-Specific Mobility Management Enhancements of the NR-NTN UE in the RRC_IDLE and RRC_INACTIVE States

Mobility management of the NR-NTN UE in RRC_IDLE and RRC_INACTIVE states reuses the procedures defined for an NR UE operating in a TN, with specific enhancements introduced for the NTN. Examples of NTN-specific enhancements include enhanced system information, enhanced tracking area management, and support for TN-NTN mobility.

Enhanced System Information. Enhanced SIB1 facilitates the overall mobility of NR-NTN UEs in the NTN. The NR-NTN UE can determine the network type, such as a terrestrial network or a non-terrestrial network, implicitly by the existence of cellBarredNTN in SIB1, as summarized in Section 7.2.2 in Chapter 7. If cellBarredNTN is set to "barred," the NR-NTN UE considers that the cell is barred for connectivity to the NTN. Furthermore, if the cellBarredNTN is absent in SIB1, the NTN UE also considers the cell to be barred. When the cell currently being monitored is barred, the NR-NTN UE performs cell reselection to another cell at the same frequency as the barred cell (3GPP 2024e). If cellBarredNTN is set to "notBarred," the cell being monitored is allowed to connect to the NTN.

Enhanced Tracking Area Management. As discussed in Section 4.5 in Chapter 4, moving NTN cells can result in massive registration update signaling if only one tracking area is allowed per cell. In an NTN, Earth-fixed TAs are defined even when NTN cells are moving. The NR-NTNB gNB may broadcast multiple Tracking Area Codes (TACs) per PLMN in an NR-NTN cell. Hence, a stationary UE does not see different TACs even as the NTN cells move, avoiding the need for registration updates when the NTN cell changes for the NR-NTN UE. Note that the TACs being broadcast in the system information may not be precisely synchronized with the illumination of beams on the ground at a given instant. In other words, the NR-NTN may be broadcasting TACs X, Y, and Z in an NTN cell at a given instant, but the beam(s) of the NTN cell may be illuminating only TACs X and Y at that instant.

Support for TN-NTN Mobility. In support of mobility between (i) the NR-NTN and (ii) NR-TN or E-UTRA-TN, the gNB may broadcast cell information in NR TN and E-UTRA TN coverage areas in SIB25. This cell information is broadcast for all three types of NTN cells: Earth-fixed, quasi-Earth-fixed, and Earth-moving cells. The cell coverage information includes a list of geographical TN areas and carrier frequency-related information. The NR-NTN UE can skip TN measurements based on the broadcast TN coverage information. To facilitate NR-NTN measurements, the NTN ephemeris is broadcast in SIB19. In an NR-NTN cell, the NTN ephemeris information includes the serving cell's NTN payload ephemeris and (optionally) the neighbor cell's NTN payload ephemeris.

9.3.2.2 Activities of an NR-NTN UE in the RRC_IDLE and RRC_INACTIVE States

In a typical scenario, in the absence of any data traffic for a few seconds, the NR-NTN UE-specific inactivity timer expires like a TN. The NR-NTN gNB, in an implementation-specific manner, asks the NR-NTN UE to enter the RRC_INACTIVE state or the RRC_IDLE state. The NR-NTN UE context is not preserved in the NG-RAN when the UE is in the RRC_IDLE state. However, when the UE is in the RRC_INACTIVE state, the radio connection is released, but the NR-NTN UE context is preserved in the last serving NR-NTN gNB. Hence, when the NR-NTN UE needs to enter the RRC_CONNECTED state, a faster transition can be made due to more rapid restoration of the radio connection made possible by the preserved NR-NTN UE context in the NG-RAN. The RRC_IDLE state and the RRC_INACTIVE state reduce the amount of signaling and processing for both the NR-NTN UE and the NR-NTN.

Figure 9.6 summarizes primary operations related to the RRC_IDLE state of the NR-NTN UE. These operations are very similar in a TN and an NTN. The NR-NTN gNB sends the NGAP UE Context Release Request to the AMF. The NR-NTN gNB then asks the NR-NTN UE to enter the RRC_IDLE state by sending the RRC release command in *Step 1* as in a TN. The NR-NTN gNB sends the NGAP UE Context Release Complete to the AMF. The AMF preserves the UE context with information such as the 5G-GUTI allocated to the NR-NTN UE and the location of the UE

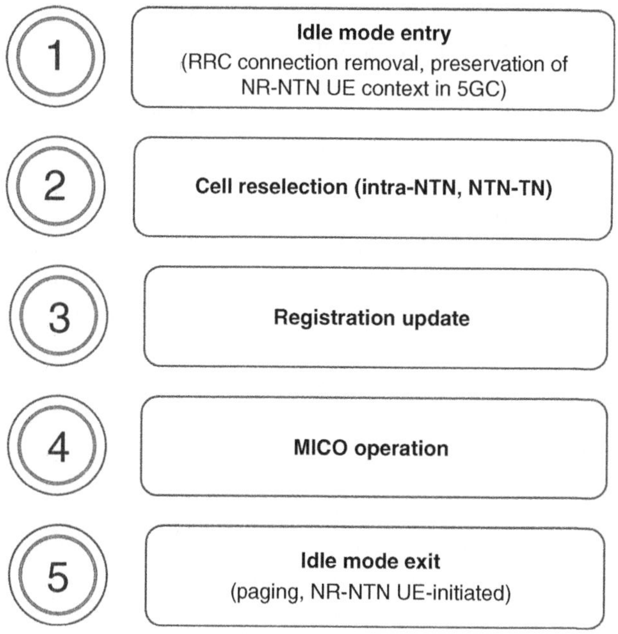

Figure 9.6 NR-NTN UE Activities in the RRC_IDLE State.

(e.g., the registration area where the NR-NTN UE is considered to be registered by the AMF; this location information is used for any paging when needed). As in a TN, the registration area for the NR-NTN UE may include one or more TAs. Note that the IP address allocated to the NR-NTN UE also remains valid in the RRC_IDLE state of the UE.

In *Step 2*, the RRC_IDLE UE typically camps on the strongest NR-NTN cell to ensure a high probability of success in receiving any incoming page message and establish a radio connection by initiating the random-access procedure for any UE-initiated data transfer. Hence, in an implementation-specific manner, the NR-NTN UE searches for strong cells and performs cell reselection to a neighbor NR-NTN cell if such a neighbor NR-NTN cell can provide a better signal than the currently monitored NR-NTN cell. The NR-NTN UE utilizes the system information to learn about the NR-NTN neighbor cells for potential cell reselection. For example, the NR-NTN UE can obtain the ephemeris of the serving cell and neighbor cells from SIB19. Note that the NR-NTN UE may find a suitable TN cell and hence may perform cell reselection to a TN cell. To save processing power and increase battery life, the NR-NTN UE may use information about TN and NTN coverage areas from SIB19 to facilitate its search for NTN and TN neighbor cells. Intra-NTN cell reselection may occur between two different types of NTNs, such as an LEO satellite-based NTN and a GEO satellite-based NTN. See 3GPP (2024e) for additional details on the NR-NTN UE operations in the RRC_IDLE state.

In *Step 3*, the NR-NTN UE carries out a registration update when necessary. For example, if the periodic registration timer expires, the NR-NTN UE executes the registration update procedure. Furthermore, when the NR-NTN UE performs cell reselection, the NR-NTN UE checks if the new NR-NTN cell belongs to the tracking area that is part of the registration area, which is equivalent to the Tracking Area Identifier (TAI) list. If the new NR-NTN cell's TA is not in the NR-NTN UE's registration area, the NR-NTN UE executes the mobility-based registration update procedure.

5G supports a new type of connectivity in the idle called Mobile-Initiated Connection Only (MICO) shown as *Step 4*. When the UE is in the MICO mode, only the UE can initiate the connection with the network; the network cannot send a page message to the UE in the case of any incoming traffic. The UE is unreachable by the network, but the UE can contact the network anytime. The MICO mode enables the UE to save power by minimizing network monitoring for paging. The MICO mode can be used in a TN and an NTN.

There may be a few reasons for the NR-NTN UE to exit the idle mode. For example, the NR-NTN UE exits the idle mode to execute the registration update procedure, carry out the random-access procedure, establish an RRC signaling connection, and exchange NAS signaling with the AMF. Specifically, the NR-NTN UE sends the Registration Request message to the AMF, and the AMF replies with the Registration Accept message. The NR-NTN UE sends the Registration Complete message to complete the registration update procedure. The AMF (especially if it is an AMF that has not allocated the current 5G-GUTI to the NR-NTN UE) allocates a new 5G-GUTI to the NR-NTN UE and provides an updated Registration Area to the NR-NTN UE. In the future, when the AMF needs to contact the RRC_IDLE UE, it sends a paging message to the gNBs that are part of the TAs of the UE-specific registration area. Upon completion of the registration update procedure, the NR-NTN UE usually re-enters the idle mode. While a TN typically has hundreds of cells in a tracking area, an NTN cell may have one or just a few cells in a tracking area.

The NR-NTN UE also exits the RRC_IDLE mode when sending user traffic. Furthermore, the NR-NTN UE exits the idle mode to respond to an incoming paging message. To exit the idle mode, the NR-NTN UE carries out the random-access procedure and establishes an RRC signaling connection. The NR-NTN gNB establishes suitable UE-specific radio bearers, such as SRBs and DRBs, to facilitate the transfer of signaling messages and user traffic.

In support of paging, the NR-NTN UE monitors a specific paging occasion (which is a specific subframe) in a paging frame as in a TN. There is one paging occasion per discontinuous reception (DRX) cycle, and the NR-NTN UE monitors for PDCCHs in multiple slots. The specific paging occasion and the paging frame that the NR-NTN UE monitors for paging are a function of the NR-NTN UE's ID (e.g., IMSI). The NR-NTN UE monitors the paging occasion during the ON period of the DRX paging cycle, which typically lasts for about two seconds. The idle mode NR-NTN UE, like in a TN, wakes up every two seconds or so and looks for a page message during a specific paging occasion based on its permanent identity (i.e., IMSI). The NR-NTN gNB sends the RRC paging message during the paging occasion, which the NR-NTN UE is expected to monitor. The paging message typically includes 5G-S-TMSI as the identity of the NR-NTN UE.

The RRC_INACTIVE state is a new state introduced by 5G originally for a TN. This state can also be used in an NTN based on the capabilities and the configuration of the network and the choices made by the service provider. In support of the RRC_INACTIVE state, a tracking area is divided into multiple RNAs, and the user mobility is managed from the perspective of RNAs. For example, as long as the user stays in the same RNA, no RNA-specific signaling needs to be exchanged between the NR-NTN UE and the RAN. The NR-NTN UE is still considered to be accessible by the 5GC as if the UE were in the RRC_IDLE state. When a signaling message or use traffic needs to be exchanged with the UE in the RRC_INACTIVE state, the RAN manages paging and the resumption of the NR-NTN UE's RRC connection. The transition from the RRC_INACTIVE state to the RRC_CONNECTED state is fast due to the preservation of the NR-NTN UE's AS context in the last serving NR-NTN gNB, which is the gNB controlling the NR-NTN cell where the NR-NTN UE entered the RRC_INACTIVE state. In general, the RRC_INACTIVE state can be more beneficial in a TN than an NTN because NTN cells are much larger than TN cells, and an RNA is likely to consist of multiple NTN cells.

In case of user inactivity, as in a TN, the serving NR-NTN gNB may ask the NR-NTN UE to enter the RRC_INACTIVE state by sending the RRC Release message with SuspendConfig containing parameters that influence the inactive state operations. Example parameters of SuspendConfig include RAN paging cycle (e.g., 32, 64, 128, or 256 frames), periodic RNA update timer (e.g., 5 to 720 min), a 40-bit full Inactive—Radio Network Temporary Identifier (I-RNTI), and 1 24-bit short I-RNTI, RNA information (i.e., the cell list constituting the current RNA and the RAN area specified as the combination of the TAC and the RAN area code) (3GPP 2024g). The RRC_INACTIVE UE looks for an RAN paging message every RAN paging cycle. Periodic RNA updates can be viewed as a keep-alive mechanism, where the NR-NTN UE exchanges signaling with the RAN to make the RAN aware of its presence in an RNA. I-RNTI is the identity assigned to the RRC_INACTIVE UE by the serving NR-NTN gNB.

To exit the RRC_INACTIVE state, as in a TN, the NR-NTN UE sends RRC Resume Request containing its Short I-RNTI and Resume cause (e.g., mobile-originated signaling, mobile-originated data, and RNA update). The NR-NTN gNB retrieves the NR-NTN UE context from the last serving gNB and replies to the NR-NTN UE with the RRC Resume message. Finally, the NR-NTN UE sends RRC Resume Complete to complete the process of the resumption of the RRC connection. The serving NR-NTN gNB switches the NR-NTN UE-specific N2 connection toward the AMF and the N3 connection toward the UPF by exchanging path switch-related signaling messages with the AMF. See 3GPP (2024e) for additional details on the NR-NTN UE operations in the RRC_INACTIVE state.

9.4 Mobility Management in the LTE-M NTN

Mobility management in the LTE-M NTN reuses the principles of mobility management used in the NR-based TN. As in a TN, mobility management is based on the LTE-M NTN UE's RRC state. Section 9.4.1 discusses mobility management for an LTE-M NTN UE in the RRC_CONNECTED state, while Section 9.4.2 describes mobility management for an LTE-M NTN UE in the RRC_IDLE state.

9.4.1 Handover in the LTE-M NTN

Handover is carried out for an LTE-M NTN UE in the RRC_CONNECTED state. Key handover principles of the LTE-M-based TN, such as neighbor cell search, measurements, measurement or handover triggers, and inter-eNB signaling for handover, are reused in the LTE-M NTN. However, NTN-specific enhancements such as location- and time-based triggers are defined for more reliable handovers similar to those of the NR-NTN. Section 9.4.1.1 discusses measurements, measurement events, and handover triggers for the LTE-M NTN. The overall signaling for the CHO in the LTE-M NTN is illustrated in Section 9.4.1.2.

9.4.1.1 Measurements and Events for Handover in the LTE-M NTN

Measurements and associated handover events used in LTE-M-based TN are reused in LTE-M NTN. Additionally, NTN-specific enhancements are introduced for the LTE-M NTN. The measurements and handover triggers for the LTE-M NTN are similar to those discussed earlier in Section 9.3.1.1 for the NR-NTN. Note that there are significant similarities among LTE, LTE-M, and NR for handover-related measurements and handover events. Similarly, there are significant similarities among LTE-M-based TN, NR-NTN, and LTE-M NTN with respect to measurements and handover events.

As mentioned in Section 9.3.1.1 for the NR-NTN, to address the NTN-specific challenge of the lack of sufficient signal strength variations between the cell center and the cell edge in an NTN, location-based handover trigger has been introduced in the LTE-M NTN as well. Additionally, location-based handover trigger has been introduced to address the challenge of ensuring reliable handover of LTE-M NTN UEs from the outgoing LTE-M NTN cell to the incoming LTE-M NTN cell in the case of quasi-Earth-fixed beams or cells.

Location-Based Handover Trigger for the LTE-M NTN The LTE-M NTN UE's location relative to the serving cell and a neighbor cell can be used to reliably estimate the need for handover from the serving cell to a neighbor target cell. When the LTE-M NTN UE is farther from the currently serving LTE-M NTN cell and closer to the neighbor NR-NTN cell, RSRPs of the serving and neighbor NTN cells measured at the LTE-M NTN UE may not be widely different. However, distances between the LTE-M NTN UE and the serving and neighbor cells can be used to determine the need for handover more reliably. This situation for the LTE-M NTN is like the NR-NTN situation illustrated earlier in Figure 9.3 in Section 9.3.1.1.

As in the case of the NR-NTN, the distance-based trigger can be implemented in the LTE-M NTN using the criteria defined by Eqs. (9.7) and (9.8), which are identical to Eqs. (9.1) and (9.2) (3GPP 2024c).

$$\text{Ml1} - \text{Hys} > \text{Thresh1} \tag{9.7}$$

$$\text{Ml2} + \text{Hys} < \text{Thresh2} \tag{9.8}$$

In Eqs. (9.1) and (9.2), Ml1 is the distance between the LTE-M NTN UE and a reference location specified as referenceLocation1 in the IE reportConfigEUTRA (e.g., the center of the serving NTN cell), Ml2 is the distance between the LTE-M NTN UE and a reference location specified as referenceLocation2 in reportConfigEUTRA (e.g., the center of the neighbor NTN cell), Hys is the hysteresis parameter specified as hysteresisLocation in reportConfigEUTRA, Thresh1 is the threshold represented by the parameter distanceThreshFromReference1 specifying the distance from a reference location configured with the parameter referenceLocation1 in reportConfigEUTRA, Thresh2 is the threshold represented by the parameter distanceThreshFromReference2 specifying the distance from a reference location configured with the parameter referenceLocation2 in reportConfigEUTRA. Ml1, Ml2, Hys, Thresh1, and Thresh2 are expressed using meters.

Note that the LTE-M NTN UE predicts the movement of the reference location based on the ephemeris information and epoch time provided in CondReconfigurationAddMod if this event is used in the case of Earth-moving NTN cells (3GPP 2024c).

Equations (9.7) and (9.8) represent the entering conditions 1 and 2, respectively. The related leaving conditions are specified below in Eqs. (9.9) and (9.10), which are identical to Eqs. (9.3) and (9.4) (3GPP 2024c).

$$\text{Ml1} + \text{Hys} < \text{Thresh1} \tag{9.9}$$

$$\text{Ml2} - \text{Hys} > \text{Thresh2} \tag{9.10}$$

The location-based trigger is used in NTN-specific events such as conditional event condEvent D1 (3GPP 2024c). In essence, conditional event D1 corresponds to the case where the distance between the LTE-M NTN UE and referenceLocation1 is above threshold1 and the distance between the UE and referenceLocation2 is below threshold2.

Time-Based Handover Trigger for the LTE-M NTN A time-based trigger for handover is very helpful in the case of quasi-Earth-fixed NTN cells, as illustrated for the NR-NTN in Figure 9.4. The LTE-M NTN UE needs to be moved to the incoming LTE-M NTN cell before the outgoing LTE-M NTN cell stops serving the currently served geographic area. Hence, the LTE-M NTN UE compares time with time parameters defined for the time-based event. Specifically, the conditional event condEvent T1 uses time-based parameters. The entering condition and the leaving condition are specified in Eqs. (9.11) and (9.12), which are identical to Eqs. (9.5) and (9.6) shown earlier for the NR-NTN in Section 9.3.1.1 (3GPP 2024c).

$$\text{Mt} > \text{Thresh1} \tag{9.11}$$

$$\text{Mt} > \text{Thresh1} + \text{Duration} \tag{9.12}$$

In Eqs. (9.5) and (9.6), Mt is the time measured at the LTE-M NTN UE, Thresh1 is the threshold specified as t_1-Threshold in reportConfigEUTRA, and Duration is the duration specified in reportConfigEUTRA (3GPP 2024c). Mt, Thresh1, and Duration are expressed in ms. The serving LTE-M NTN eNB can provide suitable handover parameters to provide LTE-M NTN UEs adequate time to transition to a different NTN cell.

9.4.1.2 Signaling Flow for Conditional Handover in the LTE-M NTN

Like LTE and 5G NR, LTE-M supports UE-assisted network-controlled handover, where the LTE-M UE in the RRC_CONNECTED state provides measurements to the serving LTE-M eNB, and the serving LTE-M eNB makes a traditional handover or CHO decision. The source and target eNBs communicate to reserve resources for the UE in the target cell(s). The overall CHO signaling flow in the LTE-M NTN is very similar to that in the NR-NTN described earlier in Section 9.3.1.4. Figure 9.7 summarizes the key steps of the CHO in the LTE-M NTN when the MME and the S-GW remain the same (3GPP 2024a). Details of these steps can be found in 3GPP (2024a).

The LTE-M NTN eNB configures the LTE-M NTN UE with suitable measurements using the RRCConnectionReconfiguration message in *Step I*. The NR-NTN eNB specifies one or more carrier frequencies that the LTE-M NTN UE should measure. Furthermore, the LTE-M NTN eNB also specifies what measurements to make, such as RSRP and RSRQ. Additionally, the LTE-M NTN eNB

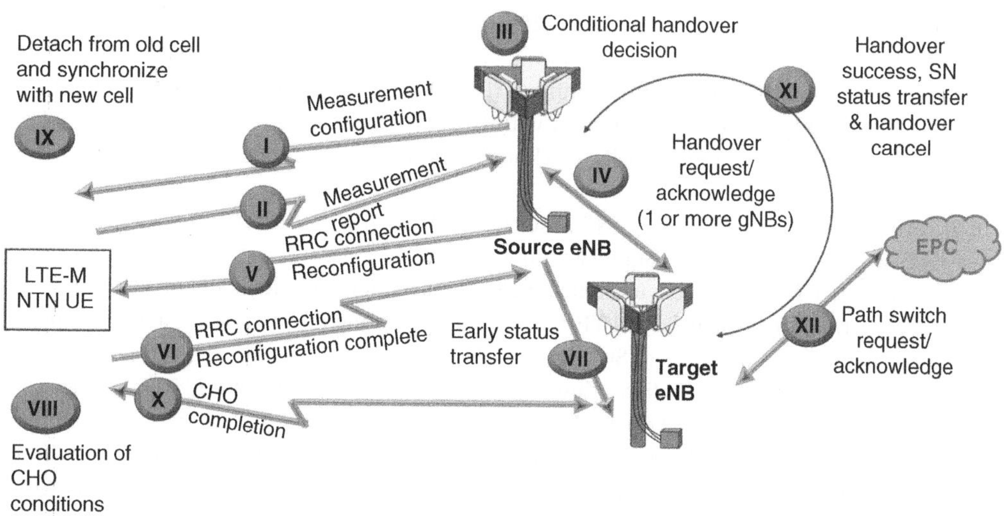

Figure 9.7 Conditional Handover in the LTE-M NTN.

configures the measurement reporting or handover events for which the LTE-M NTN UE needs to send a Measurement Report message. The LTE-M NTN eNB specifies the periodicity of measurement reports in case of periodic reporting. Note that both time-based and location-based events are available for configuration for the CHO in the LTE-M NTN.

The 3GPP has defined several intra-RAT and inter-RAT handover events originally for a TN, which can potentially be reused in an NTN. Both LTE and LTE-M can make use of these events. A1 to A6 are intra-RAT handover events, while B1 and B2 are inter-RAT handover events. See Section 3.2.2 in Chapter 3 for a summary of these events; see 3GPP (2024a) for details. Note that these A1 to A6 and B1 and B2 events are similar for LTE, LTE-M, and 5G NR.

In *Step II*, the LTE-M NTN UE makes measurements of the serving cell and neighboring cells. The LTE-M NTN UE can use PCIs of LTE-M NTN cells to determine the RSRP and RSRQ of these cells. When the reporting criterion for the measurement reporting is met (e.g., the occurrence of an event such as A3 or A5), the LTE-M NTN UE sends the MeasurementReport message containing the cell identity (i.e., PCI) and RSRP and RSRQ of the serving and relevant neighbor cell(s).

In *Step III*, the serving LTE-M NTN eNB processes the LTE-M NTN UE's measurement report and decides to use the CHO instead of the traditional handover as in a TN.

In *Step IV*, the serving LTE-M NTN eNB sends a Handover Request message to one or more target LTE-M NTN eNBs using the X2 interface to request the CHO for the identified CHO candidate cells. The Handover Request message contains information such as the target PCI and QoS parameters (e.g., a QCI) for established EPS bearers. The target LTE-M NTN eNB executes an implementation-specific admission control algorithm and evaluates the QoS needs of the current EPS bearers of the LTE-M NTN UE against the availability of radio resources. If the target LTE-M NTN eNB decides to accept the CHO request, it responds to the serving LTE-M NTN eNB with the Handover Request Acknowledge containing the CHO configuration for the CHP candidate cell. The Handover Request Acknowledge message may include parameters such as C-RNTI allocated to the LTE-M NTN UE in the CHO candidate cell, a dedicated random-access preamble (to avoid contention with other UEs using the PRACH in the new CHO candidate cell), and system information. This message includes the list of EPS bearers admitted by the target LTE-M NTN eNB. If packet forwarding is intended for the LTE-M NTN UE, the X2-U tunnel information is also provided in this message.

In Step V, the currently serving LTE-M NTN eNB sends the RRCConnectionReconfiguration message to the LTE-M NTN UE. This message contains the configuration of CHO candidate cells provided by target LTE-M NTN eNBs and CHO execution conditions determined by the source LTE-M NTN eNB.

After receiving the RRCConnectionReconfiguration message containing CHO-specific information, the LTE-M NTN UE in *Step VI* replies with the RRCConnectionReconfigurationComplete message to the serving LTE-M NTN eNB to acknowledge the reception of the RRCConnectionReconfiguration message.

In *Step VII*, the source LTE-M NTN eNB sends the EARLY STATUS TRANSFER message to the target LTE-M NTN eNBs if early data forwarding is used.

In *Step VIII*, the LTE-M NTN UE evaluates the CHO execution conditions. If one or more CHO execution conditions are satisfied, the LTE-M NTN UE selects the relevant CHO candidate cell as the new serving cell and proceeds to Step IX.

In *Step IX*, the LTE-M NTN UE discontinues communications with the serving LTE-M NTN cell and synchronizes with the new serving LTE-M NTN cell.

In *Step X*, the LTE-M NTN UE completes the CHO procedure by sending the RRCConnection-ReconfigurationComplete message to the target LTE-M NTN gNB associated with the new serving

LTE-M NTN cell. After successfully completing the CHO procedure, the LTE-M NTN UE releases previously stored CHO configurations.

In *Step XI*, the target LTE-M NTN eNB (which is now the new serving LTE-M NTN eNB for the LTE-M NTN UE) sends the HANDOVER SUCCESS message to the source LTE-M NTN eNB (which is the previously serving eNB for the UE) on the X2 interface to convey to the source LTE-M NTN eNB that the LTE-M NTN UE has successfully accessed the target LTE-M NTN cell. The source LTE-M NTN eNB replies with the SN STATUS TRANSFER message. Note that late data forwarding may be initiated as soon as the source LTE-M NTN eNB receives the HANDOVER SUCCESS message from the target (i.e., new serving) LTE-M NTN eNB. The source LTE-M NTN eNB sends the HANDOVER CANCEL message to the other candidate target LTE-M NTN eNBs (if any) to cancel CHO for the LTE-M NTN UE so that any reserved resources can be released.

In *Step XII*, the LTE-M NTN eNB-core network connectivity is modified. For example, the LTE-M NTN UE-specific S1-MNE connection is changed so the MME can exchange UE-specific signaling messages with the new serving LTE-M NTN eNB. Furthermore, the S1-U connection is modified to enable the S-GW to exchange user traffic with the new serving LTE-M NTN eNB. To modify S1-MME and S1-U connections, the target LTE-M NTN eNB (i.e., new serving LTE-M NTN eNB) sends the Path Switch Request to the MME. The MME exchanges E-RAB update messages with the S-GW to provide the S-GW with information about the target eNB side of the S1-U tunnel and to obtain information about the S-GW side of the S1-U tunnel from the MME. Next, the MME responds to the new serving LTE-M NTN eNB with the Path Switch Request Acknowledge message to inform the LTE-M NTN eNB about the change in the network's UE-specific S1-MME and S1-U connections.

9.4.2 Activities of an LTE-M NTN UE in the RRC_IDLE State

In a typical scenario, in the absence of any data traffic for a few seconds, the LTE-M NTN UE-specific inactivity timer expires as in a TN. The LTE-M NTN gNB, in an implementation-specific manner, asks the LTE-M NTN UE to enter the RRC_IDLE state. The LTE-M NTN UE context is not preserved in the NG-RAN when the LTE-M NTN UE is in the RRC_IDLE state. However, when the LTE-M NTN UE is in the RRC_INACTIVE state, the radio connection is released, but the LTE-M NTN UE context is preserved in the last serving LTE-M NTN gNB. Hence, when the LTE-M NTN UE needs to enter the RRC_CONNECTED state, it can make a faster transition due to more rapid restoration of the radio connection made possible by the preserved NR-NTN UE context in the NG-RAN. The RRC_IDLE state reduces the amount of signaling and processing for both the LTE-M NTN UE and the LTE-M NTN.

Figure 9.8 summarizes primary operations related to the RRC_IDLE state of the LTE-M NTN UE (3GPP 2024b). These operations are very similar in a TN and an NTN.

To initiate entry of the LTE-M NTN UE into the RRC_IDLE state, the LTE-M NTN eNB sends the S1AP UE Context release Request to the MME, and the MME works with the S-GW to release the S1-U/Access Bearer(s) by exchanging Release Access Bearers Request and Release Access Bearers Response messages. After releasing the S1-U/Access Bearer(s), the MME responds to the LTE-M NTN eNB with the UE Context Release Command. The LTE-M NTN eNB then asks the LTE-M NTN UE to enter the RRC_IDLE state by sending the RRC Connection Release command in *Step 1*. The LTE-M NTN eNB removes the access stratum UE context. The MME still preserves the UE context with information such as the GUTI allocated to the LTE-M NTN UE and the location of the LTE-M NTN UE (e.g., a given tracking area to be used for any needed paging). The IP address allocated to the UE also remains valid in the RRC_IDLE state of the LTE-M NTN UE.

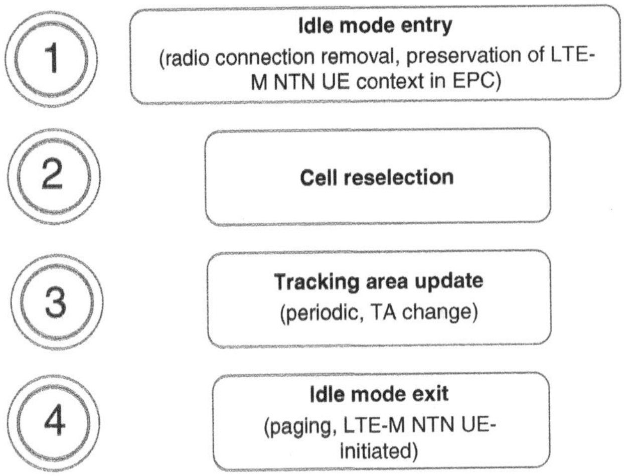

Figure 9.8 Activities of an LTE-M NTN UE in the RRC_IDLE State.

In *Step 2*, the RRC_IDLE UE typically camps on the strongest cell to ensure a high probability of success in receiving any incoming page message and establishes a radio connection by initiating the random-access procedure for any UE-initiated data transfer. Hence, in an implementation-specific manner, the LTE-M NTN UE searches for strong cells and performs cell reselection to a neighbor cell if such neighbor cell can provide a better signal compared to the currently monitored cell.

In support of cell reselection in the NTN, NTN-specific measurements or triggers related to timing and location are available for use. For example, the LTE-M NTN eNB broadcasts the timing and location-related information about the serving cell in SIB3 and SIB31. The LTE-M NTN gNB broadcasts timing information on when the LTE-M NTN cell will stop serving the current geographic area for quasi-Earth-fixed cells. The LTE-M NTN UE can use this cell stop time to determine when to start measuring neighbor cells. The LTE-M NTN eNB also broadcasts location information related to the reference location of the serving cell and a distance threshold to the reference location in support of location-based trigger for cell reselection. The LTE-M UE can use both location and time to decide about measurements for cell reselection.

Cell reselection principles of a TN are reused in the LTE-M NTN. For example, intra-frequency reselection is based on the ranking of cells, and inter-frequency reselection can be based on absolute priorities of carrier frequencies or redistribution priority. It may utilize layer-specific cell reselection parameters that are common to all neighboring cells on a given carrier frequency. Furthermore, the LTE-M NTN cell can provide an NCL containing cell-specific cell reselection parameters for special intra-frequency and inter-frequency neighboring cells. Additionally, the LTE-M NTN cell can provide exclude-lists to prevent the LTE-M NTN UE from reselecting to specific intra-frequency and inter-frequency neighboring cells. While cell reselection can be speed dependent and useful in a TN, its utility for an NTN may not be high, given that Earth-fixed TAs are used, and an NTN cell may broadcast multiple TAs.

In *Step 3*, the LTE-M NTN UE carries out a TAU when necessary. For example, if the periodic TAU timer expires, the LTE-M NTN UE executes the TAU procedure. Furthermore, when the LTE-M NTN UE performs cell reselection and finds that none of the TAs broadcast by the NTN cell is part of the TAI list conveyed to the LTE-M NTN UE by the MME, the LTE-M NTN UE executes the TAU procedure.

The LTE-M NTN UE may exit the idle mode for various reasons. For example, to execute the TAU procedure, the LTE-M NTN UE exits the idle mode, carries out the random-access procedure, establishes an RRC signaling connection, and exchanges non-access stratum TAU signaling with the MME. Specifically, the LTE-M NTN UE sends the TAU Request message to the MME, and the MME replies with the TAU Accept message. The LTE-M NTN UE sends the TAU Complete message to complete the TAU procedure. The MME (especially if it is an MME that has not allocated the current GUTI to the LTE-M NTN UE) allocates a new GUTI to the LTE-M NTN UE and provides an updated TAI list to the LTE-M NTN UE. The MME has now become aware of the UE's latest location and has updated the TAI list for the UE in its database. In the future, when the MME needs to contact the RRC_IDLE UE, it sends a paging message to the eNBs that are part of the TAs of the UE-specific TAI list. Upon completion of the TAU procedure, the LTE-M NTN UE usually re-enters the idle mode.

The LTE-M NTN UE also exits the RRC_IDLE mode when sending IoT data. Furthermore, the LTE-M NTN UE exits the idle mode to respond to an incoming paging message.

To exit the idle mode, as in a TN, the LTE-M NTN UE carries out the random-access procedure and establishes an RRC signaling connection. The LTE-M NTN eNB establishes suitable UE-specific radio bearers, such as Signaling Radio Bearers and Data Radio Bearers, to facilitate the transfer of signaling messages and user traffic. For user traffic, the S1-U/Access Bearers are also established.

In support of paging, as in a TN, the LTE-M NTN UE monitors a specific paging occasion (which is a specific subframe) in a paging frame. The specific paging occasion and the paging frame that the LTE-M NTN UE monitors for paging are a function of the UE's ID (e.g., IMSI). The LTE-M NTN UE monitors the paging occasion during the period of the DRX paging cycle, which typically lasts for about two seconds. An LTE-M NTN UE wakes up every two seconds or so and looks for a page message in a specific subframe of a specific 10-ms paging frame based on its permanent identity (i.e., IMSI). The LTE-M NTN eNB sends the RRC paging message during the paging occasion, which the LTE-M NTN UE is expected to monitor. The paging message typically includes S-TMSI as the identity of the LTE-M NTN UE.

9.5 Mobility Management in the NB-IoT NTN

While handover in the RRC_CONNECETD state of the UE is similar to handover in LTE, LTE-M, and 5G NR, NB-IoT does not support UE-assisted network-controlled handover. Specifically, to reduce the complexity and processing requirements on the NB-IoT UE, the eNB does not configure the NB-IoT UE to make frequent measurements of neighbor cells or to report measurements based on events to the serving NB-IoT eNB. Mobility management principles used in the NB-IoT-based TN are reused in the NB-IoT NTN.

When the NB-IoT NTN UE in the RRC_CONNECTED state moves from one NB-IoT NTN cell to another, its radio link with the currently serving NB-IoT NTN cell experiences a Radio Link Failure (RLF). The NB-IoT NTN UE enters the RRC_IDLE state and performs cell reselection to the strongest neighbor cell. An RRC connection is established between the NB-IoT NTN UE and the new NB-IoT NTN cell (which may belong to the previously serving eNB or the new eNB). The NB-IoT NTN can carry out or resume data transfer upon establishing suitable SRBs and DRB(s).

The idle mode activities, such as cell reselection and TAU, are performed by the NB-IoT NTN UE like an LTE-M UE, as shown earlier in Figure 9.8. For increased power savings, the PSM may be activated like in a TN. Furthermore, the MME can configure a long periodic TAU timer value for the NB-IoT UE like in a TN.

As in a TN, the NB-IoT NTN uses intra-frequency reselection based on the ranking of cells (which may use cell-specific offsets per configuration by the NB-IoT NTN eNB). Inter-frequency reselection utilizes the ranking of frequencies with support for frequency-specific offsets in NB-IoT in a TN and an NTN. Furthermore, for both a TN and an NTN, blind redirection is supported for load balancing and a dedicated offset for the frequency, where the NB-IoT UE is being redirected to.

9.6 Feeder Link Switchover in the NTN

Feeder link switchover is a unique aspect of an NTN. Feeder link switchover is supported for 5G NR, LTE-M, and NB-IoT. A feeder link switchover is a procedure where the feeder link is changed from a source NTN Gateway to a target NTN Gateway for a given NTN payload. The feeder link switchover is considered to be a Transport Network Layer procedure. A change of the serving NTN payload for a given UE is called a service link switch. When the feeder link switchover occurs, NTN UEs would need to be transferred from one NTN-gNB to another NTN-gNB when two NTN-GWs are connected to two different NTN-gNBs, which is expected to be a common scenario.

Consider Figure 9.9, which illustrates two types of feeder link switchover: hard and soft. The network operator can choose the specific type of feeder link switchover based on factors such as equipment capabilities, service requirements, and performance goals.

In the case of a hard feeder link switchover, an NTN payload connects to a single NTN-GW in a given instant. Hence, there is a certain transition period during the radio communication does not occur on the feeder link for a given NTN payload. For example, in Figure 9.9, the NTN payload communicates with the NTN-GW1/NTN-gNB1 from t_1 to t_2. Then, the feeder link switchover occurs during the transition period t_2 to t_3. The feeder link is reestablished for the NTN payload

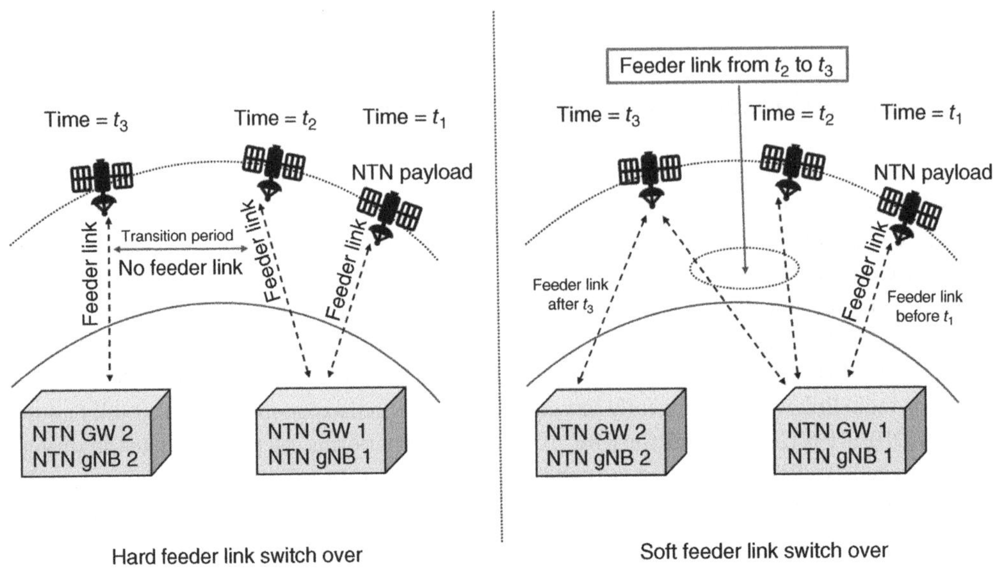

Figure 9.9 Feeder Link Switchover in the NTN.

at time t_3. The NTN payload communicates with the NTN-GW2/NTN-gNB2 beginning time t_3. The NTN UEs that experience the service link switch due to the hard feeder link switchover would also experience a period of service interruption between t_2 and t_3. The exact impact on the NTN user experience or NTN device depends on the types of active services and the length of the feeder link switch transition period.

When a soft feeder link switchover occurs, an NTN payload connects to two NTN-GWs at a given instant. Hence, the radio communication continues to occur on the feeder link for a given NTN payload. For example, in Figure 9.9, the NTN payload communicates with the NTN-GW1/NTN-gNB1 from t_1 to t_3. The feeder link switchover occurs during the transition period t_2 to t_3. The NTN payload communicates with both the NTN-GW1/NTN-gNB1 and the NTN-GW2/NTN-gNB2 during the transition period t_2 to t_3. The feeder link between the NTN payload and the NTN_GW1/NTN-gNB1 is removed at time t_3. The NTN payload communicates with only the NTN-GW2/NTN-gNB2 beginning time t_3. The NTN UEs that experience the service link switch as a result of the soft feeder link switchover could continue their communications even during the feeder link switchover due to the continuous existence of the feeder link via the suitable implementation of soft feeder link switchover.

To facilitate the feeder link switch over, the NTN control function determines the instant for the feeder link switch over. The transfer of the contexts of NTN UEs from one gNB/eNB to another is performed via NG/S1-based or Xn/X2-based handover based on the gNBs'/eNBs' implementation of NTN gNBs/eNBs and configuration information provided to the gNBs/eNBs by the NTN control function (3GPP 2024a,d).

9.7 Discontinuous Coverage in the NR-NTN or IoT-NTN

The satellite constellation may not be complete in an NTN that uses NGSO satellites, such as LEO satellites. In such a sparse satellite constellation case, the target service area may not be covered continuously as the satellites move into and out of the visibility of NTN UEs. Hence, such NTN UEs experience discontinuous coverage.

The 3GPP has introduced an enhancement for NTN UEs experiencing discontinuous coverage. The NTN eNB broadcasts NTN payload assistance information in the form of ephemeris parameters, the start time, and/or carrier frequency of upcoming NTN payload coverage to enable NTN UEs to predict when an upcoming NTN payload will provide coverage (3GPP 2024a). The NTN UE, in an implementation-specific manner, utilizes such information and predicts out-of-coverage and in-coverage periods. The NTN UE in the RRC_IDLE state can save precious battery power by avoiding the NTN cell search during a period of no coverage. The NTN eNB, in an implementation-specific manner, may initiate a UE Context Release Request procedure toward the serving MME if it detects that the NTN UE is experiencing discontinuous coverage during a given period.

9.8 Major Takeaways

Mobility management principles for a TN are reused in an NTN along with NTN-specific enhancements. The NTN UE's RRC state influences mobility management. Here are the major takeaways of this chapter.

1) Major NTN mobility challenges include moving NTN cells, signal strength variations, feeder link switchover, massive handover, TN-NTN mobility, and discontinuous coverage.
2) Key handover principles of the NR-based TN, such as neighbor cell search, measurements, measurement or handover triggers, and inter-gNB signaling for handover, are reused in the NR-NTN. Furthermore, NTN-specific enhancements, such as location and time-based triggers, are defined for more reliable handover.
3) To address the NTN-specific challenge of the lack of sufficient signal strength variations between the cell center and the cell edge in an NTN, a location-based handover trigger has been introduced. New events such as Event D1 and conditional Event condEvent D2 in the NR-NTN and the conditional event condEvent D1 in the LTE-M NTN utilize the location-based trigger in the NR-NTN.
4) To address the challenge of ensuring reliable handover of NR-NTN UEs from the outgoing NR-NTN cell to the incoming NR-NTN cell, a time-based handover trigger has been introduced. The key idea for the time-based trigger is that the NR-NTN UE needs to be moved to the incoming NTN cell before the outgoing NTN cell stops serving the currently served geographic area. A new event, such as the conditional Event content T1 in the NR-NTN and LTE-M NTN, utilizes the time-based trigger in the NR-NTN.
5) Mobility management of the NTN UE in RRC_IDLE (and RRC_INACTIVE) state reuses the procedures defined for a TN with NTN-specific enhancements such as enhanced system information, enhanced tracking area management, and support for TN-NTN mobility.
6) Handover types supported in the NTN include traditional handover, DAPS handover, RACH-less handover, CHO, NTN-NTN handover, TN-NTN handover, and satellite switch with re-sync.
7) Major steps in the CHO in the NR-NTN and the LTE-M NTN include configuration of the NTN UE for measurements by the NTN gNB/eNB, reporting of measurements by the NTN UE, determination of the use of the CHO by the serving gNB/eNB, retrieval of CHO configuration for target CHO neighbor cells by the serving gNB/eNB, configuration of the NTN UE with CHO configurations and CHO execution conditions by the serving gNB/eNB, implementation of CHO by the NTN UE, release of radio resources at nonselected CHO cells, and update of radio-core connections.
8) Mobility management principles used in the NB-IoT-based TN are reused in the NB-IoT NTN. When the NB-IoT NTN UE in the RRC_CONNECTED state moves from one NB-IoT NTN cell to another, RLF occurs, the NB-IoT NTN UE enters the RRC_IDLE state, and the NB-IoT NTN UE performs cell reselection to the strongest neighbor cell.
9) A feeder link switchover is a procedure where the feeder link is changed from a source NTN Gateway to a target NTN Gateway for a given NTN payload. When the feeder link switchover occurs, NTN UEs are transferred from one NTN-gNB to another NTN-gNB when two NTN-GWs are connected to two different NTN-gNBs.
10) In the case of a sparse satellite constellation, an NTN UE experiences discontinuous coverage. The NTN UE in the RRC_IDLE state can save battery power by avoiding the NTN cell search during a period of no coverage by exploiting the coverage information broadcast by the NTN eNB.

References

3GPP (2024a). *Evolved Universal Terrestrial Radio Access (E-UTRA) and Evolved Universal Terrestrial Radio Access Network (E-UTRAN): Overall description.* TS36.300. https://portal.3gpp.org/desktopmodules/Specifications/SpecificationDetails.aspx?specificationId=2430 (accessed 6 September 2024).

3GPP (2024b). *Evolved Universal Terrestrial Radio Access (E-UTRA); User Equipment (UE) procedures in idle mode.* TS36.304. https://portal.3gpp.org/desktopmodules/Specifications/SpecificationDetails.aspx?specificationId=2432 (accessed 6 September 2024).

3GPP (2024c). *Evolved Universal Terrestrial Radio Access (E-UTRA); Radio Resource Control (RRC); Protocol specification.* TS36.331. https://portal.3gpp.org/desktopmodules/Specifications/SpecificationDetails.aspx?specificationId=2440 (accessed 6 September 2024).

3GPP (2024d). *NR and NG-RAN overall description.* TS38.300. https://portal.3gpp.org/desktopmodules/Specifications/SpecificationDetails.aspx?specificationId=3191 (accessed 6 September 2024).

3GPP (2024e). *User Equipment (UE) procedures in idle mode and in RRC Inactive state.* TS38.304. https://portal.3gpp.org/desktopmodules/Specifications/SpecificationDetails.aspx?specificationId=3192 (accessed 6 September 2024).

3GPP (2024f). *NR; Medium Access Control (MAC) protocol specification.* TS38.321. https://portal.3gpp.org/desktopmodules/Specifications/SpecificationDetails.aspx?specificationId=3194 (accessed 6 September 2024).

3GPP (2024g). *Radio Resource Control (RRC); Protocol specification.* TS38.331. https://portal.3gpp.org/desktopmodules/Specifications/SpecificationDetails.aspx?specificationId=3197 (accessed 6 September 2024).

3GPP (2024h). *Procedures for the 5G System (5GS); Stage 2.* TS23.502. https://portal.3gpp.org/desktopmodules/Specifications/SpecificationDetails.aspx?specificationId=3145 (accessed 6 September 2024).

Tripathi, N.D. and Reed, J.H. (2019). *5G Cellular Communications: Journey and Destination.* The Wireless University. https://thewirelessuniversity.com/ (accessed 6 September 2024).

10

Evolution of the NTN in 5G-Advanced and 6G

The 3GPP introduced the NTN in Release 17. Enhancements were made to the NTN specifications in Release 18. The evolution of the NTN continues in Release 19. This chapter provides a brief overview of target NTN enhancements in Release 19. While the NTN-related specifications work carried out by the 3GPP is formally part of 5G and 5G-Advanced, the NTN is expected to play an even more key role in 6G. A brief introduction to the 6G vision and requirements is given, followed by the usage scenarios that the International Telecommunication Union (ITU) has defined for 6G. The ITU-specified performance goals for 6G are also summarized. An overview of 6G technology enablers is also given. The NTN's role in realizing the vision of 6G is explained. Finally, examples of potential NTN research directions are given.

10.1 Evolution of the NTN

Cellular technologies have evolved from the first generation (1G) to the fifth generation (5G) in about five decades. The 3GPP defined the first release of 5G, so-called 5G Phase 1, in Release 15. The 3GPP introduced new features and capabilities in Release 16 to Release 18. The 3GPP features beginning with Release 18 are considered part of 5G-Advanced. Release 19 completion is targeted for 2025. Initial 3GPP-based 6G specifications are expected to be defined in Release 21 (Qualcomm 2022; Chen and Jain 2024).

The 3GPP introduced the NTN as a 5G feature in Release 17. The 3GPP made enhancements to the NTN in Release 18. The Release 19 workshop held by the 3GPP in June 2023 identified the NTN as one of the candidate areas for normative specifications in Release 19. The R19 specifications work is expected to be carried out in 2024 and 2025. R19 addresses enhancements such as support for a regenerative NTN payload. Section 10.2 describes the NTN enhancements in R19.

The NTN is expected to play an essential role in enabling 6G to meet 6G goals. Various organizations, including the ITU and the Next G Alliance (NGA), have initiatives to define 6G vision and requirements. Section 10.3 lists examples of key 6G organizations and summarizes 6G vision and requirements and the role of the NTN in 6G.

3GPP-based Non-Terrestrial Networks in 5G and 6G: Expanding the Frontiers of Wireless Communications, First Edition.
Nishith D. Tripathi and Jeffrey H. Reed.

Companion website: www.wiley.com/go/tripathi5g

10.2 NTN Enhancements in Release 19

As in Release 17 and Release 18, the 3GPP has two parallel work streams related to the NTN, NR-NTN, and IoT-NTN. Section 10.2.1 discusses R19 enhancements in the NR-NTN, often known as NR-NTN Phase 3. Section 10.2.2 discusses R19 enhancements in the IoT-NTN, often known as IoT-NTN Phase 3.

10.2.1 NR-NTN Enhancements in Release 19

The NR-NTN enhancements in Release 19 aim to enhance downlink coverage, enhance uplink capacity, support the Multimedia Broadcast Service (MBS), support selected 5G network functions (NFs) on the NTN platform, including a regenerative payload, and support reduced capability (RedCap) UEs (3GPP 2024a). Figure 10.1 lists key enhancements targeted by the 3GPP for the NR-NTN Phase 3 in Release 19 (3GPP 2024a).

10.2.1.1 Enhanced Downlink Coverage

One NR-NTN enhancement goal is enhanced downlink coverage for handheld devices such as smartphones with a −5.5 dBi antenna gain (3GPP 2024a). This enhancement considers the NTN deployment constraints, which include payload power limitation, a large satellite footprint, and limited feeder link bandwidth. The DL coverage enhancements at both the link and system levels are considered. The link-level enhancements aim to improve the link margin of physical channels such as the Physical Downlink Control Channel (PDCCH) and the Physical Downlink Shared Channel (PDSCH) so that these channels can be operated with reduced EIRP in the FR1 frequency band. The system-level enhancements aim to support efficient, dynamic, and flexible power sharing between beams or different beam sizes across the satellite footprint for the FR1 and the FR2 frequency bands. Both GSO and NGSO satellites are considered. Furthermore, energy-saving techniques defined for 5G NR up to Release 18 are considered to be the baseline over which enhancements are designed.

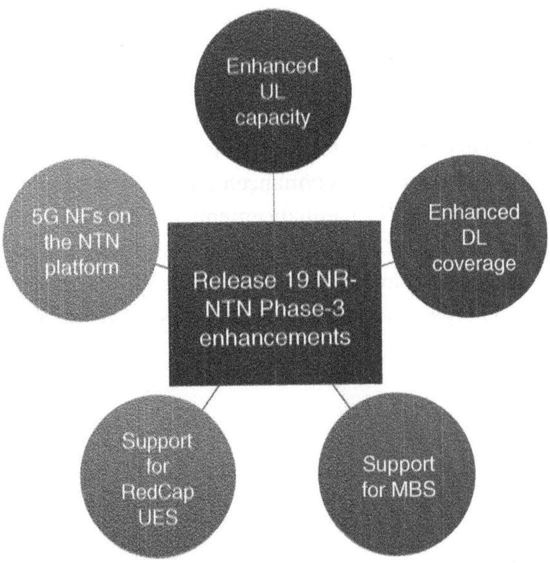

Figure 10.1 NR-NTN Enhancements in Release 19.

10.2.1.2 Enhanced Uplink Capacity

This enhancement aims to optimize uplink capacity through multiplexing techniques. Recall from Chapter 3 that an NTN cell typically covers a large area and many devices. When the amount of spectrum available to an NTN is limited, it is challenging to support high capacity or throughput per device. Fast access to and prompt release of radio resources of the NTN cell are required to maximize resource utilization and achieve relatively higher capacity or throughput in the cell. Furthermore, finer granularity of resource multiplexing can help meet diverse QoS requirements of different users. Enhanced uplink capacity makes more radio resources available to a device to improve Voice over NR[1] (VoNR) services, especially in coverage-limited situations. Techniques such as Orthogonal Cover Codes (OCCs) across OFDM symbols, across slots, and within an OFDM symbol are potential technology enhancements (3GPP 2024a).

10.2.1.3 Support for MBS

The MBS is a value-added service for the NR NTN because a large number of users can be served very efficiently compared to unicast transmissions, where a large number of devices download the same content. In general, the terrestrial MBS specifications can be leveraged in the NTN. However, enhancements are needed when the intended service area is smaller than the coverage of the entire NTN cell (3GPP 2024a). For example, a notification about the service area of a broadcast service is needed. Furthermore, suitable signaling between the NG-RAN and the 5GC is also required.

10.2.1.4 Support for 5G NFs on the NTN Platform

One or more NFs can be placed on the NTN platform. For example, the NR-NTN gNB can be placed on the NTN platform to implement a regenerative payload. One or more 5GC NFs, such as the UPF, can also be placed on the NTN platform. The placement of an NF on the NTN platform leads to new architecture options for deployment, increasing flexibility. A regenerative payload with onboard 5G NFs facilitates real-time connectivity (i) between two UEs and (ii) between the network and the UE via the NTN platform, both with ISL and without ISL (3GPP 2024a).

10.2.1.5 Support for RedCap UEs

This enhancement supports RedCap devices, including handheld devices such as smartwatches and IoT devices in the FR1 frequency band. RedCap enables the NR-NTN to offer enhanced service capabilities using a wider channel bandwidth than IoT devices without significantly increasing the device complexity. Full-duplex FDD RedCap UEs, eRedCap UEs, and HD-FDD RedCap UEs and eRedCap UEs are supported (3GPP 2024a). The RedCap and eRedCap UEs support GNSS capabilities and simultaneous GNSS and NR-NTN operation.

10.2.2 IoT-NTN Enhancements in Release 19

Figure 10.2 lists two key enhancements targeted by the 3GPP for the IoT-NTN Phase 3 in Release 19 (3GPP 2024b). These two enhancements include support for Store & Forward and uplink capacity enhancements.

---

1 VoNR is usually provided to a user via the NR-based NG-RAN, the 5GC, and the IP Multimedia Subsystem (IMS). VoNR utilizes voice over IP (VoIP) to transport packets from one endpoint (e.g., a smartphone) to the other endpoint.

Figure 10.2 IoT-NTN Enhancements in Release 19.

10.2.2.1 Store & Forward (S&F) Using a Regenerative Payload

The support for Store and Forward (S&F) operation by the NTN platform with a regenerative payload offers several benefits (3GPP 2024b). For example, the S&F operation allows delay-tolerant and non-real-time IoT services available in the areas (e.g., middle-of-the-ocean area or remote areas) illuminated by the NTN platform without requiring an NTN-GW in the direct satellite visibility. The S&F operation enables cost-effective and immediate deployment of IoT-NTN services with sparse LEO satellite constellations and reduced NTN-GW infrastructure. Note that specific non-3GPP solutions natively support S&F services (3GPP 2024b). Suitable enhancements in the E-UTRAN and the UE are made in support of the S&F operation. Furthermore, suitable enhancements related to the S1AP and feeder link switch are needed.

10.2.2.2 Uplink Capacity Enhancements

The increasing popularity of IoT devices requires an increase in uplink capacity to support a massive number of IoT devices. Furthermore, several types of IoT devices, such as low-cost and low-complexity IoT devices and medium-cost and medium-complexity smart wearables, possess different capabilities. Multiplexing of UEs using OCCs on a single 3.75 kHz or 15 kHz subcarrier may be investigated to increase the uplink capacity. For example, NPUSCH Format 1 and NPRACH may benefit from OCCs. Multitone support for 15 kHz SCS may also be investigated by the 3GPP. Furthermore, the reduction in uplink and downlink signaling to complete an Early Data Transmission (EDT) procedure can help improve the overall efficiency of resource utilization, contributing to the capacity increase. For example, transmission of Msg3 transmission without msg1/Random-Access Response (RAR) may be studied. Additionally, the overhead reduction of msg4/RRCEarlyData-Complete may also be studied.

10.3 NTN and 6G

The NTN will play an essential role in the Sixth-Generation (6G) wireless technology. Section 10.3.1 describes the key organizations for 6G. Section 10.3.2 summarizes the vision and requirements of 6G based on the work carried out by various alliances and organizations. Section 10.3.3 defines potential performance requirements or goals for 6G. Technologies that may become part of 6G are briefly introduced in Section 10.3.4. Section 10.3.5 discusses the potential role of the NTN in 6G.

10.3.1 Key Organizations for 6G

Research on 6G started even when 5G specifications were being developed. For example, the University of Oulu in Europe started performing research in 6G by launching the 6G Flagship project. Figure 10.3 lists examples of organizations or alliances expected to play a key role in defining the vision and requirements for 6G and 6G specifications. While individual companies such as equipment manufacturers and device manufacturers are not explicitly mentioned in Figure 10.3, they

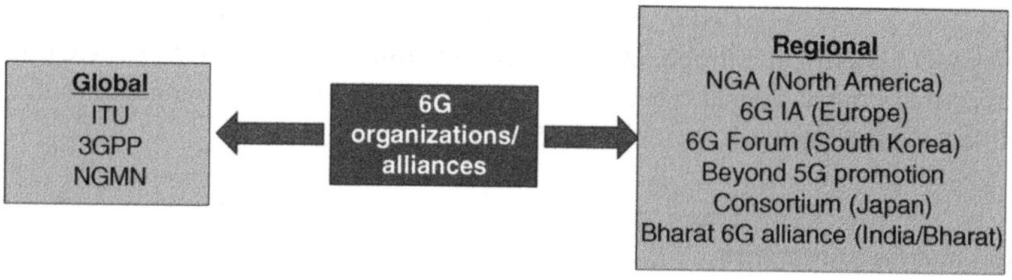

Figure 10.3 Key 6G Organizations.

will continue to play a critical role both as part of one or more of these alliances or organizations as well as separately.

The organizations or alliances are divided into two categories to facilitate discussions—global and regional. Global organizations include significant participation from multiple continents and countries. Regional organizations typically have significant participation from a specific region or country. Examples of key global organizations for 6G include the International Telecommunication Union (ITU), the Third-Generation Partnership Project (3GPP), and the Next-Generation Mobile Network (NGMN) Alliance. Examples of key regional organizations for 6G include the Next G Alliance (NGA) in North America, 6G Smart Networks and Services Industry Association (6G-IA) in Europe, 6G Forum in South Korea, Beyond 5G Promotion Consortium in Japan, and Bharat 6G Alliance in India/Bharat.

The ITU has typically defined use cases and performance requirements for different generations of wireless technologies, such as International Mobile Telecommunication (IMT-2000) for 3G, IMT-Advanced for 4G, and IMT-2020 for 5G. The ITU has started identifying use cases and potential performance goals (ITU 2023), and Sections 10.3.2 and 10.3.3 reflect the work done by the ITU in this regard.

The 3GPP is expected to lead the development of specifications for 6G. The 3GPP's work on 5G enhancements in Release 19 is expected to end in 2024. Release 20 may also focus on enhancements to 5G, but studies related to 6G could begin in Release 20. The actual specification work on 6G is expected to start in Release 21 (Qualcomm 2022; Chen and Jain 2024).

The NGMN Alliance is an alliance of the world's leading mobile network operators. The NGMN Alliance had developed a white paper on 5G vision and requirements and has also developed a 6G vision paper (NGMN Alliance 2021). The NGMA Alliance has also published a white paper on 6G use cases and analysis (NGMN Alliance 2022).

The NGA initially published a 6G vision paper containing a roadmap for 6G along with six audacious goals for 6G (NGA 2022a) and has since published multiple white papers covering various aspects of 6G (NGA 2024a). The NGA has six working groups to address diverse topics. These WGs include the 6G National Roadmap, Technology, Spectrum, Societal and Economic Needs (SEN), Green G, and Applications.

The 6G Smart Networks and Services Industry Association (6G-IA) in Europe represents European Industry and Research for next-generation networks and services. Its primary objective is to contribute to Europe's leadership in 5G, 5G evolution, and SNS/6G research (6G-IA 2024). 6G-IA published its vision for 6G in 2021 (6G-IA 2021).

Examples of country-specific alliances include 6G Forum in South Korea (6G Forum 2024), Beyond 5G Promotion Consortium in Japan (Beyond 5G Promotion Consortium 2024), and Bharat 6G Alliance in India (Bharat 6G Alliance 2024).

Consortia such as Hexa-X in Europe (led by Nokia Bell Labs), 6G SHort range extreme communication IN Entities (6G-SHINE) in Europe, 6G-Research and Innovation Cluster (6G-RIC) in Germany, and Eurecom in France have also been conducting research in 6G. Universities around the globe have also been doing research on advanced 5G and 6G.

10.3.2 6G: Vision and Requirements

6G is expected to be different from all previous generations of cellular technologies from the perspectives of vision and requirements. The 3GPP considers the requirements suggested by organizations such as the ITU and the requirements identified by its working groups to develop specifications. The ITU has traditionally defined use cases and target services for generations of cellular technologies. For example, the ITU defined usage scenarios of eMBB, URLLC, and mIoT for 5G. Figure 10.4 illustrates a wheel diagram showing six usage scenarios envisioned by the ITU for 6G (ITU 2023).

Three of the six usage scenarios are extensions of IMT-2020 (i.e., 5G) scenarios, and three are new scenarios. The extension of the 5G's eMBB usage scenario is immersive communication, the extension of the 5G's mMTC usage scenario is massive communication, and the extension of the 5G's URLLC usage scenario is Hyper Reliable and Low-Latency Communication (HRLLC). Three new usage scenarios are Ubiquitous Connectivity, Integrated AI and Communication, and Integrated Sensing and Communication. Furthermore, there are four overarching aspects that act as design principles commonly applicable to all usage scenarios. These aspects include sustainability, connecting the unconnected, ubiquitous intelligence, and security/privacy/resilience.

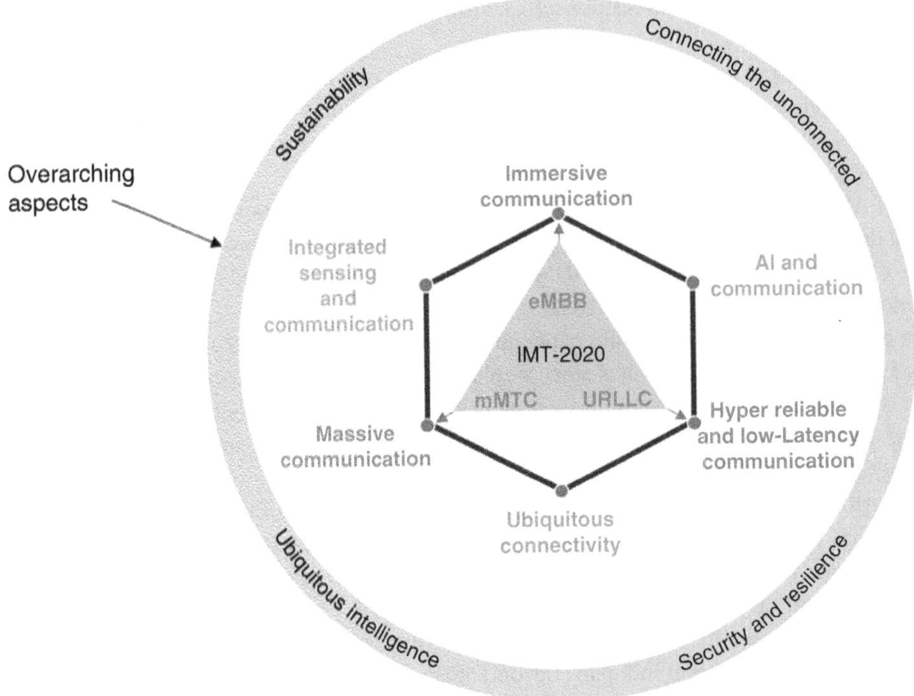

Figure 10.4 IMT-2030 Usage Scenarios and Overarching Aspects.

10.3.2.1 6G Usage Scenario: Immersive Communication

This usage scenario provides users with a rich and interactive video or immersive experience. Interactions with machine interfaces are also covered. Diverse service environments, such as hot spots, urban, and rural, are covered. Examples of use cases for this usage scenario include communication for immersive eXtended Reality (XR), remote multisensory telepresence, and holographic communications. In addition to the standalone voice services, mixed audio, video, and other data traffic with suitable time synchronization is supported.

10.3.2.2 6G Usage Scenario: Massive Communication

This usage scenario provides connections for a massive number of devices or sensors for various use cases and applications. Examples of use cases for this usage scenario include expanded and new applications in smart cities, transportation, logistics, health, energy, environmental monitoring, agriculture, and areas requiring a variety of IoT devices (i) without batteries or (ii) with batteries that last for the life of the device.

10.3.2.3 6G Usage Scenario: HRLLC

This usage scenario supports more stringent requirements on reliability and latency than the URLLC usage scenario of 5G and typically involves time-synchronized operations. Examples of use cases include communications in an industrial environment with full automation, control, and operation. This usage scenario supports applications such as machine interactions, emergency services, telemedicine, and electrical power transmission and distribution monitoring.

10.3.2.4 6G Usage Scenario: Ubiquitous Connectivity

This usage scenario provides connectivity to bridge the digital divide and addresses uncovered or scarcely covered geographic areas, such as rural, remote, and sparsely populated areas. Both IoT and mobile broadband communication are supported.

10.3.2.5 6G Usage Scenario: Integrated AI and Communication

This usage scenario supports distributed computing and AI applications. Examples of use cases include automated driving, autonomous collaboration between devices for medical assistance applications, offloading of heavy computation operations across devices and networks, and creation of and prediction with digital twins.

10.3.2.6 6G Usage Scenario: Integrated Sensing and Communication

This usage scenario supports applications and services that require sensing capabilities and offers wide-area multidimensional sensing that provides spatial information about unconnected and connected objects, including their movements and surroundings. Examples of use cases include (i) navigation; (ii) activity detection and movement tracking, such as posture/gesture recognition, fall detection, and vehicle/pedestrian detection; (iii) environmental monitoring, such as rain/pollution detection; and (iv) provisioning of sensing data/information on surroundings for AI, XR, and digital twin applications.

Table 10.1 summarizes typical capabilities required for the six usage scenarios of 6G (ITU 2023).

10.3.3 6G Performance Goals

To support the use cases associated with the six usage scenarios described above, the 6G system would need to meet specific performance goals. Figure 10.5 specifies capabilities identified by the ITU for 6G (ITU 2023). Figure 10.5 shows the range of values given for specific capabilities

Table 10.1 Required Capabilities for 6G Usage Scenarios.

6G Usage Scenario	Required Capabilities
Immersive Communication	Enhanced spectrum efficiency, consistent service experiences, the balance between higher data rates and increased mobility in different environments, high reliability, and low latency for responsive and accurate interaction with real and virtual objects, larger system capacity to simultaneously connect numerous devices
Massive Communication	Very high connection density, different data rates, low power consumption, mobility, extended coverage, high security, and high reliability
HRLLC	Enhanced reliability, low latency, precise positioning, and high connection density
Ubiquitous Connectivity	Extensive coverage
Integrated AI and Communication	High area traffic capacity, high user-experienced data rates, low latency, high reliability, data acquisition, preparation, and processing from different sources, distributed AI model training, model sharing and distributed inference across IMT systems, and computing resource orchestration and chaining
Integrated Sensing and Communication	High-precision positioning, sensing-related capabilities such as range/velocity/angle estimation, object and presence detection, localization, imaging, and mapping

Source: Adapted from ITU IMT-2030 Framework-2023.

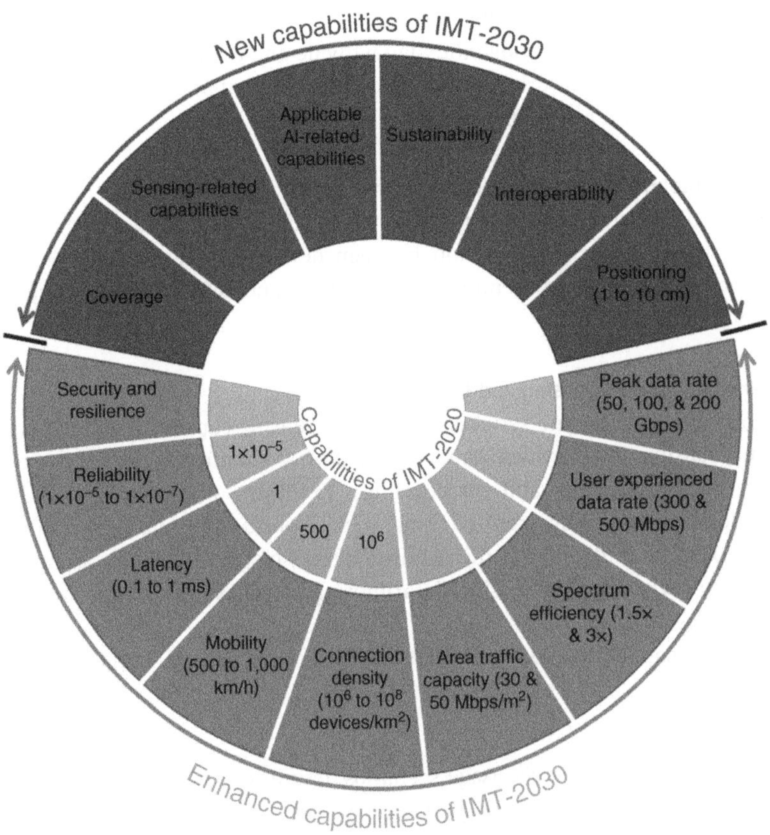

Figure 10.5 Performance Targets for IMT-2030 Capabilities.

targeted for IMT-2030 systems. Note that some capabilities have been enhanced relative to 5G (i.e., IMT-2020), while others are new capabilities. Not all usage scenarios would not need the 6G system to achieve all the performance targets specified in Figure 10.5. Some values of performance targets would be applicable or relevant to some use cases, while other values of performance targets would apply to another set of use cases. These capabilities and associated performance targets are briefly discussed later (ITU 2023).

Peak Data Rate. This is the maximum achievable data rate under ideal conditions per device (ITU 2023). Example values for the target peak data rate include 50, 100, and 200 Gbps.

User-Experienced Data Rate. This achievable data rate is available ubiquitously across the coverage area to a mobile device (ITU 2023). Example values for the target user-experienced data rate include 300 and 500 Mbps.

Spectrum Efficiency. This is the average data throughput per unit of spectrum resource and per cell (ITU 2023). Example values for the target spectrum efficiency include 1.5 and 3 times greater than that of IMT-2020.

Area Traffic Capacity. This is the total traffic throughput served per geographic area (ITU 2023). Example values for the target area traffic capacity are 30 and 50 $Mbps/m^2$.

Connection Density. This is the total number of connected and/or accessible devices per unit area (ITU 2023). Example values for the target connection density are 106 to 108 $devices/km^2$.

Mobility. This is the maximum speed at which a defined QoS and seamless transfer between radio nodes (which may belong to different layers and/or radio access technologies) can be achieved (ITU 2023). Example values for the target mobility are 500–1,000 km/h.

Latency. This is the latency or delay over the air interface and refers to the contribution by the radio network to the time from when the source sends a packet of a certain size to when the destination receives it (ITU 2023). Example values for the target latency are 0.1–1 ms.

Reliability. This is the capability of successfully transmitting a predefined amount of data within a predetermined time duration with a given probability (ITU 2023). Example values for the target reliability over the air interface are 1×10^{-5} to 1×10^{-7}.

Coverage. This is the ability to provide access to communication services for users in a desired service area. It is formally defined as the cell edge of a single cell through link budget analysis (ITU 2023).

Positioning. This is the ability to calculate the approximate position of connected devices. Positioning accuracy is defined as the difference between the calculated horizontal/vertical position and a device's actual horizontal/vertical position (ITU 2023). Example values for the target positioning accuracy are 1–10 cm.

Sensing-Related Capabilities. These are the abilities to provide functionalities in the radio interface, including range/velocity/angle estimation, object detection, localization, imaging, and mapping (ITU 2023). These capabilities could be measured in terms of metrics such as accuracy, resolution, detection rate, and false alarm rate.

Applicable AI-Related Capabilities. These are abilities to provide certain functionalities throughout IMT-2030 to support AI-enabled applications (ITU 2023). Examples of these functionalities include distributed data processing, distributed learning, AI computing, AI model execution, and AI model inference.

Security and Resilience. Security refers to the preservation of confidentiality, integrity, and availability of information (e.g., user data and signaling) and protection of networks, devices, and systems against cyberattacks (e.g., hacking, distributed denial of service, and man-in-the-middle attacks) (ITU 2023). Furthermore, resilience refers to the capabilities of the networks and systems to continue operating correctly during and after a natural or man-made disturbance (e.g., the loss of a primary source of power).

Sustainability. Environmental sustainability is the network and devices' ability to minimize greenhouse gas emissions and other environmental impacts throughout their life cycle (ITU 2023). The factors that can help achieve environmental sustainability include increased energy efficiency, minimum energy consumption, and an increased period of the use of resources (e.g., optimizing for equipment longevity, repair, reuse, and recycling). Energy efficiency can be used to quantify sustainability and refers to the number of information bits transmitted or received per unit of energy consumption in bits/Joule. Improvement in energy efficiency with the capacity increase helps minimize overall power consumption.

Interoperability. It is the radio interface based on member inclusivity and transparency that enables functionality between different entities of the system.

10.3.4 6G Technology Enablers: A Concise Overview

Like 5G, the 3GPP is expected to play a leading role in developing 6G specifications. While the technologies that will be part of 6G will be known only when the 3GPP develops the 6G specifications, candidate 6G technology enablers identified by the NGA are mentioned in Figure 10.6 (NGA 2022a, 2022b). Details of these technology enablers can be found in NGA (2022a, 2022b, 2023).

The NGA has identified five major categories of technology enablers for 6G. These include component technologies, network O&M and service enablement technologies, radio technologies, system and network architectures, and trustworthiness technologies. Below is a brief description of these technology enablers.

Component Technologies. Semiconductor technologies, RF subsystems for diverse frequencies, devices, modems, and holographic displays are examples of component technologies that would be needed in 6G.

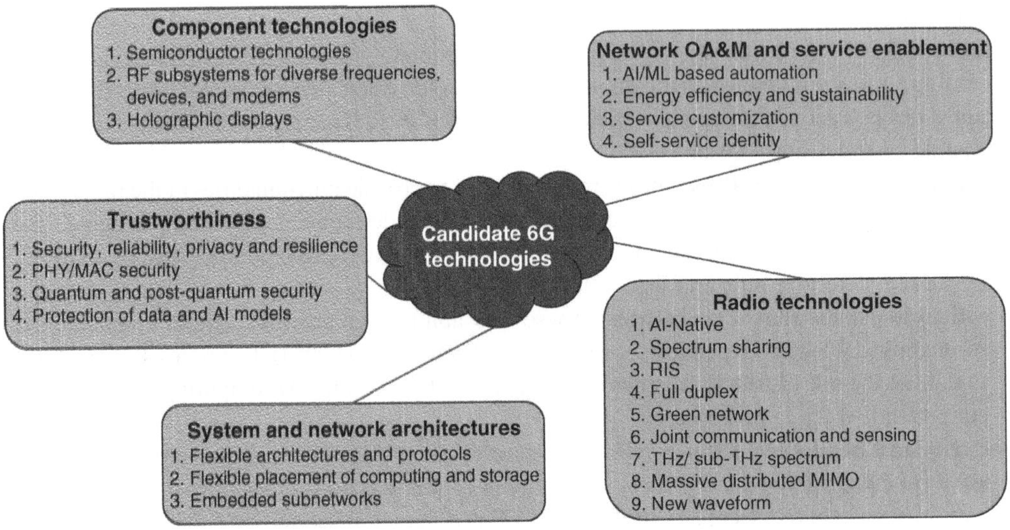

Figure 10.6 Potential 6G Technology Enablers.

Radio Technologies. On the radio interface, 6G may use technologies such as AI, spectrum sharing, Reflective/Reconfigurable Intelligence Surface (RIS), full duplex, green network, joint communication and sensing, THz/sub-THz spectrum, massively distributed MIMO, and a new waveform.

System and Network Architectures. 6G systems may be characterized by flexible architectures and protocols, flexible placement of computing and storage, and embedded subnetworks.

Network Operations, Administration, and Maintenance (OA&M) and Service Enablement. OAM and service enablement may utilize AI/ML-based automation, energy efficiency and sustainability, service customization, and self-service identity.

Trustworthiness. Technologies that address various aspects of trustworthiness, such as security, reliability, privacy, and resilience, PHY/MAC security on the radio interface, quantum and post-quantum security, and data and AI model protection, would be important in 6G.

Sections 10.3.4.1–10.3.4.5 take a closer look at these 6G technologies. See NGA (2022b) for more details about these 6G technologies.

10.3.4.1 Component Technologies

6G will need enhancements in component technologies such as semiconductors, circuits and subsystems, antennas, and packaging and displays to realize its full potential in meeting service requirements of target use cases (NGA 2022a).

Semiconductors A semiconductor technology needs to operate at least 3–5 times faster than the carrier frequency to realize a radio with acceptable range, power dissipation, and link margin characteristics (NGA 2022a). For a system operating in the 100–300 GHz spectrum, the semiconductor technology needs to operate with a speed of 0.5–1.5 THz. Candidate technologies for these speeds include silicon and III-V semiconductors with SiGe and InP aiming to provide speeds greater than 1 THz. Transceiver design depends on semiconductor technologies that are selected based on the Figures of Merit (FOMs) such as the carrier frequency, the peak data rate, and energy efficiency and factors such as the form factor, cost, reuse of intellectual property, and product reliability. Example components of the transmitter portion of a transceiver include inverse fast Fourier transform (IFFT), digital-to-analog converter (DAC), low pass filter (LPF), mixer, and power amplifier (PA), and example components of the receiver portion of a transceiver include low noise amplifier (LNA), mixer, LPF, variable gain amplifier (VGA), analog-to-digital converter (ADC), and fast Fourier transform (FFT). For a time division duplex (TDD)-based transceiver, an RF switch is also needed to switch the antenna connectivity between the transmitter and the receiver. In a transceiver employing an active antenna array (which is common for millimeter wave systems), a large-power PA is replaced by multiple low-power Pas connected to an antenna element or a group of antenna elements. The transmitter portion of the transceiver is expected to be more challenging than the receiver portion of the transceiver from the perspective of semiconductor technologies.

As the transceiver consists of different processing blocks, which have different requirements, different semiconductor technologies are required in a given transceiver (NGA 2025a). For example, digital baseband in 6G that includes IFFT/FFT and ADC/DAC may be realized in high-performance and highly scaled Complementary Metal-Oxide-Semiconductor (CMOS) technology. The analog baseband that includes the mixer may be realized using advanced RF-compatible Si technology. The RF Front End that includes power amplifiers needs to generate an adequate amount of transmit power with high efficiency even at mmW, sub-THz, and THz RF frequencies.

Hence, III-V materials complement or replace silicon-based technologies based on the frequency band, such as FR3 (e.g., 7–24 GHz), FR2 mmW band, and D-Band (e.g., 110–170 GHz). Thermal management is expected to become even more important in 6G at relatively high frequencies due to dense stacking of amplifiers in arrays.

In the receiver portion of the transceiver, LNAs are typically implemented in Silicon-Germanium (SiGe) and Silicon-on-Insulator (SOI) technologies, because they provide both low noise figure (NF) and a high gain. Bulk CMOS LNA is considered to be a low-cost LNA solution. At higher mmW and D-band frequencies, LNA designs may balance NF and gain or prioritize gain over NF. LNAs using GaAs metamorphic High-Electron-Mobility-Transistors (mHEMT) and InP Heterojunction Bipolar Transistor (HBT) technologies provide a low NF and a high gain. The SiGe LNAs provide good NF-gain balance. The CMOS LNA is a low-cost and fully integrated solution.

In support of the AI-based air interface in AI-native 6G, the AI-based digital baseband would need to include new hardware, often referred to as an AI accelerator, that can perform large matrix–vector and matrix–matrix multiplications by employing passive parallelism and large cache memory (NGA 2025a).

To realize efficient transceivers, especially at high mmW and D-band frequencies, new methods for dense heterogeneous integration such as 3D integration and 2.5D integration may be explored in 6G (NGA 2025a). 3D integration stacks multiple layers of semiconductor materials to provide benefits such as a compact footprint and large bandwidths. The 3D integration is realized using wafer-to-wafer or die-to-wafer bonding in combination with Through-Silicon-Via-based interconnections. Thermal dissipation is expected to be a major challenge faced by 3D integration. In the case of 2.5D integration on a Si-interposer, various dies are connected through a Si substrate processed in a CMOS fab. Low-loss dielectrics can be exploited to minimize substrate losses while supporting high interconnection density. The addition of passives on the interposer increases the overall functionality.

Circuits and Subsystems Important considerations related to RF circuits and subsystems in 6G include ADCs, LNAs, PAs, carrier frequency, local oscillator (LO) frequency generation, link budgets, high-order modulation (HOM), beamforming, integrated sensing and communication (ISAC), non-terrestrial network (NTN), and electronic design automation (EDA) tool (NGA 2025c). Several insights discussed in NGA (2025c) are briefly summarized later.

Higher frequencies such as sub-THz and THz may incur high costs in terms of component performance, power, and capital from a hardware perspective (NGA 2025c). The hardware and power costs associated with the midband FR3 bands are expected to be moderate. As 6G is expected to be AI native, it is important to evaluate the tradeoffs and design challenges of incorporating AI in RF circuits and subsystems.

In a typical 5G system, frequency range 1 (FR1) utilizes a traditional architecture of high-power PAs, acoustic filters, and uniformly spaced antenna elements, while frequency range 2 (FR2) typically utilizes phased arrays with analog phase shifting and regularly spaced antenna elements (NGA 2025c). As some of the midband frequency range 3 (FR3) important for 6G is occupied by satellite systems, suitable filtering and transceiver and power amplifier designs would be needed to minimize interference between 6G and satellite systems. The lower portion of the FR3 band could reuse the FR1 approach for the circuits and systems. The upper portion of the FR3 band may use an FR1 approach or an FR2 approach.

In sub-THz and THz bands, the number of beams could be in thousands, requiring a large look-up table (LUT) memory (NGA 2025c). Beam management is also expected to be challenging due to very narrow beams, user mobility, shorter channel coherence time, and the limited data rates supported by control interfaces in the transceiver circuitry, like serial peripheral interface (SPI). If beam weights are not retrieved from an LUT but are computed in real time, faster interfaces like USB and Ethernet can be more effective than the SPI in transmitting the optimal weights from the processing node to the beamformer (NGA 2025c). The processing node may be a traditional computing platform, such as a central processing unit (CPU) or a field programmable gate array (FPGA) or an emerging computing platform such as a graphics processing unit (GPU), tensor processing unit (TPU), or compute in memory (CIM) module.

To overcome the LO frequency generation challenge at mmW, sub-THz, and THz frequencies, 6G may utilize several frequency multipliers such as frequency doublers and triplers to achieve multiplying ratio of 2, 3, 4, 6, 9, and so on (NGA 2025c). A tripler may be mixer based or harmonic based. A mixer-based tripler utilizes the second harmonic at the source of a differential pair and mixes it with the fundamental frequency to generate the third harmonic. A harmonic-based tripler may generate the third harmonic through proper biasing or by exploiting a short circuit at the second harmonic frequency. Drain-harmonic shaping techniques may help achieve wider bandwidth and high power.

Link budgets need to be derived for new 6G frequency bands. To enhance the uplink link budget, the receive antenna gain can be quite helpful although the devices effective isotropic radiated power (EIRP) cannot be increased much (NGA 2025c). The downlink link budget benefits from the larger EIRP of the gNB. If a suitable antenna array with high receive beamforming gain is available at the device, the downlink link budget can be increased further.

At sub-THz and THz frequencies, due to large channel bandwidths, the noise contribution to the error vector magnitude (EVM) increases from the signal path and the ADC (NGA 2025c). Additionally, the Integrated Phase Noise (IPN) from the frequency synthesizers used for LO frequency generation is also large. The IPN increases as the frequency and the signal bandwidth increase.

At the midband FR3 band, large antenna arrays are not feasible in a UE's limited physical area due to the relatively large wavelengths and the corresponding larger antenna array sizes. Hence, the potential for analog RF beamforming is limited, and the output power is supported by one or a limited number of PAs (NGA 2025c).

In support of ISAC, the sensing transmitter and receiver operate at the same frequency, causing in-band leakage, where the transmit signal causes interference to the receiver. The ISAC receiver requires a wide dynamic range to avoid being blocked by the leakage. The ISAC transmitter needs to minimize the noise in the transmit signal. When the device implements the ISAC transceiver, the in-band leakage worsens due to limited space in the small form factor device. Enhancements in system, circuit, and algorithm designs become important to support ISAC in a 6G transceiver (NGA 2025c).

The NTN is expected to be quite important in 6G to meet the ubiquitous connectivity requirement. Low-band NTN deployments cannot support high data rates in a handheld device due to limited channel bandwidths in low-frequency bands. High-band NTN deployments can support higher data rates in a very small aperture terminal (VSAT) device due to high antenna gain. Support for handheld NTN devices at high frequencies is challenging. For example, 6G needs to accommodate a bigger antenna array in an NTN device. High-power and high-efficiency transmitters and very low-noise receivers are also needed in 6G (NGA 2025c).

Electronic Design Automation (EDA) is expected to play a key role in 6G (NGA 2025c). The EDA tool may exploit a neural network to generate better models for transistors (e.g., GaN and other III-V devices) to reflect memory and load pull effects. The EDA tool can also help with automated layout generation, such as classical structures for passive elements to realize a required frequency response. The EDA tool can help design signal conditioning via digital pre/post-distortion (DPD) techniques and interference cancellation. The EDA tool facilitates built-in-self-test (BIST) and other signal path optimizations. For example, a load modulated amplifier (e.g., Doherty or load modulated balanced amplifier) can be optimized by considering signal statistics, frequency bands and bandwidths, memory effects, bias conditions, power levels, phase control, and variable parameter components. Joint electrical-thermal simulations can be performed by the EDA tool to address issues such as low-power efficiencies and high-power consumption. Furthermore, the EDA tool can process numerous variables such as antenna elements, number of beams, modulation scheme, duty cycle, frequency, and bandwidth selection to carry out joint optimization at the phased array level.

The ADC/DAC is one of the important components in the transceiver. Key parameters that characterize an ADC's performance are sampling rate and effective number of bits (ENOB) representing the sampled analog signal. The ADC typically adheres to the Nyquist sampling requirement and samples the signals at a frequency greater than twice the signal bandwidth. Hence, operating at sub-THz and THz frequencies requires a very high ADC sampling rate due to large signal bandwidths available at these frequencies. Furthermore, ADC power dissipation increases almost linearly as a function of frequency and ENOB. Additionally, the cost of ADCs increases significantly as the sampling rate and ENOB increase.

The LNA in the receiver amplifies the received signal without adding much noise. In general, an SNR degrades by 3 dB when a frequency increases by a factor of 20 (NGA 2025c). Hence, a significant SNR degradation is expected for sub-THz and THz frequencies.

The power amplifier efficiency degrades by 6% when the frequency doubles (NGA 2025c). Hence, PAs at sub-THz and THz frequencies are expected to operate inefficiently. Furthermore, the high peak-to-average power ratio (PAPR) for OFDM systems influences the PA's saturation point, creating challenges in supporting a high-order modulation scheme. PAPR reduction techniques will become important in 6G.

Antennas and Packaging To meet the use-case requirements, 6G needs to package components such as integrated circuits, passive elements (e.g., filters, inductors, capacitors, and matching and tuning/switching elements), and antennas into subassemblies and modules in a limited space (NGA 2022a, 2025b). Hence, new packaging, redistribution, and interconnect technology would need to be evaluated to support 6G transceivers. Antenna design, antenna packaging, and antenna testing-related insights discussed in NGA (2022a, 2025b) are briefly summarized later.

A 6G smartphone would likely support one or more new 6G frequency bands, and some frequency bands supported in pre-6G technologies. 6G may utilize an antenna-on-display to address the limited real estate in a handheld device. The antenna and the radio front-end components may be codesigned. For example, heterogeneous integration with an interposer may be utilized to derive benefits from silicon, III-V, and emerging technologies such as GaN-on-Sic and GaN-on-Si, especially at high frequencies.

The need to overcome large path losses in sub-THz and THz frequency bands in 6G leads to the beam explosion problem, where thousands to tens of thousands of beams need to be managed (e.g., using large codebooks). 6G may use innovations such as memory expansion and implementation of ML-based algorithms in CIM systems. Furthermore, thousands of elements in sub-THz and THz antenna arrays require complex routing feedlines to minimize stray currents and short feedlines to minimize power loss. Antenna systems are likely to be quite complex in 6G due to the need to consider numerous factors, including gain, efficiency, return loss, and isolation. Hence, 6G may exploit ML for optimization of antenna systems.

The use of a RIS in 6G introduces additional hardware challenges because numerous small reflective antenna elements need to be designed. Furthermore, a RIS may require special mechanical and esthetic features, including weather resistance, light transparency for installation on glass, and esthetic compatibility with the surrounding physical environment. Special RIS components such as varactors, PIN diodes, and RF Micro-Electromechanical System (MEMS) need to be low cost and low power.

As device enclosure and protective encasing technologies influence signal transmissions, enclosure materials suitable for sub-THz and THz frequencies need to be determined. Thermal challenges are also expected to be nontrivial due to the need for overall circuit and amplifier efficiency and circuit miniaturization at such high frequencies. Note that antenna array sizing, spacing between elements, and radio frequency integrated circuit (RFIC)-to-antenna routing are expected to cause significant loss for the transmit signal power and receive signal strength. Additionally, solutions for verification and conformance testing, production testing, and calibration are also needed in 6G.

A typical 5G smartphone may utilize eight to ten individual antennas for FR1 frequency bands and two or three antenna arrays for FR2 frequency bands (NGA 2025b). A 6G smartphone may have up to 14 individual antennas for frequency bands below 10 GHz and three or four antenna arrays for frequency bands above 10 GHz. Furthermore, 6G devices may have two to four antennas for Wi-Fi (two to four), one Global Navigation Satellite System (GNSS) antenna, and one Near-Field Communication (NFC) antenna.

As the center portion of the smartphone is occupied by one or more Printed Circuit Boards (PCBs) and a large battery, antenna integration is feasible only in the top and bottom portions of the smartphone for FR1 frequency bands (NGA 2025b). The frame or chassis is usually metallic to create a strong smartphone body. Some of the FR1 antennas are integrated into such a frame or chassis using separate electrically isolated sections. Antenna elements are fed from an advanced feeding circuit integrated on a PCB or Flexible Printed Circuits (FPC). Such feeding circuits contain tunable antenna apertures and impedance matching circuits. Isolation between two antennas operating in a given frequency range determines how much energy from one antenna is absorbed by the other antenna. An isolation target greater than 12 dB, corresponding to the energy loss of around 6%, is often utilized for smartphone antennas (NGA 2025b). Achieving the target isolation between the antennas is more challenging at low frequencies. As more antennas share the same small physical space in 6G, achieving a 12-dB isolation between the antennas is quite challenging.

At higher frequencies such as mmW frequencies, a compact "Antenna in Package (AiP)" solution is often implemented. Multiple packaged antenna modules are needed to provide full spatial coverage. Antenna on/in Display (AoD) may be used in 6G to support higher frequencies. The AoD is often implemented on a transparent film with a clearance to the metallic display panel to achieve the target bandwidth and radiation efficiency. To ensure transparency of the display, thin metal grids or transparent metals such as Indium Tin Oxide (ITO) are used. However, such materials increase the signal loss and reduce power efficiency. The AoD needs to ensure that the display's touch sensitivity does not degrade.

6G would likely require testing enhancements. For example, as there may be thousands of beams in sub-THz and THz frequency bands, characterizing the antenna beam patterns through over-the-air (OTA) testing can be quite expensive and time consuming. The required number of tests is a function of the number of beams, the number of polarizations, the number of frequency points per frequency band, and the number of frequency bands to be tested.

6G may need fast calibration and beam pattern measurement techniques. 6G may need more BIST capabilities to reduce the testing costs while achieving accuracy targets. 6G may utilize

AI-based test automation and novel OTA testing techniques. For example, AI/ML may be used to recommend the number of samples and test configurations based on knowledge of historical data.

Displays To provide an immersive user experience, 6G is expected to introduce new displays. Immersive displays can enhance engagement, realism, and connectivity through multisensory experiences. Experiences such as augmented reality (AR), virtual reality (VR), and extended reality (XR) are expected to be better in 6G compared to 5G. Display-related insights discussed in NGA (2025d) are briefly summarized later.

6G would likely exploit hardware innovations in displays to provide an immersive user experience while achieving goals such as efficiency, small size, and low power consumption. For example, the use of multi-lens arrays and double-pass pancake lenses instead of pancake lenses can enhance efficiency and reduce the display (and hence device) size. Micro-light emitting diode (LED) and micro-organic light emitting diode (OLED) displays provide benefits such as high efficiency, small form factors, and increased brightness. Waveguide-based AR displays enhance transparency, efficiency, and usability compared to traditional reflective optics. Furthermore, the use of micro-LED for AR smart glasses results in ultrabright and energy-efficient displays with high pixel density. Additionally, advanced light engine technologies reduce power consumption and enhance AR display quality.

Hardware innovations for 6G would need to expand the overall field of view (FOV) through design and manufacturing advancements. Furthermore, support for 3D immersive experiences increases data rates and latency by a factor of 10 to 100. Additionally, immersive experiences lead to a significant amount of power consumption due to display and processors such as a CPU and a GPU. VR devices such as Meta Quest 3 and Apple Vision Pro (AVP) have shown significant advancements in visual fidelity compared to previous generations of VR devices. 6G is expected to exploit such display enhancements. 6G would also need to reduce the size and increase the usability of VR devices for user comfort and enhanced immersive experience.

6G may also implement software innovations for an immersive user experience. For example, 6G may apply AI-driven compression and real-time scene adaptation to increase rendering speeds and reduce memory consumption. Foveated rendering that prioritizes rendering detail in the user's focal area while reducing peripheral detail can reduce both the processing requirements and bandwidth requirements. Furthermore, 6G may utilize new compression techniques such as Multiview High Efficiency Video Coding (MV-HEVC), MIV, and AI-based point cloud compression to optimize the delivery of immersive content. Video compression techniques such as Gaussian splatting enhance real-time rendering.

10.3.4.2 Radio Technologies

Figure 10.7 summarizes candidate radio technologies that may become part of 6G specifications.

Both 4G LTE and 5G utilize orthogonal frequency division multiplexing (OFDM) waveform on the radio interface. OFDM is still a leading candidate waveform in 6G. However, the 3GPP may evaluate other waveforms for 6G. Examples of candidate waveforms for 6G include orthogonal time–frequency space (OTFS), peaky frequency shift keying (FSK), nonorthogonal frequency division multiple access (NOMA), and low-complexity waveforms.

6G may be sharing native, where spectrum is shared using one or multiple spectrum sharing mechanisms. Examples of new frequency bands that 6G may use include mid-band spectrum between 7 and 24 GHz (which is being discussed for 5G-Advanced as well), sub-THz spectrum ranging from 100 to 300 GHz, and THz spectrum ranging from 300 GHz to 3 THz.

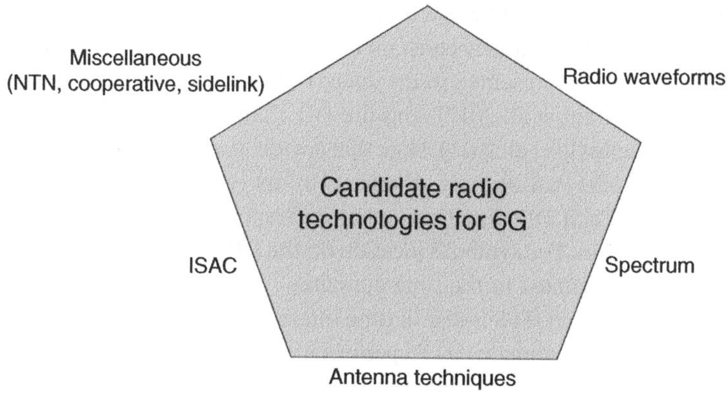

Figure 10.7 Candidate 6G Radio Technologies.

5G supports high-performance advanced antenna techniques. 6G is expected to continue making further advancements in antenna techniques. For example, 6G may explore reconfigurable/reflective intelligent surface (RIS), massively distributed multiple input multiple output (MIMO), cell-free MIMO, holographic beamforming, and orbital angular momentum (OAM).

Integrated Sensing and Communication (ISAC) is expected to be a unique technology of 6G. ISAC utilizes radio signals for both communications and sensing. ISAC facilitates support for sensing-based use cases such as weather and pollution monitoring, geofencing, gesture control, health monitoring, security monitoring, and RF optimization.

While the NTN is introduced by 5G, 6G is expected to further expand the NTN capabilities. UE cooperative communications where multiple UEs could collaborate to enable enhanced performance for a given UE may be explored by 6G.

The radio interface technologies mentioned in Figure 10. are briefly explained later.

Peaky FSK, OTFS, NOMA, and low-complexity waveforms are briefly summarized later.

Peaky FSK Communication is quite challenging when the average signal-to-noise ratio (SNR) is low. Hence, a waveform that can concentrate power in both the time domain and the frequency domain can provide a more reliable radio link compared to the case where power is distributed in the frequency domain at a given instant. A peaky FSK waveform is such a waveform.

Impulsive FSK (I-FSK) is one way of realizing a peaky FSK. I-FSK increases peak transmit SNR through frequency-concentrated transmission with a low duty cycle when the average SNR is low (NGA 2023). In challenging radio environments with high delay spread and Doppler spread, I-FSK has the potential to perform better than OFDM. Both the transmitter and the receiver have the knowledge of the time–frequency resources of the I-FSK transmission.

Wideband time frequency coding (WTFC) is a peaky FSK waveform like I-FSK. However, in WTFC, the exact transmission resources are known only to the transmitter (NGA 2023). Hence, it is possible to encode information to correspond to the specific resources used by the transmitter. In one example implementation of WTFC, if the transmitter uses one group of radio resources, it implies that the transmitted signal is carrying one piece of information. In contrast, if the transmitter uses another group of radio resources, it implies that the transmitted signal is carrying a different piece of information.

OTFS OTFS, in conjunction with equalization, converts a time-varying and fading radio channel into a time-independent radio channel with a relatively constant channel gain during the designed transmit time interval. The radio channel is represented in the delay-Doppler (DD) domain. OTFS was proposed by Hadani and others (Hadani et al. 2017), and the DD channel model of OTFS is based on the framework developed by Bello (Li et al. 2021). Note that a wide-sense stationary uncorrelated scattering (WSSUS) channel model can be fully described by its response in the time–frequency domain or the DD domain. Each DD domain symbol corresponds to a delayed and phase-rotated pulse train in the time domain. The symbol's location on the DD domain grid dictates how the pulse train is delayed and phase-rotated in the time domain.

OTFS provides a near-constant gain over an OTFS transmit time interval (TTI) or frame, because the information is spread across the entire DD domain. OTFS performs well even when the Doppler spread is high. OTFS does not need the transmitter to adapt to the dynamic radio channel conditions because the radio channel becomes static due to OTFS processing.

NOMA Orthogonality of radio resources minimizes interference, leading to high spectral efficiency. However, orthogonality cannot be achieved under all circumstances. For example, during random access, when multiple devices are attempting to send signals to the base station, signals arriving at the base station are not orthogonal due to the lack of synchronization. In another example, if the same radio resources can be utilized by different users in the same cell in the same beam, spectral efficiency can be increased if interference due to nonorthogonality can be eliminated or mitigated using interference mitigation techniques.

For example, in NOMA, multiple UEs can transmit data using the same time–frequency-space radio resources in the same beam, and the base station can implement advanced signal processing techniques such as multiuser detection to distinguish information for different UEs. NOMA is computationally complex. NOMA needs to provide a significant performance gain to justify the increased complexity of signal processing in 6G. 6G may exploit AI/ML techniques to enable NOMA to extract information for multiple devices (NGA 2023).

Low-Complexity Waveforms While some 6G use cases require a waveform with high spectral efficiency, some use cases prefer a simple waveform to reduce cost, processing power requirements, and power consumption. Hence, 6G may explore low-complexity waveforms such as Binary Phase Shift Keying (BPSK), Differential Phase Shift Keying (DPSK), and On–Off Keying (OOK) (NGA 2023).

In one example configuration identified in (NGA 2023), significant cost reduction is mentioned for a digital transceiver chain with a low-complexity waveform. It is observed in NGA (2023) that each digital transceiver chain can have its own antenna in massive MIMO deployments (NGA 2023).

Spectrum Recall that two key frequency ranges (FRs) for 5G are FR1 from 410 MHz to 7.125 GHz and FR2 from 24.250 GHz to 71.000 GHz, with FR2-1 covering 24.250 GHz to 52.600 GHz and FR2-2 covering 52.600 GHz to 71.000 GHz. Additionally, specific to the NTN, FR2-NTN includes 17.300 GHz to 20.200 GHz for the downlink and 27.500 to 30.000 GHz for the uplink.

6G may utilize these 5G FRs as well as new frequency bands. New candidate frequency spectrum for 6G includes upper midband spectrum, sub-THz spectrum, and THz spectrum.

Upper Mid-band Spectrum For 6G, the upper mid-band frequency spectrum from 7.125 to 24.25 GHz, often referred to as FR3, may be used. This frequency range is evaluated by the 3GPP for

5G-Advanced as well. FR3 is often viewed as a good balance between capacity/throughput and coverage. FR3 can support higher capacity and throughput compared to FR1, because it has more spectrum available compared to FR1. Furthermore, FR3 offers larger coverage than FR2 due to smaller propagation path losses.

Sub-THz Spectrum and THz Spectrum The candidate sub-THz spectrum for 6G includes the frequency range 100–300 GHz, and the candidate THz spectrum covers the frequency range 300 GHz to 3 THz (NGA 2024). Examples of use cases for sub-THz spectrum and THz spectrum include sensing and short-distance and high-speed communications such as video transfer.

Spectrum Sharing Access to a large amount of spectrum is important to achieve high data rates for new 6G services such as immersive video calls, holograms, and digital twins or replicas. As it is challenging to find a large amount of unoccupied spectrum, spectrum sharing is expected to play an important role in 6G. Examples of spectrum sharing methodologies include the geographic isolation method, geographic isolation method, database/server and sensing method, temporal sharing method, incumbent informing capabilities (IIC) method, multi-RAT spectrum sharing (MRSS) method, listen-before-talk (LBT) method, spatial sharing method, and propagation-based isolation method. See NGA (2024d) for a detailed look at these spectrum sharing methodologies.

The geographic isolation method defines exclusion zones where secondary services cannot use the shared spectrum. However, the primary service can use the shared spectrum in these exclusion zones.

The database/server and sensing method allow secondary users to utilize a database/server to seek permission to use the shared spectrum.

The temporal sharing method allows different services to use the given spectrum at different instants in the same geographic area.

In the IIC method, the primary service informs a suitable database or server about the period when it is using or plans to use the spectrum so that other services do not use the spectrum during those periods.

The MRSS method allows two RATs or air interfaces to share the spectrum through close coordination.

According to the LBT method, a transceiver does not transmit unless it finds the target spectrum unoccupied based on observation of the spectrum.

In a spatial sharing method, transmissions are confined to specific geographic regions.

The propagation-based isolation method exploits the isolation created by propagation path loss to enable spectrum sharing between two systems.

One or multiple spectrum sharing methodologies can be used in a given deployment. Furthermore, AI/ML algorithms can be exploited to facilitate spectrum sharing. IN one example, xApps in O-RAN can be used to manage spectrum sharing.

Antenna Techniques Candidate antenna techniques and architectures that may be considered for 6G include the reconfigurable/reflective intelligent surface (RIS), massively distributed Multiple Input Multiple Output (MIMO), cell-free MIMO, holographic beamforming, and orbital angular momentum (OAM) (NGA 2023). These techniques and architectures are briefly explained below.

Reconfigurable Intelligent Surface A reconfigurable intelligent surface (RIS), often referred to as a reflective intelligent surface, intelligent reflective or reconfigurable surface, smart surface, or large intelligent surface, is one of the potential antenna technologies for 6G (NGA 2023).

In a traditional approach, the base station considers the prevailing radio environment to determine transmission parameters of antenna techniques such as single-user MIMO and multiuser MIMO. In contrast, the RIS changes the properties of the radio environment.

In one implementation approach, the RIS can be placed between the transmitter and the receiver. Such RIS placement creates an additional propagation path between the transmitter and the receiver. If there is an obstacle between the transmitter and the receiver, the receiver usually receives a weak non-line-of-sight signal. However, due to the additional signal path created by the RIS, the receiver experiences a stronger signal compared to the case where the RIS is not used.

A RIS can be constructed using unit cells on a metasurface. In general, properties of unit cells can be modified to mimic propagation mechanisms such as reflection, refraction, diffraction, focusing, refraction, collimation, and modulation. A RIS controller can be used to modify properties of the RIS to influence the operation of the RIS. For example, the RIS controller can modify RIS design parameters so that the RIS reflects the incoming or incident signal in a target direction.

Massively Distributed MIMO In a typical deployment, the base station utilizes three cells or sectors with 120° azimuth coverage per sector, with one antenna structure per sector. With the introduction of the concept of transmission/reception points (TRPs), there is more flexibility to place antenna structures in a sector. 6G could broaden the scope of TRPs via massively distributed MIMO, where numerous antenna structures are widely distributed in a given sector or cell.

Massively distributed MIMO can significantly enhance coverage by reducing the probability of NLOS propagation paths between the base station and the UE. Furthermore, the placement of MIMO antenna structures closer to UEs implies increased SINR and increased potential for the use of high-order MIMO spatial multiplexing. 6G would need to design a solution to support backhaul between the base station's baseband processor and distributed MIMO antenna structures.

Cell-Free MIMO In a typical deployment scenario, a UE with a dedicated radio connection has a dedicated signaling connection with a base station in a specific cell. Furthermore, cells transmit cell-specific signals. Additionally, handover between cells is managed through cell-specific and device-specific signaling. In a multi-TRP scenario, a UE communicates with multiple TRPs, where TRPs belong to one or more cells.

Cell-free MIMO, often referred to as user-centric MIMO, creates a user-specific cluster of TRPs for the user device, where the UE exchanges traffic with a set of TRPs (NGA 2023). A central processing unit (CPU) connects to multiple access points (APs), which are equivalent to TRPs. The set of TRPs communicating with the UE is updated as the UE moves from one place to another. As the user moves from one location to another, the cluster of TRPs is dynamically modified. To make cell-free MIMO a reality, 6G would need to address challenges such as clustering, mobility, propagation delays, coordination among TRPs, and RF planning and design methodologies (NGA 2023).

Holographic Beamforming Holographic beamforming (HBF) creates beams by applying principles of optical holography. In one possible implementation of HBF (Pivotal Commware 2019), an RF port, located in the center of the back portion of the antenna, connects to an RF distribution network. An RF wave, also known as a reference wave, traverses from the central feed point toward the edges along the RF distribution network. The HBF antenna, through suitable tuning of its elements, maps the reference wave into the object wave that is associated with the target beam pattern. This mapping is created through a hologram. The hologram represents the impedance pattern of the HBF antenna (Pivotal Commware 2019).

The required overall impedance pattern of the HBF antenna is determined using the reference wave and the desired object wave associated with the target beam pattern. Note that a beam can be formed in the target direction by choosing a suitable hologram (i.e., an appropriate impedance pattern). An HBF antenna can be visualized to be a leaky antenna, where the RF energy is leaked into free space for radiation. Potential benefits of HBF relative to active antenna arrays are lower weight, lower cost, thinner design, and reduced power consumption (NGA 2023).

Orbital Angular Momentum Orbital angular momentum (OAM) can be exploited to design a new type of antenna system. OAM describes the helical phase pattern of a wavefront (NGA 2023). An OAM mode is an important parameter that specifies the amount of phase front twisting. As different OAM modes are spatially orthogonal, they can carry information in parallel to increase throughput and spectral efficiency. These multiplexing gains can be achieved when transmitter-receiver separation is short, there is line-of-sight between the transmitter and the receiver, and the transmit antenna array and the receive antenna array are aligned. Due to such constraints, OAM is suitable for fronthaul and backhaul transmissions, which utilize static point-to-point transmissions.

In one OAM implementation approach, a uniform circular array (UCA) is used at the transmitter and the receiver (NGA 2023). Multiple omni or directional antennas are uniformly placed on a circle at equal angles at both the transmit antenna array and the receive antenna array. Different OAM modes are created at the transmitter by applying different counterclockwise rotations on the transmit signal. A reciprocal clockwise rotation to the received signal is applied at the receive antenna array for different OAM modes (NGA 2023).

ISAC Integrated Sensing and Communications (ISAC) or Joint Communications and Sensing (JCAS) is one of the most widely discussed candidate 6G technologies. A typical 4G or 5G cellular communication system focuses on wireless communications for UEs. Sensing is typically carried out by a radar system. Sensing involves the detection of an object and its characteristics (e.g., position, speed, and shape) or characterization of the environment (e.g., rain and pollution). The 3GPP has carried out a study on ISAC, where more than 30 use cases have been identified (3GPP 2024c). Examples of ISAC use cases include intrusion detection, traffic safety, UAV detection and tracking, health monitoring, and environmental monitoring.

Availability of large channel bandwidths in millimeter wave and higher spectrum bands creates opportunities for high-resolution sensing in 5G and 6G. ISAC supports both communications and sensing using a given radio channel. In other words, separate communication and sensing systems are not needed; one system performs both sensing and communication functions. As one system is carrying out both sensing and communication functions, the spectrum is utilized more efficiently. Furthermore, sensing can be used to enhance the performance of communications. For example, sensing can be used to manage beams, allocate resources, and enhance physical layer security.

There are several types of sensing topologies. In monostatic sensing, the same entity performs transmission of sensing signals and reception of sensing signals (e.g., the signal reflected from an object). In bi-static or multi-static sensing, one entity performs transmission of sensing signals, and a different entity performs reception of sensing signals. In the context of a 5G or 6G system, the entity participating in sensing is typically a base station or a UE. A special sensor could also be utilized in an implementation-specific manner to facilitate sensing.

6G needs to address several challenges to make ISAC a reality. For example, 6G would need to have adequate flexibility in the system design to achieve the target tradeoff between sensing performance and communications performance. The time scale for sensing can be significantly smaller than the communication time scale. For example, the sensing signal reflected from an object may need to be detected within microseconds of the sensing signal transmission. Radio resources would need to be allocated to sensing needs and communications needs. 6G would also need to address the challenge of NLOS propagation, because sensing typically assumes LOS propagation. Channel modeling is needed to facilitate the system design for various sensing topologies and scenarios (e.g., frequency bands, channel bandwidths, clutter clusters, and radar cross-sections [RCSs]). 6G would need to work on various ISAC aspects such as ISAC waveform, design, parametrization, frame structure, and time–frequency–space resource allocation to meet diverse communication and sensing performance requirements.

Miscellaneous Technologies The NTN is expected to be one of the essential components of 6G (NGA 2024b). Indeed, the NTN will enable 6G to support the ITU's IMT-20230 usage scenario of "Ubiquitous Connectivity" and the overarching aspect of "Connecting the Unconnected" shown in Figure 10.4. See Section 10.3.5 on the NTN's role in 6G.

A sidelink is the link between two UEs. The sidelink allows direct transfer of data between two UEs such that the user traffic does not need to traverse through the radio and core networks. While the sidelink has been specified in LTE and 5G specifications, 6G could make significant enhancements to the sidelink (NGA2024b). 6G could explore supporting a new physical layer design for low-complexity and low-power devices. 6G may define enhanced discovery mechanisms that facilitate the discovery of neighbors for mesh or multi-hop connections. 6G may also enhance mobility support for mesh networking scenarios. An efficient and simple routing protocol for mesh networking may be needed in 6G. 6G may also consider load balancing in the mesh network. As UAVs may communicate via the sidelink, precise routing in an urban environment may require routing protocol enhancements in 6G. Furthermore, in 6G, the sidelink may be integrated with other network architectures such as IAB, smart repeaters, the NTN, Wi-Fi, and Bluetooth. 6G may enhance the synchronization procedure for multi-hop communications. 6G may also support scheduling enhancements in the sidelink to increase efficiency and reduce interference. Bem management enhancements may be defined in 6G to optimize sidelink performance. Cross-band sidelink may also be considered in 6G, where a relay UE may use a low frequency band to connect to a base station and a high frequency band to connect to a remote UE.

In UE cooperative communications, multiple nearby UEs cooperate with an active UE so that the active UE achieves better performance compared to the case when such cooperation is absent (NGA 2024b). The active UE and the cooperating helper UEs together form a virtual UE, and the virtual UE communicates with the base station. Physical resources of all the physical UEs can be leveraged by the virtual UE to communicate with the base station. UE cooperative communications aim to enhance one or more target KPIs and can be exploited in several use cases. For example, one use case may aim to save power by offloading paging and cell search procedures. Another use case is data sharing, where multiple UEs receive the same data or transmit the same data to the base station for enhanced reliability or resilience. The coverage use case could use different radio access technologies for different UEs. Cooperating UEs could form a subnetwork in another use case. Aggregation of radio resources can help increase throughput. Computing can also be distributed among cooperating UEs.

10.3.4.3 System and Network Architectures

While 5G wireless communication has begun influencing other industries, 6G is expected to accelerate and broaden such influence. Key aspects of 6G system and network architectures include network topologies, network adaptability, distributed cloud and computing, AI in network and devices, and protocol stack enhancements. These aspects are discussed in NGA (2022a, 2022b) and briefly summarized later.

Network Topologies 6G may utilize diverse and integrated network topologies to support diverse use cases and deployment scenarios. For example, 6G may integrate an NTN with satellites, an aerial network with drones and aircraft, and a terrestrial network. The integrated network would need to consider factors such as long distances, high speeds, and different spectrum regimes. While 6G will continue using the traditional hub-and-spoke cellular communication architecture, diversification started in areas such as sidelink, relays, and integrated access and backhaul (IAB) is expected to expand (NGA 2022a). For example, sidelink and mesh topologies may be expanded.

While the NTN is introduced by the 3GPP in 5G, it is expected to play a crucial role in 6G by providing reliable, broadband, and cost-effective coverage in remote and rural areas and for in-flight communications. Furthermore, a 6G network operator could exploit sidelink communications and ad-hoc and dynamic mesh networks to cost-effectively deploy new dense networks to support multisensory XR, V2X, industrial automation, swarm communications, and local source and sink applications (NGA 2022b). Such deployments could yield additional benefits such as increased network resilience, load balancing, and scalable deployment.

Network Adaptability While 5G is designed to be flexible, 6G is expected to be even more flexible due to the expected applications to multiple industry verticals. Increased **network disaggregation** and **virtualization** may be exploited in 6G to facilitate the mix and match of network components from different vendors. 6G may implement embedded **subnetwork connectivity** so that service continuity and discovery of new services are feasible even during the mobility of devices across subnetworks. To expand the ecosystem to various industry verticals, 6G may support new **open network frameworks** and application programming interfaces (APIs). Key aspects of network adaptability—disaggregation, subnetworks, enhanced application framework—are briefly explained later.

Disaggregation Disaggregation may be realized using various dimensions (NGA 2022b): (i) Disaggregation of open hardware and software platforms, (ii) disaggregation through the separation of control, management, and user/data plane functions, (iii) disaggregation of the RAN into components such as a radio unit, a distributed unit, and a central unit. The standard organizations, such as the 3GPP and the O-RAN ALLIANCE, have worked on disaggregation in 5G. Disaggregation and distributed cloudification would likely require optimal placement of network functions (NFs). Some NFs can be placed at the edge, while other NFs can be placed at a regional or centralized location based on the use case. Furthermore, different use cases may benefit from various levels of disaggregation.

5G supports 1:1 connectivity between the RAN and the core network for a given device. For example, a gNB connects to one AMF for one non-access stratum (NAS) connection for a UE. 6G may support 1:N connectivity between the RAN and the core network to support multiple distributed NAS connections. Different NAS security contexts for a UE can then be defined in 6G to provide additional resilience in case one NAS connection is compromised.

Subnetworks A subnetwork is a network consisting of several devices and one or more access points, which is connected to an overlay network such as a 6G system, 5G system, EPS, and WLAN (NGA 2022b). The access point includes the gateway functionality. Note that the subnetwork can operate autonomously even if no overlay network is available. The access point serves as a 6G network while communicating with a 6G UE and as a 6G UE while communicating with the overlay network. Key characteristics of a subnetwork include (NGA 2022b) (i) support for extreme communication requirements such as a latency less than 100 μs and the service availability 99.999%, (ii) low transmit power, (iii) star or tree topology, and (iii) limited mobility between subnetworks.

A subnetwork can be viewed as a wireless network that replaces a legacy wired network, yielding benefits such as lower cost, increased deployment flexibility, and the need to maintain cables. For example, a subnetwork can be applied to avionics to replace a wired network on an aircraft, avoiding cables to connect sensors, controllers, and actuators. An in-vehicle subnetwork can replace the controller area network (CAN) bus and automotive Ethernet, reducing the overall weight and thereby reducing fuel consumption. An in-body subnetwork can be implemented on the surface such as a wristband to help maintain a suitable glucose level in diabetic patients or in an implant such as a brain implant to enable movements in patients with motor disabilities (NGA 2022b).

Enhanced Application Framework 6G would need to identify service requirements for new 6G services and determine how to expose the requirements to the 6G network. An extensive open and standardized service framework facilitating suitable interactions of these new services with the capabilities of the 6G CN and 6G RAN (e.g., dynamic QoS indicators and user policy changes) would be needed in 6G to ensure that the required QoE is met (NGA 2022b).

6G may need to support following types of services (NGA 2022b): (i) services to enable applications to configure, train, deploy, and manage AI/ML models in the 6G system; (ii) services to automate and assist applications and users in establishing end-to-end trust, security, and data privacy with each other; (iii) services that provide near real-time secure customization and personalization of applications, devices, and networks for individual users based on the user context; (iv) services that enable applications and users to incorporate and/or use lower-level services supported in the 6G network such as integration and use of network-based AI/ML, computing/data services in the distributed cloud platform, and/or network-based data storage services; and (v) services that provide just-in-time scheduling and deployment of application components across the 6G distributed cloud and communications fabric.

Distributed Cloud and Computing Some use cases have very stringent end-to-end latency requirements, necessitating low communication latency and placement of computing near the edge. A distributed cloud framework distributes computing and data resources or services in devices, local network nodes, and local data centers. Close coordination among communication, computing, and data enables a distributed cloud in a wide-area system. For example, 6G may make computing plane and data plane resources and services integral components of a 6G system in addition to traditional communication plane services and resources (NGA 2022a).

Indeed, "Distributed Cloud and Communications System" is one of the six audacious goals identified by the NGA (NGA 2022b). A node in the 6G system may have computing, data, and communication capabilities. 6G may utilize both in-node computation and distributed computation to optimize resource usage and power consumption while meeting the target task's requirements. Unique features of a 6G distributed cloud and communications system relative to 5G edge computing include (i) built-in computing and data services, (ii) the ability to enable a large-scale distributed cloud in a heterogeneous and ubiquitous computing environment, and (iii) the incorporation of device compute.

6G transforms a mobile communication system from a communication-centric system to a communication-computing-data-converged system. Ubiquitous computing enables new use cases and applications, including immersive XR and sensing applications. It also helps address compute limitations of certain types of user devices such as wearables. For example, compute-intensive ML training/inference can be offloaded in a distributed or collaborative fashion in a 6G system.

AI in Networks and Devices 6G is expected to be AI native. AI may be exploited in both the device and the network. Examples of AI applications in 6G networks and devices include AI-based PHY/MAC, AI-assisted mobility, AI-optimized resource allocation, AI for orchestration, and AI for security. 6G also needs to consider explainability, predictability, and freedom from bias while applying AI. 6G may exploit ambient intelligence (AmI), where various sensing technologies are used to sense the environment, and such sensing is combined with AI. AmI enables the development of embedded, context-aware, personalized, adaptive, and anticipatory user interfaces (NGA 2022a).

AmI enables the design and implementation of cyber-physical systems such as service robots with multisensory applications (NGA 2022b). AmI considers complex input from a given environment for decision-making and actuation. AmI exploits three main technologies (NGA 2022b): (i) Seamless and pervasive computing that integrates microprocessors/cloud servers into objects such as furniture, clothing, vehicles, and roads; (ii) seamless and ubiquitous communication that enables such objects to communicate with each other, users, and cloud servers; and (iii) an intelligent user interface that enables human users or machine users of the AmI to control and interact with the environmental factors in a customized manner.

AI/ML is expected to be a key component of 6G. 6G is expected to be more complex than 5G because of its expected role in serving multiple industry verticals and addressing numerous use cases. Increasing complexity in 6G necessitates novel approaches to optimize RAN and CN functions. AI/ML can facilitate network planning and optimization of complex 6G networks in areas such as coverage optimization, handover robustness, and traffic forecast (NGA 2022b). AI/ML can enhance network operations by adapting operational parameters of the radio protocol stack to prevailing traffic characteristics. AI/ML can also help with predictive maintenance, failure diagnostics, and forecasting of security attacks. Note that AI/ML can be integrated into 6G at various levels. For example, AI/ML can (i) replace an individual function or module, (ii) optimize two or more functions jointly, and (iii) create an end-to-end AI-native system. It is important to evaluate the performance of an AI/ML function using a set of target KPIs. Continuous learning can help enhance the performance of an AI/ML function. Furthermore, a fallback mechanism can be used if an AI/ML function does not perform satisfactorily.

While AI/ML can provide significant benefits in 6G, it does have some challenges that need to be addressed. For example, AI/ML needs additional computational processing. Furthermore, AI/ML increases power consumption. AI/ML may also increase signaling overhead. These costs must be weighed against functional or performance improvements. Development of explainable AI techniques is needed so that their operation such as prediction can be understood by system designers. Depending on the AI/ML function, certain interactions between the device and the network may be needed during one or more stages in the AI/ML lifecycle, such as training, inference, and performance evaluation. For example, in the case of federated learning, the network and the device may coordinate with each other to exchange model updates and to generate a new global model. For effective AI/ML usage, 6G would need to address the challenge of the availability of relevant datasets. Furthermore, 6G would need to ensure the confidentiality, security, and privacy of AI/ML data.

Protocol Stack Enhancements 6G may exploit closer coordination between applications and the communication protocol stack to enhance overall user experience (NGA 2022b). Cross-layer optimizations in the radio protocol stack can help an immersive application adapt its QoS requirements to radio conditions, thereby enhancing the user experience.

As 6G is expected to offer new extreme-throughput and real-time services such as holographic and MR communications, stricter requirements need to be supported by 6G, including stricter QoS requirements, stricter security requirements, increased processing requirements, increased memory requirements, and increased energy consumption. Protocol stack enhancements play a key role in meeting such requirements even when the radio environment is changing quickly.

To enable large-scale adoption of immersive services, the 6G protocol stack needs to be energy efficient and spectrally efficient while meeting the user experience performance goals (NGA 2022b). For example, 6G may jointly optimize upper layers of the radio protocol stack and the physical layer to leverage synergies, simplify buffer management, and reduce overheads. Security in the 6G protocol stack may be incorporated in an energy-efficient way. Awareness of packet-level QoS by the radio protocol stack and application awareness of radio conditions can be used to adapt QoS to address any user experience disruptors and seamless mobility challenges while reducing the frequency and need of interactions between the UE and the 6G network.

10.3.4.4 Network OA&M and Service Enablement

Key topics related to Network OA&M and service enablement in 6G include (NGA 2022a) (NGA 2022b) (i) service management/orchestration with AI/ML-based automation and data management, (ii) public safety, (iii) technology enablers for convergence of business services, and (iv) energy-efficient green network (NGA 2022a). These topics are described in NGA (2022a, 2022b, 2024c) and briefly summarized below.

Service Management/Orchestration with AI/ML-based Automation and Data Management While 5G adds AI/ML to the network as a bolt-on solution, 6G is expected to be AI-native and hence could transform OA&M from a manual operation to an intelligent and zero-touch operation. 6G OA&M can enable diverse services, provide a better user experience, and efficiently and cost-effectively operate the 6G system. In 6G, the management plane and control plane may need to be closely coordinated and jointly designed.

A 6G OA&M system may implement end-to-end and closed-loop system automation using suitable observations or measurements, derivation of analytics, and decision-making without human intervention. 6G may incorporate self-configuring, healing, optimizing, and protecting (self-CHOP) enhancements to achieve increased flexibility, extensibility, finer granularity, and adaptability.

In support of AI/ML-based automation, the 6G OA&M system would need to provide relevant capabilities during various stages of the AI/ML workflow, such as model training, validation, testing, deployment, and evaluation. The OA&M system may support multiple learning methods.

The 6G OA&M system is expected to exploit the ongoing trend of network virtualization and softwarization, where target functions are implemented using several software-defined components in the cloud infrastructure. The OA&M system may utilize an architectural framework that enables the 6G system to be programmable, open, and interoperable in a multi-vendor deployment.

In support of ubiquitous computing, the 6G OA&M system may include capabilities to allocate computing, networking, and storage resources to computing services. The OA&M system can orchestrate and control the required resources.

The 6G OA&M system may support user personalization, where users can customize their QoE. The OA&M system can orchestrate the services and networks to automatically adapt them to meet the configuration requirements defined by a user's preference.

The 6G OA&M system needs to perform data management such that relevant data is available where needed. Depending on the use case, the OA&M system may need to support data collection and distribution to facilitate local data processing and analytics. Furthermore, the OA&M system may need to support data sharing or data federation when real-time data movement for computing is not feasible.

The 6G OA&M system may exploit digitalization to significantly enhance the OA&M capabilities. For example, the OA&M system could utilize a digital twin to develop a real-time and holistic vision by gaining insights and executing prompt and accurate actions. The OA&M system in 6G may address various aspects of digitalization of a data service, such as data encryption; compression; transmission; and extract, transform, and load (ETL).

In support of diverse use cases and industry verticals, the 6G OA&M could enable different management capability exposure levels and permissions as a function of the application and the consumer type. Furthermore, the 6G OA&M system may provide dynamic management capability exposure to reflect changes in application layer adaptations, network conditions, and availability.

Public Safety 6G has the potential to integrate multiple technologies to provide clear situational awareness to the first responders and support staff. 6G can facilitate secure discovery of network peers, facilities, and clients, provide secure proxy services, enable secure and dependable geolocation, and assure the QoE in support of public safety. 6G can assist with rapid network recovery and minimize the impact after a disaster has occurred. As physical resources may be limited after a disaster, 6G needs to allocate these resources effectively. 6G needs to facilitate monitoring of the progress of the overall emergency response.

6G could harden the network through a suitable design to enhance services for public safety. For example, 6G may define multiple methods to restore a given network function. Network functions could potentially be recreated automatically from the cloud to restore the services seamlessly. Furthermore, the network context and data could be backed up in real time in the cloud with minimal or no manual configuration by the operator.

A 6G ecosystem for public safety in emergencies and disasters needs to consider *sustainability* and *security* (NGA 2022b). From the perspective of public safety, sustainability incorporates multiple areas beyond the green network. For example, sustainability requires high business operational efficiency and needs to validate preparedness for emergency and disaster surge demands. Sustainability needs to keep first responders and the public safe during disasters. Sustainability helps install an isolated or minimally interconnected network with limited resources and restore services.

In support of security for public safety, a 6G network providing local coverage may need to be set up securely and efficiently using limited physical resources. Authentication of users, applications, and new local infrastructure in an isolated environment needs to be carried out. Integration of an isolated network with the main network needs to be designed. 6G may also determine which secure means and tools to use to audit and assess the isolated network and its operation during the emergency. The presence of bad actors in the isolated or main network also needs to be detected.

Technology Enablers for the Convergence of Business Services 6G may define a shared foundational orchestrator with categorized components for service convergence. For example, IoT service enablement platforms provide management of connectivity, devices, and sensors, where operations such as activation, onboarding, and performance monitoring of IoT devices are carried out, and sensor data is used to derive advanced analytics for control decisions and usage analysis.

Technology enablers for converged business services would need to define (i) the form of the operational user interface (UI) such as determination of the degree of autonomy, (ii) the method of interface between services and network, and (iii) a framework for how various business services may converge to fully utilize service offerings.

In support of ubiquitous service enablement, the OA&M system in 6G may incorporate a digital composable autonomous collaboration platform, advanced tools for online privacy and safety, analytics and adaptive technologies in network management and orchestration, portable cloud infrastructure for "anywhere placement," and mobile devices that can enable access to content and activities anytime and anywhere (NGA 2022b).

Examples of technology enablers for service enablement in the 6G OA&M system are (NGA 2022b) (i) IoT sensors and biometrics for identity management, (ii) autonomous logistics for auto-sending logistics to the cloud, (iii) blockchain and distributed data for security, (iv) computer vision to interface with IT for troubleshooting and other purposes, (v) voice recognition for the user interface, (vi) robotics for QoE in using OA&M system and services, (vii) AR/VR/MR/XR for QoE in using OA&M systems and services, and (viii) dynamic network slicing for end-to-end QoS.

Energy-Efficient Green Network The 6G OA&M system may include the capabilities to operate 6G networks with higher energy efficiency. For example, 6G could exploit tools or technologies to monitor energy consumption, derive analytics, and implement controls to reduce overall energy consumption. Examples of such technologies include zero-touch network automation and AI-based energy-efficient network operation and deployment.

Sustainable 6G is expected to utilize sustainable data centers and digital infrastructure. Indeed, green data centers are expected to become quite important in 6G, because their share of network energy consumption is projected to grow by a factor of five in the coming years (NGA 2022b).

While a green data center could theoretically be operated fully on renewable energy, it is challenging to do so because of the intermittent availability of renewable energy sources. Hence, innovations are needed to make a data center green. For example, grid-interactive batteries and intelligent workload management can be exploited. Grid-interactive batteries can be charged with renewable energy sources and can then be used to protect against voltage and frequency anomalies, optimize energy usage, and manage the flow of energy. These batteries enable higher penetration of renewables and provide emergency backup power when needed (e.g., during blackouts or disasters). Intelligent workload management could defer low-priority jobs to times with lower emissions.

While traditional backup power systems often use diesel, sustainable methods such as hydrogen fuel cells and other advanced battery technologies could be used in 6G. Additionally, data center cooling could also be made more sustainable in 6G. Note that data center cooling consumes a significant amount of energy and water. Example approaches to reduce water consumption in data centers include (i) geothermal temperature regulation such as building underground data centers, (ii) geothermal heat pumps, (iii) use of river water or seawater, and (iv) closed loops with water filtering and cooling powered by renewable energies (NGA 2022b). A 6G OA&M system can help reduce energy and water consumption by considering ambient metrics and weather usage predictions and facilitating (a) the monitoring of the environment and the data center system, (b) optimization of temperature, and (c) analysis of water and energy usage, and (d) execution of intense operations such as backup during cool days.

10.3.4.5 Trustworthiness

Trustworthiness is one of the six audacious goals identified by the NGA for 6G. Details of the NGA's work on trustworthiness in 6G can be found in NGA (2022a, 2022b, 2022c). Trustworthiness includes multiple aspects such as security, reliability, resilience, and privacy.

Figure 10.8 shows eight key technology areas of trustworthiness identified by the NGA for 6G (NGA 2022b). These technology areas include (i) PHY/MAC security; (ii) quantum security; (iii) confidential computation and storage for proprietary AI models for network clouds; (iv) data provenance; (v) secure identities, data, and protocols; (vi) automated closed loop security creation; (vii) security assurance and defense; and (viii) service availability enablers.

These trustworthiness technology areas are described in detail in (NGA 2022b) and briefly summarized later.

PHY/MAC Security In the 5G NR radio protocol stack, security is implemented at the Packet Data Convergence Protocol (PDCP) layer (NGA 2022b). In 6G, the security of the radio interface can be strengthened by applying suitable techniques at PHY and MAC layers. MAC security can help enhance security through radio resource allocation through scheduling.

In 6G, PHY security can use anti-jamming and authentication measures. For example, PHY security can reduce the data interception probability at eavesdroppers using beamforming and MIMO precoding. Randomness of the radio channel can potentially be used in 6G to dynamically generate encryption keys and authenticate users. PHY security in 6G may aim to detect jammers. For example, some channels that are unprotected in 5G can be protected in 6G to counteract jammers. MIMO techniques can be exploited for the detection, localization, interference mitigation, and avoidance of jammers. 6G may also address the challenge of rogue base stations.

Quantum Security Security algorithms used in 5G are quite robust in generating effective keys such as ciphering/encryption keys and integrity protection keys (NGA 2022b). However, due to the arrival of quantum computing, those algorithms become vulnerable. Note that the security of a cryptographic method is a function of the key length and mathematical complexity of the algorithm

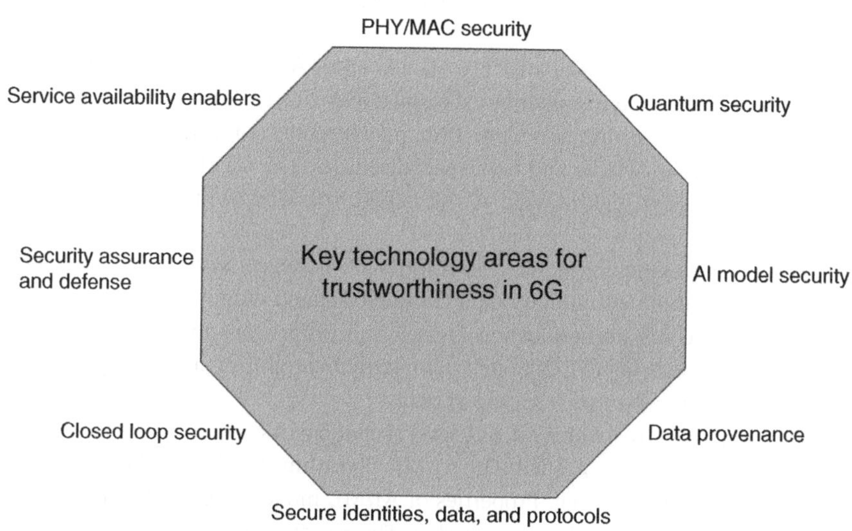

Figure 10.8 Key Technology Areas for Trustworthiness in 6G. *Source:* (NGA 2022B) NGA (2022). 6G Technologies. July 2022. https://nextgalliance.org/white_papers/6g-technologies/ (accessed 18 April 2025).

used to generate keys for encryption or integrity protection. Quantum computing significantly reduces the time required to carry out a given mathematical function. Hence, it can potentially make traditional cryptographic techniques vulnerable to attackers who possess powerful quantum computers. Note that powerful quantum computers are not yet widely available. However, in the 6G deployment timeframe, they could become commercially available. Hence, 6G would need to support Post-Quantum Cryptography (PQC) security framework. PQC is also called quantum-resistant cryptography.

The PQC security framework aims to develop cryptographic systems that are secure against both quantum and classical computers and can interoperate with existing communication protocols and networks (NGA 2022b). The National Institute of Standards and Technology (NIST) has extensively worked in the area of quantum-resistant public-key cryptographic algorithms. NIST has already released the first three finalized post-quantum encryption standards by August 2024 (NIST 2024). Internet Engineering task Force (IETF) is also addressing various aspects of PQC solutions in its working groups (IETF 2025).

Quantum cryptography enables the cryptographic key exchange securely using principles from quantum mechanics principles such as the no-cloning principle. 6G would need to identify how the PQC algorithms can complement or affect the existing key sharing and authentication methods. Hybrid key sharing techniques based on quantum and classical cryptography would likely be needed in 6G.

6G may need to address the following PQC areas (NGA 2022b): (i) identification and evaluation of post-quantum protocol candidates; (ii) implementation of post-quantum validation tools; (ii) cryptography design based on code, lattices, multivariate cryptography, and isogeny; (iii) development of hardware that can accelerate the PQC algorithm performance; and (iv) design of a reliable quantum communication network.

A mechanism called *entropy as a service* utilizes the cloud architecture to distribute random numbers through a quantum random number generator from multiple sources of entropy. Such a mechanism enables entropy-poor or constrained devices, such as IoT devices, to generate strong and unpredictable keys. *Quantum entangled networks* may become a reality in the 6G timeframe. A quantum entangled network makes use of quantum teleportation, which uses the property of entanglement from quantum mechanics to securely transmit data between endpoints.

AI Model Security To realize the vision of AI-native 6G, 6G will apply AI to many aspects of the 6G system. AI models may be implemented in a distributed cloud (NGA 2022b). Furthermore, AI models may use proprietary data or non-proprietary data. Data may be collected from various components of the system, including user devices and even special sensors. The 6G cloud infrastructure would need to support data collection, data storage, AI/ML algorithms, and AI model management and distribution.

From the security perspective, 6G may need data isolation in 6G between services and confidential compute resources. For example, proprietary data could be isolated within confidential computer hardware, where more secure authentication is used and proprietary models are utilized within trusted execution environments (TEEs). Such confidential computing avoids exploitation of data by malicious actors or opportunistic stealing of data.

To make confidential computing trustworthy in 6G, several challenges need to be addressed. For example, efficient distributed systems are needed. The AI/ML algorithms need to be able to operate with adversarial data during model training. Properties of AI/ML models need to be checked to ensure that the AI/ML models are operating as intended. The validity of data could potentially be evaluated using adversarial AI/ML techniques.

6G would likely exploit sensing technologies to enhance the network performance and support diverse use cases. Hence, AI/ML frameworks may need to get secure access to sensing data generated by sensors. 6G would then need to address issues related to data integrity, anonymization/privacy, data bias, and compliance with regulations such as General Data Protection Regulation (GDPR)) (NGA 2022b).

Data Provenance Data is valuable in wireless networks and will become even more important in 6G due to the availability and processing of diverse data in a 6G system. Data provenance[2] "means the lineage of a data record is understood, including place and time of origin, modifications made to the data, the purpose of the data, who recorded it, and for whom it was produced" (NGA 2022b). Secure data provenance in a 6G system provides confidence to a third-party service provider about the authenticity and reliability of the data that they receive.

As 6G is expected to be AI native, it is critical that AI/ML algorithms have access to correct and trustworthy data. The use of correct data by AI/ML algorithms will enable them to obtain a correct view of the state of the network and its operations. Data should be generated from the relevant data. One approach to realize a secure chain of trust is to design a 6G system with a hardware root of trust that provides integrity and to utilize mechanisms that provide data and code security over possibly untrusted components. In one approach to achieve a chain of trust, 6G may utilize blockchain or distributed ledger technology (DLT) systems to decentralize security and privacy controls.

To ensure the provenance of generated data (i.e., root of trust), 6G could explore physical measurements that can verify that the hardware systems are free of counterfeit components with vulnerabilities or poor-quality components (NGA 2022b). Such root of trust is important to address, because counterfeit or faulty components may cause issues such as data loss, data modification, disruption, failures, and loss of system availability. Secure enclaves such as a trusted platform module (TPM) and a trusted execution environment (TEE) are examples of trust technologies that can help protect data integrity and provide proof-of-data ownership tied to silicon and hardware. Furthermore, blockchain and attestation mechanisms can help track data access rights securely in the distributed data brokerage framework.

The provenance of stored data (i.e., chain of trust) is also important. 6G would need to store data securely in centralized and edge clouds. Hence, 6G needs to ensure cloud data provenance. For example, 6G needs to provide tamper-proof records, enable the transparency of data accountability in centralized and edge clouds, and provide for both privacy and availability of provenance data (NGA 2022b).

The TPM can enable data provenance in 6G by providing secure storage of keys, confidential data, certificates, cryptographic measurements of system components, cryptographic functions, and key generation (NGA 2022b). DLTs and blockchain can help develop trust-based secure 6G networks. For example, blockchain can be quite effective in addressing the challenges such as exclusive peer-to-peer transactions without centralized third parties, fraudulent replication of digital asset/value or transaction, establishing trust with pseudonymity, transparent and immutable record keeping, provenance and auditable enabled DLT, and digital signing and execution of legal agreements between parties using digital or software-enabled contracts.

Secure Identities, Data, and Protocols Secure identities, data, and protocols are essential elements of a secure and privacy-preserving 6G system.

2 Provenance refers to place or source of origin.

Secure identity management is relevant to devices and network components. 5G uses a privacy-preserving subscription concealed identifier (SUCI) instead of a subscription permanent identifier (SUPI) to identify a subscriber in clear-text signaling to increase subscriber privacy compared to 4G LTE. 6G may explore expanding such an approach for enhanced identity management to network components (e.g., network functions, virtual machines, and containers). 6G may also dynamically change paths to software components to prevent malicious actors from carrying out side-channel attacks.

Identity management in 6G may utilize secure identities for network components as well. Secure identities may be used in conjunction with microservices, cloud-native applications, virtual machines, containers, and virtualization technologies. 6G may establish authenticated and encrypted channels between the network components that constitute a 6G network function (NF). Furthermore, NFs themselves may also be connected by authenticated and encrypted channels. Suitable orchestration is needed to facilitate the processing of data and storage of keys for identifiers. Relevant hardware and software components need to be verified through a chain of trust to a trusted anchor using a secure environment (e.g., a TPM or a TEE). Messages that help verify authenticity need to be transported via reliable and trustworthy links.

Data security is also important in 6G. Specifically, secure and privacy-preserving mechanisms are needed in 6G to collect, process, analyze, and store the data for further analysis and/or decision making within or outside the 6G network.

Increased degree of network softwarization in 6G also has implications for security. For example, a variety of APIs will be used in 6G. Hence, an unintended exposure of an API of a software function becomes vulnerable to attacks. Furthermore, 6G will have centralized and decentralized NFs. In support of secure decentralization, decentralized mechanisms for identification, including decentralized identifiers (DIDs) and self-sovereign identity (SSI), are helpful. These mechanisms enable users and organizations to own digital identities without a central authority. Additionally, 6G may also introduce new secure mutual authentication between network entities and cross-domain identity management to distribute governance across various trust domains. 6G may also support mechanisms to verify, validate, and authorize the required NF instances to achieve resilience and trustworthiness.

Closed Loop Security A comprehensive security architecture implements the NIST-identified functions of identification, protection, detection, response, and recovery (NGA 2022b). Automated closed-loop security is an important aspect of resilient security architecture. Automated closed-loop security mechanisms utilize building blocks such as threat analysis, static threat mitigation, dynamic threat detection, and threat response (NGA 2022b).

Threat analysis involves the identification of static and dynamic security requirements and methods to counter external and internal threats. Different use cases may have different security requirements. As the threat surface can change due to new system features or sophistication of attacks, dynamic threat analysis also needs to be carried out in 6G.

Some threats are static, while others are dynamic. For example, organizational, implementational, or architectural threats are often static and can be addressed through static threat mitigation techniques. For example, attestation of compliance with structured design techniques and process mechanisms can mitigate static threats.

Dynamic threats could be due to the exposure of the system to malicious actors. Such dynamic threats need to be detected first. One approach to detect dynamic threats is to observe a change in system behavior or radio environment such as a change in system performance. Key performance metrics could be analyzed in real time, potentially by exploiting AI/ML algorithms, to detect any anomalies resulting from dynamic threats.

It may be feasible to automate responses to some threats. For example, programmed responses to known threats can be exploited. To develop responses, an active threat-and-response simulation could first be carried out to characterize the attitude and impact of threats and to determine conditions that affect system performance by unknown threats. In one example, electromagnetic threats could be identified and localized in an environment first, followed by the implementation of a potential solution such as instantiating redundant network nodes to alleviate the threat. Another solution such as interference nulling may also be evaluated for its effectiveness. In another example, denial-of-service attacks can be identified by detecting activity signatures that reflect the slowing of traffic flows. If and when needed, human intervention could be invoked even in an automated system.

Security Assurance and Defense 6G could utilize a security assurance framework called the network equipment security assurance scheme (NESAS). NESAS is developed by the 3GPP and the Global System for Mobile Communications Association (GSMA) and utilizes a variety of security test cases to evaluate the security of 5G equipment. While NESAS is defined for 5G, it could be extended to 6G. NESAS defines security requirements and an assessment framework for secure product development and product lifecycle processes. Security assurance is becoming even more important in 6G due to the expansion of the 6G ecosystem to critical infrastructure such as power grids, financial institutions, and governmental networks.

6G is expected to be even more complex than 5G. Hence, AI/ML could play a key role in developing automation techniques, including those for security assurance and defense. For example, AI/ML methods are exploited to analyze and verify compliance with security requirements. Explainability of AI/ML techniques is also important to gain confidence that the resulting automation can be trusted to achieve the design objectives.

6G could also make use of recommendations from the European Union Agency for Cybersecurity (ENISA). ENISA considers security aspects such as governance and risk management, security of human resources, security of systems and facilities, operations management, incident management, business continuity management, monitoring, auditing, testing, and threat awareness (NGA 2022b).

6G could benefit from enhanced security assurance and defense due to the increased availability of data-driven approaches and the AI/ML frameworks. Indeed, security assurance and defense in 6G can exploit automation across network operations and frameworks and AI/ML techniques to provide an effective defense against threats. Deployment profiles of security assurance and defense techniques depend on the factors such as the type of the use case, the type of use (e.g., public network use or for private/enterprise/vertical use), the degree of control that the service provider or the end user can exercise on the network configuration, and the time scale of the defense mechanisms (NGA 2022b).

Service Availability Enablers Service availability is closely tied to network reliability and resilience with the radio network playing a critical role in ensuring target service reliability goals. Example enablers for 5G-enhanced mobile broadband (eMBB) to meet eMBB service availability goals include network node pooling, geographical redundancy, multiple frequency bands, and multiple cells. These enablers provide an abundance of radio resources to achieve high data rates needed for eMBB services.

New uses cases in 6G would likely require new service availability goals and new mechanisms. For example, 6G could utilize multiple radio access technologies (RATs) for a given device to improve performance and/or redundancy, where multiple radio links with multiple RATs are established. Some 6G use cases such as robots facilitating automation would benefit from such redundancy.

6G would need to be flexible just like 5G to meet service availability goals of diverse use cases and industry verticals, including automation in enterprises and factories, and wide area use cases for utilities, transportation, and public safety. 6G would need flexibility in configuring radio bearers to meet the performance goals of initial use cases while continuing to evolve to address new requirements of new use cases.

To enhance service availability, 6G may exploit data-driven techniques that increase the internal observability of the network state. 6G may also use AI/ML to generate metrics for performance that are exposed to the service layer for feedback and control. Furthermore, 6G may use automation and AI/ML techniques to predict risks to service availability.

To ensure high service availability, 6G would need to address the following challenges (NGA 2022b): coverage provisioning. Peak load handling, dependability of hardware and software, avoiding single points of failure (SPoF) in equipment and deployment profile across RAN, transport, core, and cloud, performance management across all network components, and continuous improvement/continuous development (CI/CD) approaches for network performance improvement, and limitations of legacy systems.

10.3.5 Role of the NTN in 6G

Recall from Section 10.3.2 that the 6G usage scenario of ubiquitous connectivity aims to provide connectivity to bridge the digital divide and addresses uncovered or scarcely covered geographic areas, such as rural, remote, and sparsely populated areas. Note that both IoT and mobile broadband communication are covered under this usage scenario. Furthermore, one of the overarching aspects is to "connect the unconnected." However, cost-effective deployment of the base stations and the associated transport network is not feasible in areas with sparse populations and remote and vast regions on land, such as mountains, deserts, and national parks. For large regions of oceans and remote islands, deployment of a terrestrial cellular network is challenging due to inadequate land infrastructure and the lack of a business case. The NTN can play an essential role in 6G in providing ubiquitous connectivity.

While the 3GPP has already started the work on the NTN in 5G, the NTN would see a more prominent role in commercial deployments in the 6G timeframe. There are two basic approaches in which an NTN can be used. In one approach, the NTN can provide direct connectivity between a wireless device and an NTN payload. In another approach, a wireless device communicates with radio equipment on the ground (e.g., an integrated base station or a radio unit of a disaggregated base station), and such radio equipment utilizes the NTN payload to connect to the remaining portion of the on-the-ground network. In the second approach, the NTN provides intra-network backhaul connectivity. The first direct connectivity approach is expected to be more helpful in achieving ubiquitous connectivity.

The NTN in 6G can also provide another benefit. Indeed, the NGA has identified NTN as one of the key technologies that meet the requirements of 6G (NGA 2023). As mentioned in Chapter 1, the NGA has defined six audacious goals for 6G systems, and one such goal includes resilience (NGA 2022a). The NTN can help achieve resilience in 6G by providing an additional mode of communication beyond terrestrial communications. If the terrestrial network is damaged or destroyed due to a disaster or a conflict, the NTN can be used to provide vital communications.

10.4 O-RAN-Based NTN Deployments

The O-RAN ALLIANCE has defined a Radio Access Network (RAN) architecture that is characterized by openness and intelligence (Tripathi and Shah 2025). Section 10.4.1 illustrates the O-RAN architecture, and Section 10.4.2 discusses O-RAN-based NTN deployments.

10.4.1 O-RAN: A Concise Introduction

Figure 10.9 illustrates the Open-Radio Access Network (O-RAN) architecture specified by the O-RAN ALLIANCE (O-RAN ALLIANCE 2024; Tripathi and Shah 2025). For a 5G RAN, the O-RAN ALLIANCE utilizes the 3GPP-defined 5G RAN architecture as a baseline and increases the degree of disaggregation and introduces new interfaces to achieve its goal of openness. Such openness enables a service provider to mix and match components from different vendors to reduce costs and achieve vendor diversity. The O-RAN ALLIANCE also introduces the concept of RAN Intelligent Controllers (RICs). The RICs house intelligent apps to manage and optimize the RAN. A non-real-time RIC hosts remote Apps (rApps) that aim to optimize RAN operations with the time scale exceeding one second. A near-real-time RIC hosts extended Apps (xApps) that aim to optimize RAN operations with the time scale between 10 ms and 1 s.

As illustrated in Figure 10.9, key O-RAN nodes include Service Management and Orchestration (SMO) such as non-real-time RAN[1] Intelligent Controller (Non-RT RIC), near-real-time RAN

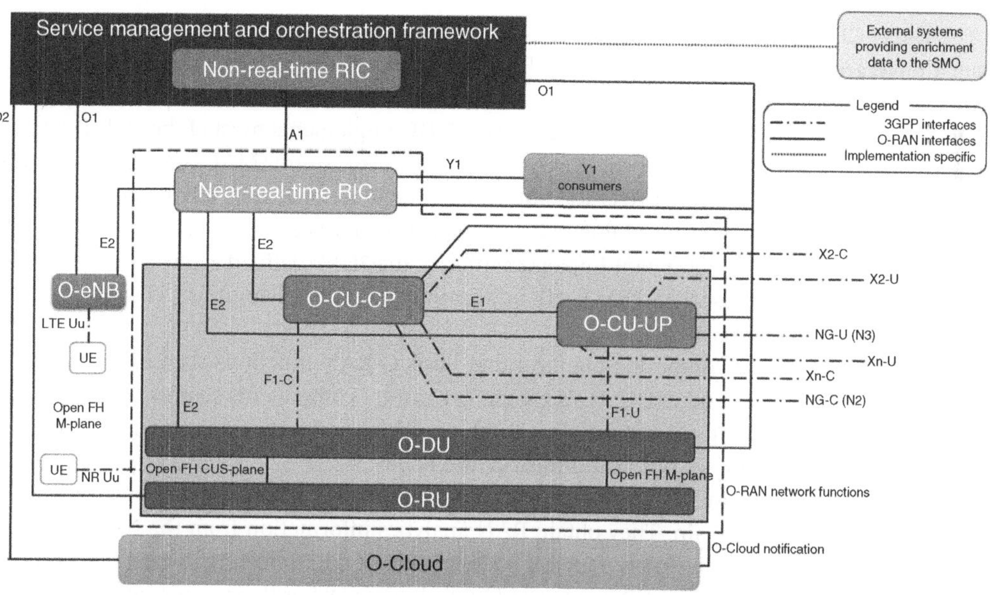

Figure 10.9 O-RAN architecture. *Source:* Reproduced from Tripathi and Shah (2025)/with permission of John Wiley & Sons.

Intelligent Controller (Near-RT RIC), O-RAN-Central Unit-Control Plane (O-CU-CP), O-RAN-Central Unit-User Plane (O-CU-UP), O-RAN Distributed Unit (O-DU), O-RAN Radio Unit (O-RU), O-RAN eNodeB (O-eNB), and O-RAN Cloud (O-Cloud). These O-RAN components are briefly summarized later (O-RAN ALLIANCE 2024; Tripathi and Shah 2025).

A 5G UE communicates with O-RAN using the 3GPP-defined 5G NR Uu interface. The O-RU, the O-DU, the O-CU-CP, and the O-CU-UP together act as a 3GPP-defined gNB and implement the 3GPP-defined NR Uu protocol stack functions while performing the O-RAN ALLIANCE-specified additional functions.

The 3GPP-defined gNB-CU-CP and gNB-CU-UP map to the O-RAN-CU-CP (O-CU-CP) and O-RAN-CU-UP (O-CU-UP), respectively, where O-CU-CP and O-CU-UP support O-RAN-defined interfaces and functions. The 3GPP-defined gNB-DU maps to O-RAN-Distributed Unit (O-DU) and O-RAN-Radio Unit (O-RU) using Functional Split 7.2x.

The SMO implements Fault, Configuration, Accounting, Performance, and Security (FCAPS) interface to O-RAN Network Functions; contains the Non-RT RIC; and performs management, orchestration, and workflow management of the O-Cloud.

The Non-RT RIC consists of the Non-RT RIC framework and rApps. The Non-RT RIC framework terminates the A1 interface and utilizes the R1 interface to expose a set of internal SMO services to rApps. The Non-RT RIC framework provides rApps with AI/ML workflow that includes model training, inference, and relevant updates. rApps enable non-real-time control and optimization of RAN and policy-based guidance to the Near-RT RIC.

The Near-RT RIC enables near-real-time control and optimization of RAN through suitable data collection and actions over the E2 interface. The Near-RT RIC may exploit AI/ML-based model training, inference, and updates. The Near-RT RIC contains xApps. xApps usually consist of one or more microservices and identify which data they consume and which data they provide at the time of onboarding. The E2 interface enables an association between the xApp and a specific RAN functionality.

The O-CU-CP implements gNB-CU-CP containing Radio Resource Control (RRC) and Packet Data Convergence Protocol-Control Plane (PDCP-C or PDCP-CP). The O-CU-UP implements gNB-CU-UP containing Service Data Adaptation Protocol (SDAP) and Packet Data Convergence Protocol-Control Plane (PDCP-U or PDCP-UP). The O-DU implements most of the gNB-DU functions and includes Radio Link Protocol (RLC), Medium Access Control (MAC), and the functional split 7.2x-based high-PHY layer. The O-RU implements the functional split 7.2x-based lower PHY layer of the gNB-DU along with suitable radio frequency (RF) processing.

The O-eNB is an O-RAN-compliant implementation of the 3GPP-defined evolved Node B (eNB) or next-generation-eNB (ng-eNB). The O-eNB utilizes the Long-Term Evolution (LTE)-based radio interface to communicate with UEs.

The O-Cloud is a cloud computing platform that hosts O-RAN functions such as the Near-RT RIC, O-CU-CP, O-CU-UP, and O-DU. It also includes software components such as Operating System, Virtual Machine Monitor, and Container Runtime. It also performs management and orchestration functions. It comprises physical infrastructure nodes.

Y1 consumers consume the Y1 services produced by the Near-RT RIC and may be within or outside of the PLMN trust domain. They obtain RAN analytics information from the Near-RT RIC.

O-RAN defines an E2 node that is a logical node terminating E2 interface toward the Near-RT RIC. E2 nodes include the O-CU-CP, the O-CU-UP, and the O-DU or their combination for 5G or LTE radio access.

10.4.2 O-RAN-Based NTNs for 5G and 6G

While O-RAN specifications currently support 4G LTE and 5G, the O-RAN ALLIANCE has already started the work on identifying the requirements to extend O-RAN to 6G (Tripathi and Shah 2025). Furthermore, the O-RAN ALLIANCE has identified the NTN as one of the important verticals where O-RAN can be utilized to yield important benefits (O-RAN ALLIANCE 2025). Important considerations for O-RAN-based NTN deployments are discussed in O-RAN ALLIANCE (2025) and (Tripathi and Reed 2024) and briefly summarized later.

10.4.2.1 Architectural/Topology Considerations

The NTN in 5G-Advanced and 6G would likely support multiple architectures such as a transparent payload architecture and a regenerative payload architecture. NTNs create collaboration opportunities between mobile network operators (MNOs) and satellite network operators (SNOs). For example, the existing network infrastructure of an MNO can be integrated with the SNO network to reduce costs, increase efficiencies, and expand the overall ecosystem. There are different possibilities to place O-RAN nodes in an O-RAN-based NTN. For example, in one candidate architecture for a transparent payload NTN, all the O-RAN nodes can be placed on the ground. In contrast, when the regenerative payload NTN architecture is used, some O-RAN entities can be placed on the NTN payload, while other O-RAN entities can be placed on the ground. The O-RAN Open Xhaul Transport Working Group focuses on defining technical requirements and integration strategies for NTN-based Xhaul transport.

10.4.2.2 Use Cases

When O-RAN network functions are integrated into an NTN, new use cases emerge. For example, use cases related to sustainability and efficiency can be envisioned, where RAN analytics can be combined with satellite data, and an RIC can use such enhanced RAN analytics to optimize beam management and handover.

10.4.2.3 Security Considerations

Integration of an NTN with O-RAN introduces unique security challenges. Examples of these challenges include data integrity attacks, unauthorized access, eavesdropping, jamming, kinetic threats, DoS attacks, and spoofing (O-RAN ALLIANCE 2025). The overall security situation becomes more complex due to frequent handovers and authentication during roaming. O-RAN WG11 has defined security requirements and controls and has considered various aspects such as access control, data protection, continuous monitoring, physical security, interference protection, regulatory compliance, and supply chain security (O-RAN ALLIANCE 2025).

10.5 NTN Research Directions

Several research areas may be explored in the NTN to enhance its scope and utility in meeting 6G requirements. Examples of these research areas include regenerative payload-related architectural enhancements, spectrum sharing, security, and multi-constellation connectivity.

10.5.1 Regenerative Payload

The 3GPP has focused on the transparent NTN payload in Release 17 and Release 18. Release 19 addresses the inclusion of a regenerative payload in the NTN architecture. A regenerative NTN payload enables the NTN to offer truly global coverage. For example, a LEO satellite X that provides middle-of-ocean coverage may not directly see an NTN-GW on the ground. In such case, the LEO satellite X with a regenerative payload can connect to a chain of other LEO satellites, and a satellite in this chain that has visibility of an NTN-GW on the ground can provide the LEO satellite X with connectivity to such NTN-GW.

The regenerative payload requires considering new challenges such as satellite connectivity and optimal routing of user-specific information on Inter-Satellite Links (ISLs). Compared to a transparent payload, a regenerative payload facilitates the implementation of higher-order spatial multiplexing because of the direct visibility of the radio environment between the wireless device and the NTN payload. However, Line-of-Sight (LOS) conditions are not conducive to high-order spatial multiplexing. Hence, research is needed to determine potential enhancements to realize high-order spatial multiplexing in an NTN. Advanced antenna configurations may be exploited to facilitate the implementation.

10.5.2 Spectrum Sharing

Spectrum availability significantly influences achievable capacity and throughput. Spectrum sharing between a TN and an NTN can make more spectrum available to both. Preliminary investigation in (Chintalapati et al. 2024) indicates that spectrum sharing between a TN and an NTN is feasible. For example, a low-power indoor TN and an NTN can share the same spectrum.

10.5.3 Enhanced TN-NTN Interworking

While 5G NTNs support mobility between a TN and an NTN, more integrated interworking can lead to enhanced system performance and user experience. For example, data may be transported simultaneously over a TN and an NTN for delay-tolerant traffic when appropriate. Load balancing between a TN and an NTN may be desirable in certain situations and configurations.

10.5.4 Security

An NTN has several unique interfaces, such as the feeder link, compared to a TN. These interfaces need to be secured. While many 5G security principles can be reused in an NTN, new enhancements may be required.

10.5.5 Multi-constellation Connectivity

An NTN UE typically transfers data via a cell of a single constellation, such as an LEO constellation or a GEO constellation. A more flexible architecture can support simultaneous connectivity between the NTN UE and multiple constellations. For example, delay-tolerant traffic may be transported over a GEO constellation, and delay-sensitive traffic can be transported over an LEO constellation. In another configuration, delay-tolerant data can be transported simultaneously over GEO and LEO constellations. In yet another configuration, traffic load balancing may be performed between a GEO constellation and an LEO constellation.

10.5.6 6G Technologies with NTNs

6G would introduce modern technologies such as RIS, massively distributed MIMO, and OAM. The suitability of new 6G technologies for an NTN should be investigated for enhanced system performance and user experience.

10.6 Major Takeaways

As in Release 17 and Release 18, the 3GPP has two parallel work streams related to the NTN, NR-NTN, and IoT-NTN. The NTN is expected to play an even more key role in 6G. Here are the major takeaways from this chapter.

- The NR-NTN enhancements in Release 19 aim to enhance downlink coverage, enhance uplink capacity, support the MBS, support selected 5G NFs on the NTN platform, including a regenerative payload, and support RedCap UEs.
- The DL coverage enhancements in the NR-NTN consider both the link and system levels. The link level enhancements aim to improve the link margin of physical channels such as the PDCCH and the PDSCH so that these channels can be operated with reduced EIRP in the FR1 frequency band and for handheld devices with a – 5.5 dBi antenna gain.
- The UL capacity enhancement in the NR-NTN aims to optimize uplink capacity through multiplexing techniques.
- The MBS for the NR-NTN focuses on the required enhancements when the intended service area is smaller than the coverage of the entire NTN cell, such as notification about the service area of a broadcast service and suitable signaling between the NG-RAN and the 5GC.
- One or more NFs (e.g., the gNB to realize a regenerative payload and the UPF for low-latency communications) can be placed on the NTN platform in the NR-NTN. The placement of NFs on the NTN platform leads to new architecture options for deployment, increasing flexibility.
- The support for RedCap devices, such as smartwatches and IoT devices in the FR1 frequency band in the NR-NTN, increases the utility of the NTN for service offerings.
- Key IoT-NTN enhancements in Release 19 support Store & Forward and uplink capacity enhancements.
- The Store & Forward operation in the IoT-NTN makes delay-tolerant and non-real-time IoT services available in areas (e.g., middle-of-the-ocean areas or remote areas) illuminated by the NTN platform without requiring an NTN-GW in the satellite's direct visibility.
- Techniques such as multiplexing of UEs using OCCs and multitone support for 15 kHz SCS can increase the uplink capacity in the IoT-NTN.
- The ITU envisions six usage scenarios for 6G or IMT-2030. Three scenarios are extensions of IMT-2020 (i.e., 5G) scenarios, including immersive communication, massive communication, and HRLLC. Three new usage scenarios are Ubiquitous Connectivity, Integrated AI and Communication, and Integrated Sensing and Communication. Four overarching aspects apply to all usage scenarios: sustainability, connecting the unconnected, ubiquitous intelligence, and security/privacy/resilience.
- The 6G performance can be quantified using perspectives such as peak data rate, user-experienced data rate, spectrum efficiency, area traffic capacity, connection density, mobility, latency, reliability, coverage, positioning, sensing-related capabilities, applicable AI-related capabilities, security and resilience, sustainability, and interoperability.

- The NGA has identified five major categories of technology enablers for 6G-component technologies, network O&M and service enablement technologies, radio technologies, system and network architectures, and trustworthiness technologies.
- Examples of NTN research areas include regenerative payload-related architectural and mobility enhancements, spectrum sharing, security, and multi-constellation connectivity.

References

3GPP (2024a). *Revised WID: Non-Terrestrial Networks (NTN) for NR Phase 3*. RP-240775. https://www.3gpp.org/ftp/tsg_ran/TSG_RAN/TSGR_103/Info_for_workplan/revised_WID_54/RAN2_5 (accessed 9 September 2024).

3GPP (2024b). *Revised WID on Non-Terrestrial Networks (NTN) for Internet of Things (IoT) Phase 3*. RP-240776. https://www.3gpp.org/ftp/tsg_ran/TSG_RAN/TSGR_103/Info_for_workplan/revised_WID_54/RAN2_5 (accessed 9 September 2024).

3GPP (2024c). *Feasibility Study on Integrated Sensing and Communication*. https://portal.3gpp.org/desktopmodules/Specifications/SpecificationDetails.aspx?specificationId=4044 (accessed 2 May 2025).

6G Forum (2024). *6G Forum*. http://www.5gforum.org/html/en/main.php (accessed 9 September 2024).

6G-IA (2021). European Vision for the 6G Network Ecosystem. June 2021. https://5g-ppp.eu/wp-content/uploads/2021/06/WhitePaper-6G-Europe.pdf (accessed 9 September 2024).

6G-IA (2024). *About the 6G-IA*. https://6g-ia.eu/ (accessed 9 September 2024).

Beyond 5G Promotion Consortium (2024). *Beyond 5G Promotion Consortium*. https://b5g.jp/en/ (accessed 9 September 2024).

Bharat 6G Alliance (2024). *Bharat 6G Alliance*. https://bharat6galliance.com/ (accessed 9 September 2024).

Chen, W. and Jain, P.H. (2024). Introduction to 3GPP Release 19 and 6G Planning. Presented at an ATIS webinar (3 April 2024). https://atis.org/webinars/introduction-to-3gpp-release-19-and-6g-planning/ (accessed 9 September 2024).

Chintalapati, R., Tripathi, N., Kumar, M., and Reed, J. (2024). Feasibility Analysis of Isolation-Based Spectrum Sharing Between Terrestrial and Non-Terrestrial Networks. *Proceedings of The IEEE 2024 Texas Symposium on Wireless and Microwave Circuits and Systems* (March 2024).

Hadani, R., Rakib, S., Tsatsanis, M., Monk, A., Goldsmith, A.J., Molisch, A.F., and Calderbank, R. (2017). Orthogonal time frequency space modulation. *IEEE WCNC'2017*, pp. 1–6, March 2017.

IETF (2025). Activity in Post-Quantum Cryptography. IETF. https://trac.ietf.org/trac/sec/wiki/PQCAgility (accessed 29 April 2025).

ITU (2023). *Framework and overall objectives of the future development of IMT for 2030 and beyond*. Recommendation M.2160. https://www.itu.int/rec/R-REC-M.2160/en (accessed 9 September 2024).

Li, S., Yuan, W., Wei, Z., He, R., Ai, B., Bai, B., and Yuan, J. (2021). A Tutorial to Orthogonal Time Frequency Space Modulation for Future Wireless Communications. *2021 IEEE/CIC International Conference on Communications in China* (ICCC Workshops), 2021.

NGA (2022a). Roadmap to 6G. February 2022. https://nextgalliance.org/white_papers/roadmap-to-6g/ (accessed 18 April 2025).

NGA (2022b). 6G Technologies. July 2022. https://nextgalliance.org/white_papers/6g-technologies/ (accessed 18 April 2025).

NGA (2022c). Trust, Security, and Resilience for 6G Systems. August 2022. https://nextgalliance.org/white_papers/trust-security-and-resilience-for-6g-systems/ (accessed 25 April 2025).

NGA (2023). 6G Radio Technology Part I: Basic Radio Technologies. November 2023. https://nextgalliance.org/white_papers/6g-radio-technology-part-i-basic-radio-technologies/ (accessed 9 September 2024).

NGA (2024a). *6G Library*. https://nextgalliance.org/6g-library/ (accessed 9 September 2024).

NGA (2024b). 6G Radio Technology Part II: Basic Radio Technologies. May 2024. https://nextgalliance.org/white_papers/6g-radio-technology-part-i-basic-radio-technologies/ (accessed 25 April 2025).

NGA (2024c). 6G Technology Management and Orchestration. February 2024. https://nextgalliance.org/white_papers/6g-technology-management-and-orchestration/ (accessed 25 April 2025).

NGA (2024d). Spectrum Access Mechanisms. August 2024. https://nextgalliance.org/white_papers/spectrum-access-mechanisms/ (accessed 1 May 2025).

NGA (2025a). 6G Component Technologies - Semiconductor Technology. March 2025. https://nextgalliance.org/white_papers/semiconductor-technology/ (accessed 18 April 2025).

NGA (2025b). 6G Component Technologies - Antenna, Packaging, and Testing. March 2025. https://nextgalliance.org/white_papers/antenna-packaging-testing/ (accessed 18 April 2025).

NGA (2025c). 6G Component Technologies - Circuits and Subsystems. February 2025. https://nextgalliance.org/white_papers/circuits-and-subsystems/ (accessed 18 April 2025).

NGA (2025d). 6G Component Technologies - Next G Displays. March 2025. https://nextgalliance.org/white_papers/next-g-displays/ (accessed 18 April 2025).

NGMN Alliance (2021). NGMN 6G Drivers and Vision. April 2021. https://www.ngmn.org/work-programme/ngmn-6g-drivers-and-vision.html (accessed 9 September 2024).

NGMN Alliance (2022). 6G Use Cases and Analysis. February 2022. https://www.ngmn.org/wp-content/uploads/220222-NGMN-6G-Use-Cases-and-Analysis-1.pdf (accessed 9 September 2024).

O-RAN ALLIANCE (2024). *O-RAN Architecture Description*. O-RAN.WG1.OAD-R003-v09.00. The O-RAN ALLIANCE. https://orandownloadsweb.azurewebsites.net/specifications (accessed 24 July 2024).

O-RAN ALLIANCE (2025). Deployments of O-RAN-based Non-Terrestrial Networks, O-RAN ALLIANCE white paper, February 2025.

Pivotal Commware (2019). Holographic Beam Forming and Phased Arrays. https://pivotalcommware.com/technology/ (accessed 2 May 2025).

Qualcomm (2022). Vision, market drivers, and research directions on the path to 6G. December 2022. https://www.qualcomm.com/content/dam/qcomm-martech/dm-assets/documents/Qualcomm-Whitepaper-Vision-market-drivers-and-research-directions-on-the-path-to-6G.pdf (accessed 9 September 2024).

Tripathi, N.D. and Reed, J.H. (2024). 3GPP-based Non-Terrestrial Networks in 5G and 6G. Tutorial presented at IEEE MILCOM 2024 (28 October 2024). https://atis.org/webinars/introduction-to-3gpp-release-19-and-6g-planning/ (accessed 9 September 2024).

Tripathi, N.D. and Shah, V.K. (2025). Fundamentals of O-RAN. IEEE Press/Wiley, https://www.wiley.com/en-us/Fundamentals+of+O-RAN-p-9781394206803.

Index

3GPP-based Non-Terrestrial Networks in 5G and 6G: Expanding the Frontiers of Wireless Communications, First Edition.
Nishith D. Tripathi and Jeffrey H. Reed.

Companion website: www.wiley.com/go/tripathi5g

Printed and bound by CPI Group (UK) Ltd, Croydon, CR0 4YY
17/09/2026
14964313-0002